Cutting Chemical Wastes

What 29 Organic Chemical
Plants Are Doing
To Reduce Hazardous Wastes

By
David J. Sarokin
Warren R. Muir, Ph.D.
Catherine G. Miller, Ph.D.
Sebastian R. Sperber

Editors
Perrin Stryker
Patricia Lone

An INFORM Report

INFORM, Inc.
381 Park Avenue South
New York, NY 10016
212/689-4040

Copyright © 1985 by INFORM, Inc.
All rights reserved

Cutting chemical wastes.

   (An INFORM report)
   Includes bibliographies.
   1. Chemical plants--United States--Waste disposal.
2. Hazardous wastes--United States. 3. Chemistry,
Organic. I. Sarokin, David. II. Series.
TD899.C5C87   1985   661'.8   85-23896
ISBN 0-918780-32-2

INFORM, Inc., founded in 1973, is a non-profit research organization that identifies and reports on practical actions for the protection and conservation of natural resources. INFORM's research is published in books, abstracts, newsletters and articles. Its work is supported by contributions from individuals and corporations and by grants from over 40 foundations.

Cover by Saul Lambert
Figures by Carolyn Cannon

Table of Contents

Preface	viii
Acknowledgements	x

PART I

1. INTRODUCTION	1
References	10
2. THE ORGANIC CHEMICAL INDUSTRY AND ITS WASTES	12
A Brief Background	12
The Chemical Wastemakers	14
How Industry Copes with its Wastes	19
References	25
3. SUMMARY OF FINDINGS	26
Findings on Hazardous Waste Generation	27
Findings on Waste Reduction Practices	28
Findings on Impacts of Waste Reduction Practices	31
Why Plants Do or Do Not Initiate Waste Reduction	32
Adequacy of Information	34
4. THE INFORM STUDY PLANTS	36
Differences in Three States	36
Regulatory Programs in the Three States	41
Criteria for Choosing the Study Plants	42
Problems in Obtaining Information for the Study	46
5. WASTE GENERATION, DISPOSAL AND TREATMENT AT THE STUDY PLANTS	53
Waste Generation	54
Methods of Waste Treatment and Disposal	63
References	74
6. WASTE REDUCTION AT THE STUDY PLANTS	75
Data on Hazardous Waste Reduction: An Inadequate Picture	76
Extent of Hazardous Waste Reduction Practices	79
Waste Reduction: Five Categories and 44 Practices Identified by INFORM	81

7. FACTORS INFLUENCING WASTE REDUCTION PRACTICES
   AT THE STUDY PLANTS ... 106
   Plant Characteristics ... 107
   The Impact of Environmental Regulations on
     Hazardous Waste Reduction ... 114
   The Impact of Cost Factors on Waste
     Reduction Practices ... 133
   The Impact of Liability on Hazardous Waste
     Reduction ... 140
   Conclusion: Waste Reduction is Usually the
     Last Choice ... 143
   References ... 148

PART II – THE PLANT PROFILES

Introduction ... 151
 Organization of the Profile Texts ... 151
 Profile Tables ... 152
American Cyanamid Company ... 157
Atlantic Industries ... 168
Bonneau Dye Corporation ... 182
Borden Chemical Company ... 183
Carstab Division ... 202
Chevron Chemical Company ... 212
CIBA-GEIGY Corporation ... 223
Colloids of California ... 255
Dow Chemical U.S.A. ... 259
E.I. Du Pont de Nemours & Company ... 279
Exxon Chemical Americas ... 310
Fibrec, Inc. ... 326
Fisher Scientific Company ... 327
Frank Enterprises, Inc. ... 342
J.E. Halma Company, Inc. ... 348
International Flavors and Fragrances Inc. ... 349
Max Marx Color and Chemical Company ... 366
Merck and Company, Inc. ... 370
Monsanto Company ... 399
Perstorp Polyols, Inc. ... 414
Polyvinyl Chemical Industries ... 423

Rhone-Poulenc Inc.	428
Scher Chemicals, Inc.	436
Shell Chemical Company	443
Sherwin-Williams Company	448
Smith and Wesson Chemical Company, Inc.	459
Stauffer Chemical Company	462
Union Chemicals Division	475
USS Chemicals	485
Appendix A-List of Hazardous Chemicals	507
Hazardous Air Pollutants	508
126 "Priority Pollutants"	509
RCRA Toxic Chemicals	510
Appendix B-Research Method and Information Sources	517
Master List	518
Government Files	520
Plant Interviews	523
Primary Sources Used for Identifying and Selecting Plants	525
Primary Sources Used for Information on Individual Study Plants	528
Authors' Biographies	532
Other Related INFORM Publications	534
INFORM's Board of Directors	535

List of Figures and Boxes

1-1.	Rate of Growth in Production of Organic Chemicals Compared to those of All Chemicals and Total U.S. Industrial Production	2
Box.	Quotes on Waste Reduction	6
Box.	Definitions Used in this Study	8
2-1.	Transformations of Ethylene	15
2-2.	Air Emissions of 14 Hazardous Organic Chemicals: A Comparison of Organic Chemical Industry Discharges with That of Other U.S. Industries	17
2-3.	Industrial Wastewater Discharges of Hazardous Organic Chemicals: A Comparison of Organic Chemical Discharges with Those of Other U.S. Industries	18
2-4.	Industrial Hazardous Solid Wastes: A Comparison of Chemical Industry Generation with Those of Other U.S. Industries	20
2-5.	Comparison of Waste Treatment, Recycling and Reduction: A Hypothetical Example for Air Emissions from a Storage Tank	24

List of Tables

1-1.	Organic Chemicals with Annual Production of more than 5 Billion Pounds	4
4-1.	INFORM's Study Plants	37
4-2.	Characteristics of the Organic Chemical Industry in INFORM's Study States and the U.S.	40
4-3.	Characteristics of INFORM's Study Plants	44
4-4.	Distribution of Study Plant Characteristics	47
4-5.	Cooperation of Plants in the Study	50
5-1.	Use and Waste Generation of Selected Hazardous Chemicals at INFORM Study Plants in New Jersey	56
5-2.	The Largest Hazardous Solid Wastestreams from INFORM Study Plants	59
5-3.	The Largest Amounts of Hazardous Air Emissions from INFORM Study Plants	60
5-4.	The Largest Amounts of Hazardous Chemicals in Wastewater from INFORM Study Plants	62

List of Tables (continued)

5-5.	Waste Treatment and Disposal Methods at INFORM Study Plants	65
5-6.	Active Class I Underground Injection Wells Used by Companies Represented in INFORM's Study	70
6-1.	Cooperation and Waste Reduction Practices Identified at Large and Small Plants in INFORM's Study	78
6-2.	Waste Reduction Practices at Study Plants: Process Changes	83
6-3.	Waste Reduction Practices at Study Plants: Product Reformulations	90
6-4.	Waste Reduction Practices at Study Plants: Chemical Substitutions	92
6-5.	Waste Reduction Practices at Study Plants: Equipment Changes	94
6-6.	Waste Reduction Practices at Study Plants: Operational Changes	98
6-7.	Waste Reduction Practices at Study Plants: Unexplained Changes	104
7-1.	Product and Process Characteristics of INFORM Study Plants	109
7-2.	Organic Chemicals in Wastewaters from Organic Chemical Industry Operations	110
7-3.	Emissions Exempted from Controls at California Plants Owned by Companies Represented in INFORM's Study	127
7-4.	Variable Fees and Restrictions at Sewage Treatment Plants Servicing INFORM Study Plants	130
7-5.	Material and Cost Savings of Waste Reduction Practices	135
7-6.	Treatment and Disposal of Hazardous Wastes: Regulations and Costs	144
B-1.	Data Sources Used for Selection of Study Plants	519
B-2.	Government Files Containing Information on INFORM Study Plants	521

Preface

We are pleased to offer this INFORM report, the result of three years of research. It explores -- through case studies of 29 plants -- initiatives taken by the U.S. organic chemical industry to reduce the hazardous wastes generated at source. It also evaluates the information available to public leaders regarding the use of chemicals in this country -- where and in what quantities they are discharged to our air, land and water resources.

The organic chemical industry is necessarily of primary importance to all government, environmental and business leaders concerned about toxic pollution problems. Its more than 1,000 plants generate over 60 percent of the toxic wastes entering our environment. We hope that the research here will make a substantial contribution to future planning and to the public debate over waste management, toxic waste problems and waste reduction as an alternative.

The 44 examples of waste reduction practices described in the 29 plant profiles in this study suggest the significant range of opportunities that waste reduction offers. These initiatives involve changes in processes and products and substitutions of chemicals as well as operational and equipment changes. These, in some cases, were achieved by simple alterations, in other cases by complex and expensive efforts. But the remarkable potential for plants to identify ways in which they might reduce or eliminate toxic wastes is clear.

Despite the extensive interest and verbal support industry and government have given to the concept of waste reduction, this report finds such initiatives still all too rare. Much more can (and inevitably must) be done in this crucial area.

While the kinds of waste reduction options documented here are interesting and exciting, the real impact of the practices could not be quantified in most cases. The information was seldom available on total plant wastes or on individual wastestreams so that reductions could be measured. What is more, the information on chemical use and discharges of wastes to air, land and water was fragmented and piecemeal at best.

Clearly, if government or the public is to have a useful picture of toxic waste problems we all face, and to assess progress in managing or reducing these problems, the specific data available to the public on industry's chemical uses and discharges must be improved.

We hope that the stories told in the 29 profiles will spur heightened initiatives and broader consideration by government and business of how waste reduction can be accelerated.

Joanna D. Underwood
INFORM
Executive Director

Acknowledgements

There is not a single person at INFORM who has not contributed significantly to the research and writing of this report and to the smooth functioning of the organization that makes the researchers' task possible.

The authors are especially indebted to Joanna Underwood and Gail Richardson, INFORM's Executive Director and Associate Director respectively, for their unbounded enthusiasm and support for this work as well as their combined wealth of experience as to what makes a research project valuable and do-able. Robbin Blaine's recognition of the critical need for research on an oft-mentioned but little understood topic, and her follow-up efforts, effectively launched this project. Our fellow researchers at INFORM, Dr. Steven Rohmann, Dr. Allen Hershkowitz, Larry Naviasky and Roger Miller deserve a warm thank you for both their insight and moral support.

Perrin Stryker, INFORM's Senior Editorial Consultant, and Pat Lone have succeeded in bringing the report to fruition through not only their editorial and production capacities, but also their incisive commentary on drafts of the report. Rebecca Cooney's artistic advice has been invaluable, and Saul Lambert's striking cover graphically conveys our theme.

Gwenn Sewell and Cary MacDonald made significant contributions during their tenure as interns. Ellen Poteet's careful reviews of text and administrative organization have contributed much more than her, thus far, brief tenure at INFORM would suggest. The care, skill and stamina of Bob Szwed, Charles Lowy and Norma Johnson Walker in typing and proofing the report are noted with due admiration.

Many outside of INFORM contributed their time, information, criticisms and assistance. Managers and plant officials in the chemical companies who recognized the importance of INFORM's work and chose to cooperate in the research, often devoted considerable effort first compiling and then reviewing information for INFORM. Their efforts were crucial to this project and especially appreciated by us.

A special note of gratitude is also due those who, under demanding time constraints, reviewed the drafts of this report and contributed greatly to its final form and content. To Robert B. Pojasek (Chas. T. Main Inc.), Hank S. Cole (Senior Scientist, Clean Water Action Project and Science Director, National Campaign Against Toxic Hazards), Mike Koenigsberger (3M Corp.) and Don Huisingh (North Carolina State University and co-author of "Proven Profit From Pollution Prevention"), thanks. We are grateful, too, to Douglas G. Bannerman, who freely gave of his time and expertise in helping to design the form and content of our meetings with study plants.

The small but expanding circle of people giving serious attention to waste reduction -- whether through research, education, political processes, environmental activism or corporate management -- were, and still are, a constant source of encouragement. Many of the members of federal and state environmental agencies, all of whom are overworked, were very generous with their time and information.

The best of intentions and research designs would have been fruitless had it not been for the generous financial support of the following foundations: The ARCA Foundation, Mary Reynolds Babcock Foundation, Geraldine R. Dodge Foundation, The Fund for New Jersey, The George Gund Foundation, The Joyce Foundation, Charles Stewart Mott Foundation, Jessie Smith Noyes Foundation, James C. Penney Foundation, The Prudential Foundation and the Victoria Foundation. We are grateful for their confidence and support.

Part I
Chapter 1

Introduction

In the forty years since World War II, daily life has been quietly transformed by the advent of the chemical age. Foods, medicines, building products, transport and communication now depend in large part on the 220 billion pounds of organic chemicals manufactured each year in the United States by the more than 1,000 organic chemical plants across the country.

For all its influence, however, the organic chemical industry is not the visible presence that other industries like the steel, coal or auto works have been. Its dominance has been almost imperceptible, yet over the past 40 years, the organic chemical industry's production has increased tenfold, a rate greatly outstripping general industrial growth, including the chemical industry as a whole (see Figure 1-1).

Unlike many other manufacturing operations, whose end products are usually immediately identifiable as consumer goods, the chemical industry's products -- currently between 60,000 and 70,000 chemicals -- remain largely unknown to the general public, but are crucial for a multitude of manufacturing processes. In addition, approximately 1,200 new organic chemical products, with potential commercial value, are created in laboratories each year.

It is not, however, only their usefulness that makes organic chemicals so important. They have the potential, as experience has shown, for being enormously

Figure 1-1. Rate of Growth in Production of
Organic Chemicals Compared to those of
All Chemicals and Total U.S. Industrial Production*
(from baseline year 1949)

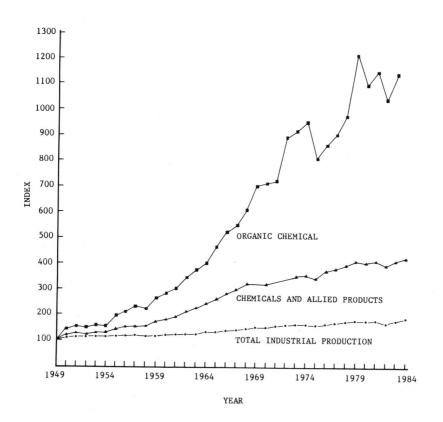

* Rate of growth is measured using production indices, with each index line showing the rate of increase in production (not the physical amount of production) for each manufacturing sector.

Sources: U.S. International Trade Commission's annual report on "Synthetic Organic Chemicals, U.S. Production and Sales," Board of Governor's of the Federal Reserve System, "Industrial Production 1957-1959 Base," and "Federal Reserve Statistical Release," July 18, 1985, and Bureau of the Census, "Statistical Abstract of the United States - 1985."

dangerous as well. PCBs, a suspected carcinogen, were found in fish in the Hudson River in New York, the Great Lakes and elsewhere; ethylene dibromide (EDB), a suspected carcinogen, appeared as a residue in cereals and grains; and dioxin, an industrial waste and one of the most potent toxic materials known, was found in schoolyards and front lawns.

Where do these hazardous substances come from? The answer lies, in large part, in the nature of the organic chemical industry itself -- in the amounts of wastes it generates, and the methods it chooses to dispose of them. PCBs and EDB are products of the organic chemical industry, as are chloroform, trichloroethylene and trichloroethane, three of the most common toxic contaminants of groundwater. The organic chemical industry is not the sole producer of these chemical wastes or solely responsible for the problems they cause, but it is by far the largest generator of them. A 1984 Environmental Protection Agency report[1] found that organic chemical plants emitted almost 40 percent of air emissions of selected chemicals from industrial sources. The industry accounts for 83 percent[2] of industrial discharges of hazardous organic chemicals to rivers. Fourteen organic chemicals are manufactured in excess of five billion pounds per year and account for about half of the total annual production of organic chemicals (see Table 1-1). Nine of these 14 high-volume chemicals are regulated as hazardous materials (see Appendix A).

The enormous volume and tremendous diversity of hazardous substances the organic chemical industry handles make its waste management practices of particular concern. There is no easy solution to the problem of disposal once hazardous wastes are generated, and the environmental lessons learned from discharging and dumping have been extremely painful and expensive. Kepone dumped into the James River in Maryland, for instance, destroyed the fish and shellfish industry in that area. Over 100 square miles of pastureland in Louisiana were contaminated by hexachlorobenzene that had evaporated from wastes dumped into local pits.[3]

Table 1-1. Organic Chemicals with Annual Production of more than 5 Billion Pounds

Chemical	1984 Production (billions of lbs)	Major End Uses
Ethylene	31.18	Polyethylene and other plastics, antifreeze
Propylene	15.47	Polypropylene and other plastics, synthetic fibers
Urea	14.30	Fertilizers, adhesives, resins
Ethylene Dichloride*	13.73	Polyvinyl chloride (PVC) plastics
Benzene*	9.86	Polystyrene plastics, nylon
Ethylbenzene*	8.61	Polystyrene plastics
Methanol*	8.28	Plastics, adhesives, fibers, solvents
Styrene	7.71	Polystyrene and other plastics
Vinyl Chloride*	7.51	Polyvinyl chloride (PVC) and other plastics
Xylene*	6.12	Polyester fibers, films, fabricated plastics
Terephthalic Acid	6.05	Polyester fibers and films
Ethylene Oxide*	5.96	Polyester fibers and films, antifreeze, additives
Formaldehyde*	5.71	Adhesives, plastics
Toluene*	5.27	Additive, industrial solvent
TOTAL	145.76 billion pounds	

* Hazardous by INFORM'S criteria (see Appendix A).
Source: "Chemical and Engineering News," June 10, 1985

By the end of 1984, the EPA had identified more than 20,000 hazardous waste dump sites throughout the U.S. and estimated the total could reach much higher. The $1.6 billion Superfund earmarked by Congress to clean up the worst of these will barely scratch the surface: containment and removal costs at Superfund sites have ranged from $10 million to more than $100 million per site.[4]

Even with strict adherence to existing regulations regarding hazardous wastes, the most widely-used methods for dealing with them -- recycling, treatment and disposal -- have major drawbacks. Air pollution control devices or wastewater treatment plants can prevent wastes from going into the air and water, but the toxic ash and sludge removed from these systems constitute hazardous solid waste problems requiring attention. Solid wastes deposited in landfills or deep wells can become water pollution problems; evaporation from ponds and lagoons can turn solid or liquid wastes into air pollution problems. Even more importantly, however, is the fact that these methods in no way address the fundamental issue of how to minimize the generation of hazardous wastes where chemicals are made and used.

Only waste reduction at source does this. It is a preventive measure that can actually help reduce the amount of waste where it is produced and used, through the modification of plant processes and products, changes in plant equipment and alterations in plant operations. As the best of all waste management practices, it has been endorsed by industry trade groups, Congress, the Environmental Protection Agency, state governments, environmental groups, the National Academy of Sciences, academic organizations, the United Nations and many others (see Box on page 6).

Despite the widespread agreement on its desirability, however, very little has been written about the extent and effects of waste reduction at source. The federal EPA's three-volume <u>Development Document</u> on proposed

The Chemical Companies Are All for It . . .

"One of the keystones in our efforts has been and will be the reduction of volume of wastes produced."
                                                    Chevron

"In our existing chemical processes, we have a formal program to reduce the quantity of waste that we are currently producing through equipment upgrading, more efficient use of raw materials, or different methods of process control."
                                                    Monsanto

. . . And So Are the Government and the Scientists

"There needs to be a concentrated effort to make changes in manufacturing processes to reduce the generation of hazardous wastes requiring disposal."
                        U.S. Environmental Protection Agency

"Elimination, rather than treatment or storage, is the optimal solution to the problems of industrial waste."
 The Industrial Waste Elimination Research Center

"The ideal solution is to reduce waste at the source by changing the industrial processes so that hazardous by-products are not produced."
           California Office of Appropriate Technology

"In-plant options are probably the most effective and economical means of managing hazardous wastes...The committee strongly recommends a major commitment, both philosophically and in funding, to approaches that prevent or eliminate hazardous materials from being discharged as wastes."
                             National Academy of Sciences

guidelines on limitations on wastewater discharges from organic chemical plants devotes hundreds of pages to discussions of treatment and disposal of toxic chemical wastes and only four pages to "in-plant source controls" -- or waste reduction. In 1982 when INFORM undertook this study, there were just six books on the subject of waste reduction, and few have been added since.[5] None offers any specific in-depth documentation of industry actions or of the impacts of those actions on the industry's wastes. Yet the subject, like the growing amounts of hazardous wastes, demands examination and attention.

INFORM, therefore, investigated 29 organic chemical plants to discover what steps are actually being taken to reduce wastes; what impact waste reduction practices have on total plant wastes; and what managerial, economic, and/or regulatory factors are stimulating or impeding hazardous waste reduction efforts. INFORM's investigation of these matters required three years of the most intensive and concentrated exploration of the organic chemical industry's waste reduction practices.

DEFINITIONS USED IN THIS STUDY

The chemical substances defined as hazardous by INFORM are those listed by the EPA as "hazardous air pollutants" under the Clean Air Act, as toxic "priority pollutants" under the Clean Water Act and as "toxic chemicals" under the Resource Conservation and Recovery Act (RCRA). Full lists of these chemicals appear in Appendix A. They include both organic and inorganic chemicals.

The term "hazardous wastes" is easily subject to misinterpretation because the words mean different things in different contexts. INFORM's use of the phrase "hazardous wastes" is broader than the conventional use of the term, referring only to solid wastes regulated under RCRA. It encompasses all wastes regardless of how they are discharged, and includes air emissions regulated under the Clean Air Act and wastewater discharges under the Clean Water Act, as well as RCRA solid wastes. Further, any type of waste stream is considered hazardous in INFORM's definition if it contains a substance appearing on any of the three hazardous substances lists regardless of whether or not it is regulated.

For example, all wastes containing the chemical toluene are defined as a hazardous waste in this study because toluene is listed on the RCRA hazardous substances list as well as on the Clean Water Act priority pollutant list. Even though air emissions of toluene are not regulated as hazardous under federal regulations (because toluene is not listed as a hazardous pollutant under the Clean Air Act), they are considered "hazardous wastes" in the context of this report.

In one sense, INFORM's use of the phrase "hazardous wastes" is more narrow than the regulatory usage, because RCRA includes in its list of hazardous wastes those substances characterized as flammable, corrosive and explosive as well as those that are toxic. INFORM has excluded substances that are solely flammable, corrosive or explosive but not toxic under RCRA.

The term "hazardous" is used in this study, rather than "toxic," so as to avoid the implication that the concentration and amount of the chemicals in the various wastes are always present in quantities sufficient to cause harm. The focus of this study is on the particular chemical substances generated as wastes and does not extend to an analysis of the extent or severity of environmental or human exposures to these substances.

The term "wastes," as used in this report, includes pollutant discharges, substances receiving destructive or containment treatment, as well as off-specification products and non-commercial co-products and by-products.

The term "waste reduction at source" means that the source of the waste is altered in some way so as to reduce the amount generated or eliminate it altogether. This is important to note because the phrase "waste reduction" is often used loosely to refer to virtually any practice that simply reduces the quantity of wastes going into landfills, even if the reduction is accomplished by treatment, recycling or by an alternative form of disposal. **This report distinguishes reduction of wastes before they are generated from recycling or treatment of wastes after they are generated.**

References

1. INFORM's calculations based on data in: Tom Lahre and Radian Corporation, "Characterization of Available Nationwide Air Toxics Emissions Data," prepared for the U.S. Environmental Protection Agency, Office of Air Quality Planning and Standards, Research Triangle Park, NC, June 13, 1984.

2. JRB Associates, *Addendum to the Report Entitled: Assessments of the Impacts of Industrial Discharges on Publicly Owned Treatment Works*, McClean, VA, February 25, 1983.

3. Michael H. Brown, *Laying Waste*, Pantheon Books, New York, 1980, p. 162.

4. "Environmental Progress and Challenges," *EPA Journal*, April 1984, p. 8.

5. The six earlier sources are:

Monica E. Campbell and William M. Glenn, *Profit from Pollution Prevention*, Pollution Probe Foundation, Toronto, 1982.

Donald Huisingh and Vicki Bailey (eds.), *Making Pollution Prevention Pay*, Pergamon Press, New York, 1982.

Office of Appropriate Technology, *Alternatives to the Land Disposal of Hazardous Wastes*, State of California, Sacramento, 1981.

Office of Technology Assessment, *Technologies and Management Strategies for Hazardous Waste Control*, Congress of the United States, Washington, DC, 1983.

Michael G. Royston, *Pollution Prevention Pays*, Pergamon Press, New York, 1979.

The 3M Company, <u>Low or Non-Pollution Technology Through Pollution Prevention: An Overview</u>, The United Nations Environment Programme, New York, June 1, 1982.

A sample of other sources:

Virginia Adamson, <u>Breaking the Barriers</u>, Canadian Environmental Law Research Foundation and The Pollution Probe Foundation, Toronto, 1984.

Gary A. Davis, <u>Measures to Promote the Reduction and Recycling of Hazardous Wastes in Tennessee</u>, Energy, Environment, and Resources Center, University of Tennessee, Knoxville, September 15, 1984.

Environmental Action Foundation, "Fact Packet: Waste Reduction," Washington, DC, October 1983.

Donald Huisingh, Helen Hilger and Sven Thesen, <u>Profits from Pollution Prevention</u>, prepared for the North Carolina Board of Science and Technology, Raleigh, May 1985.

National Research Council, <u>Reducing Hazardous Waste Generation</u>, National Academy Press, Washington, DC, 1985.

Chapter 2

The Organic Chemical Industry and Its Wastes

If the organic chemical industry's function could be summed up in one word, it would be "transformation" -- transformation of raw materials from petroleum and natural gas production (and to a lesser extent, from the farming, forestry and mining industries) into the endless variety of organic chemical products available today. To those not familiar with organic chemistry and its function of transformation, this chapter may be helpful in clarifying many of the hazardous waste reduction initiatives described in this report.

A Brief Background

To begin with, organic chemicals are those containing carbon. Chemicals without carbon -- metals, iron oxide (rust), sodium chloride (salt) and ammonia, for example -- are called inorganic. Carbon, alone among the chemical elements, readily links with itself to form chains and rings in an impressive variety of organic chemical compounds. Carbon also has the capacity to bind in almost endless combinations with other elements such as oxygen, hydrogen, nitrogen and chlorine. The enormous diversity of carbon has made this element the basis for the chemistry of all living organisms -- hence, the name "organic." In recent decades, scientists have learned to mimic some of this chemistry in the laboratory and even expand upon its diversity. Millions of organic chemicals have been developed in research laboratories and no limit has yet been glimpsed of the number of additional compounds that may be discovered.

Industrially-produced organic chemicals play a role in all other industrial operations. The production of organic chemicals is necessary for -- and is often the major component of -- pharmaceutical products, cosmetics, plastics, food additives and processing materials, photographic films, clothing, agricultural chemicals, construction materials, automotive products, electronic components, and the list goes on and on. Materials which are not themselves part of the organic chemical industry -- steel, glass, paper, wood, concrete, ceramics -- all rely on organic chemicals for their manufacture.

Transformation is at the heart of chemistry and is what distinguishes the chemical industry from other basic manufacturing efforts. The common materials that make up a great deal of the material man-made world are fabricated, extruded, molded but not fundamentally transformed to different substances as they move through the industrial world and to the ultimate consumer.

Only manufactured chemicals are fundamentally transformed to different substances during the course of their industrial life cycles. Indeed, the industrial chemist's skill at "customizing" these transformations to produce materials with desirable properties is the feature that has enabled the organic chemical industry to virtually reshape the face of modern society with an abundance of synthetic materials, many of which were never before available.

Organic chemicals are found in virtually every facet of daily life. About 75 percent of the 10 billion pounds of textiles produced annually in the U.S. are synthetic organic chemicals like polyester and nylon.[1] From one percent to eight percent of the weight of clothing is the organic chemical dye used to give the fabric its color. Whether synthetic or natural, all fabrics are treated with a wide variety of organic chemicals to impart desired properties such as flame resistance, soil resistance, antistatic behavior, color fastness, or "wicking" to carry away moisture. In rubber manufacturing, 76 percent of the 6.8 billion pounds produced in 1978 was synthetic rubber made from styrene, butadiene, acrylonitrile and other organic chemicals.[2]

Organic chemicals can also be extremely versatile. Polyvinylpyrrolidone (PVP), for instance, has a long string of credits: it is a binder for medicine tablets and retards crystallization. Its non-pharmaceutical applications include its use as an emulsifying agent; in paper manufacturing to increase wet strength; as an additive to pressure-sensitive adhesives to control tackiness and strength; in the plastics industry to modify viscosity; as a component of many cleansers and soaps to minimize soil redeposition; as a plant protection chemical and as a clarifying agent in wine and beer.

Ethylene is the largest volume chemical produced by the organic chemical industry -- more than 31 billion pounds (about 140 pounds per person) are produced annually in the U.S. It is a starting material for transformations into other organic chemicals (see Figure 2-1) and the list of final products which stem from ethylene includes aspirin, phonograph records, adhesives, pipes, floor tiles, toys, pesticides, flame retardants, dry cleaning fluid, prescription drugs and gasoline additives.

The staggering variety of chemicals in production is not only a key feature of the industry itself, but is also a key to the types of enormous quantities of hazardous wastes it generates.

The Chemical Wastemakers

Since World War II, organic chemical production has increased tenfold, from less than 20 billion to over 220 billion pounds a year. Chemical waste has surpassed even that enormous figure. Air emissions of potentially toxic organic chemicals exceed five billion pounds per year.[3] Wastewater discharges of toxic chemicals

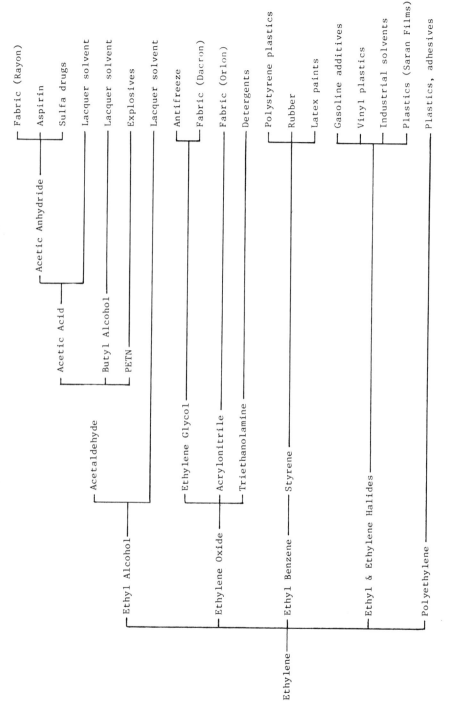

Figure 2-1. Transformations of Ethylene

to rivers and sewage treatment plants add another 412 million pounds.[4] The federal EPA has estimated that more than 580 billion pounds of hazardous solid wastes are generated annually — more than one ton for every person in the United States.[5]

The organic chemical industry is by far the largest producer of hazardous wastes, which are the by-products of the production, distribution, and use of tens of thousands of chemicals in a wide range of applications. The statistics available on wastes generated and disposed of by organic chemical plants demonstrate ample room for waste reduction.

The organic chemical industry legally releases millions of pounds per year of hazardous chemicals into the atmosphere, as only two organic chemicals (benzene and vinyl chloride) are currently regulated under the federal Clean Air Act. Air emissions from organic chemical plants account for virtually 100 percent of the discharges of five hazardous chemicals released in amounts of 50 million pounds or more (see Figure 2-2). In addition, releases of two other high volume hazardous chemicals, formaldehyde and toluene, alone add up to more than one billion pounds per year of air emissions. Of 86 toxic chemicals for which data was available on industrial air emissions, 72 were found to be emitted from organic chemical plants and half of these were emitted only from organic chemical plants.[6]

The organic chemical industry is also the major industrial source of hazardous chemicals in wastewater. As Figure 2-3 illustrates, organic chemical plants account for 83 percent (112 million pounds per year) of hazardous organic discharges to the nation's waterways, and almost 80 percent (154 million pounds per year) of hazardous organic discharges to municipal sewage treatment plants. Injection of contaminated wastewaters into underground wells is another major means of disposal. The industry accounts for 11.5 billion gallons (51 percent) of all wastewaters disposed of by this method. In wastewater streams tested for

Figure 2-2. Air Emissions of 14 Hazardous Organic Chemicals: A Comparison of Organic Chemical Industry Emissions with That of Other U.S. Industries

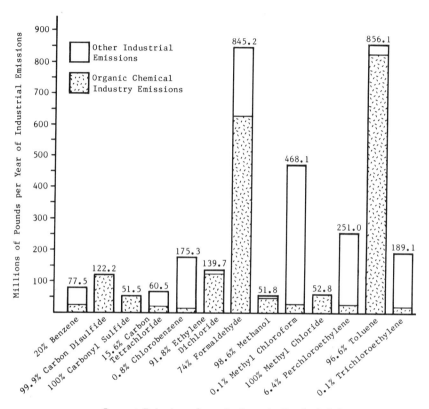

Percent Emissions from the Organic Chemical Industry

* The 14 chemicals shown here are those from EPA's study of 86 chemicals which both have emissions of more than 50 million pounds per year and are hazardous by INFORM's criteria.

Source: Tom Lahre and Radian Corporation, "Characterization of Available Nationwide Air Toxics Emissions Data," prepared for the U.S. Environmental Protection Agency, Office of Air Quality Planning and Standards, Research Triangle Park, NC, June 13, 1984.

Figure 2-3. Industrial Wastewater Discharges of Hazardous Organic Chemicals: A Comparison of Organic Chemical Industry Discharges with Those of Other U.S. Industries

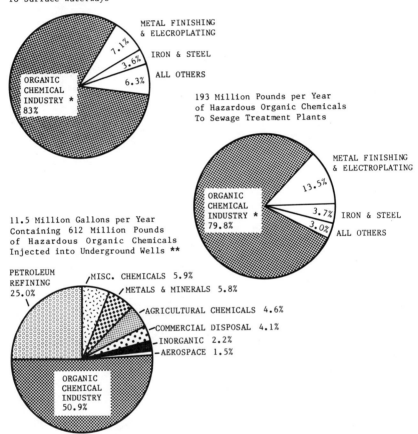

* Includes plastics manufacturing operations. Source: JRB Associates, "Addendum to the Report Entitled: Assessment of the Impacts of Industrial Discharges on Publicly Owned Treatment Works," prepared for the U.S. Environmental Protection Agency, Office of Water Enforcement, Washington, DC, February 1983.

** Source: U.S. Environmental Protection Agency, "Report to Congress on Injection of Hazardous Wastes," Washington, DC, May 1985.

129 hazardous chemicals, the organic chemical industry was found to be discharging 106, more than twice the number found in most other industrial wastewaters.[7]

As for solid wastes, a category that includes wastes that are landfilled or incinerated, separate figures for the organic chemical industry (as distinct from inorganic chemical production) have not been reported by the U.S. EPA. The chemical industry as a whole accounts for 68 percent of the 580 billion pounds per year of hazardous solid waste generated in the U.S. (see Figure 2-4). The data does not detail the quantities or types of hazardous chemicals in the wastestream.

How Industry Copes with its Wastes

Mismanagement of wastes by industry in the past has left an inheritance of hazardous waste sites and contaminated air and water resources across the nation. In confronting the dangers of improper waste disposal, the organic chemical industry has four options. These are widely acknowledged to form a hierarchy of the most to least desirable means -- in terms of environmental protection and economic benefits -- for managing hazardous wastes.

1. Waste reduction at source is the best of all waste management alternatives. Waste reduction practices minimize the generation of hazardous wastes at the source through modification of products, changes in plant processes, and alterations in plant operations. To the extent that the original generation of wastes is minimized, the need for the use of other management options is reduced. (See Figure 2-5 for comparative benefits of a waste reduction practice.)

2. Recycling, reuse and recovery practices, another widely accepted option, make constructive use of materials formerly discarded as wastes. These materials may be reused directly in the process from which they originated or may find use in an entirely different process. Acid wastewaters from one manufacturing operation, for instance, may find use as an acid raw

Figure 2-4. Industrial Hazardous Solid Wastes:
A Comparison of Chemical Industry Generation
with Those of Other U.S. Industries

580 Billion Pounds per Year *

Hazardous Solid Wastes Regulated by the
Resource Conservation and Recovery Act (RCRA)

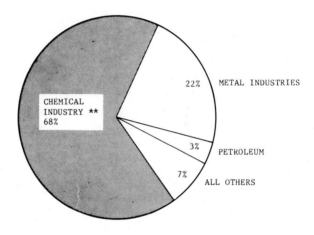

* Total quantity of wastestream, amount of hazardous component unknown.

** Includes both organic and inorganic chemical operations, separate figures are not available.

Source: Westat, Inc., "National Survey of Hazardous Waste Generators and Treatment, Storage and Disposal Facilities Regulated under RCRA in 1981," prepared for the U.S. Environmental Protection Agency, Office of Solid Waste, Washington, DC, April 1984.

material for another operation. (Recovering the energy value of wastes by burning them as a supplemental fuel is generally considered a form of waste recovery, although it is also a form of waste treatment.) Recycled materials may be used by the plants originating them or may be sold to other plants.

Recycling, however, has been difficult to regulate in practice. Waste sludges have been reused as fertilizers, and contaminated waste oils have been used on roads to reduce dust. These may pose a greater hazard than if they were disposed of in carefully engineered landfills. A highly-contaminated Superfund site in Lowell, Massachusetts was formerly a commercial recycling operation.

3. Treatment of hazardous wastes, on or off of plant sites, is an extremely common industrial means of waste management. Treatment practices, such as incineration, chemical decomposition or biological treatment, destroy some of the hazardous materials present in a waste, converting them into innocuous substances like carbon dioxide or water. Some treatment practices (particularly for metal wastes, which cannot be destroyed) simply convert the form of a waste. For instance, treatment methods remove metals from wastewater and concentrate them into sludge, which must then be disposed of.

Increasing recognition, however, is being given to the fact that many treatment technologies, in solving one waste management problem, merely create others.[9] Air pollution control devices or wastewater treatment plants can prevent wastes from going into the air and water, but the toxic ash and sludges removed from these systems constitute enormous hazardous solid waste problems requiring attention. Solid wastes deposited in landfills or deep wells can become water pollution problems; evaporation from ponds and lagoons can turn solid or liquid wastes into air pollution problems. Moreover, new waste management facilities, such as landfills to bury wastes or incinerators to destroy them, are facing increasing community opposition when they are proposed.[10]

4. Disposal of wastes containing hazardous chemicals is the option most fraught with controversy. It is, however, still routine in many forms: toxic chemicals are regularly emitted to the atmosphere, discharged to waterways, or buried in sanitary landfills. Generally, this is permitted by regulatory agencies where the quantities of wastes are presumed to be small enough or ephemeral enough (that is, they quickly decompose) so as not to present a danger to the environment or to public health. But it also occurs because of the continuing absence of standards and regulations affecting many chemical discharges. More persistent or larger quantities of wastes, or wastes more stringently regulated, must be handled by a disposal method considered "secure," that is, one which is intended to contain and isolate the waste. Such methods include carefully engineered landfills designed to prevent leakage, and underground injection wells which dispose of toxic liquids a mile or more below ground.

A history of leaks at hazardous waste landfills and underground wells have led environmental groups to question the wisdom of continuing use of these disposal practices, arguing that, at best, they merely delay rather than prevent chemical leaks. Industry maintains that when properly sited, constructed and operated, landfills can prevent contaminants from leaking into the groundwater.[11] Amendments to RCRA passed by Congress in 1984, however, deemed landfills the "least favored" means of waste disposal and require EPA to prohibit disposal of some hazardous wastes, such as solvents, in this manner, and to minimize the overall use of landfills.

From all the foregoing, it is clear that enormous potential exists for waste reduction initiatives in the organic chemical industry. To tap that potential, however, requires two of the most important "transformations" of all. One is changing industry's wastemaking habits into concrete waste reduction measures; and the other is codifying and expanding the fragmented information on total uses and discharges of chemicals, and on those waste reduction practices that are in place. The summary of INFORM's findings, which are presented in the following chapter, includes the status and the promise of these two transformations.

Figure 2-5. Comparison of Waste Treatment, Recycling and Reduction:
A Hypothetical Example for Air Emissions from a Storage Tank

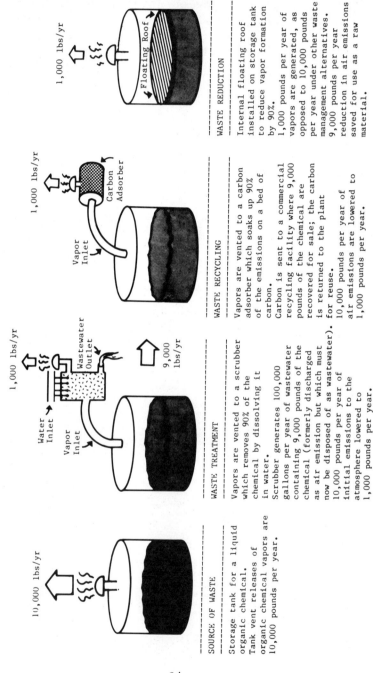

SOURCE OF WASTE

Storage tank for a liquid organic chemical. Tank vent releases of organic chemical vapors are 10,000 pounds per year.

WASTE TREATMENT

Vapors are vented to a scrubber which removes 90% of the chemical by dissolving it in water. Scrubber generates 100,000 gallons per year of wastewater containing 9,000 pounds of the chemical (formerly discharged as air emission but which must now be disposed of as wastewater). 10,000 pounds per year of initial emissions to the atmosphere lowered to 1,000 pounds per year.

WASTE RECYCLING

Vapors are vented to a carbon adsorber which soaks up 90% of the emissions on a bed of carbon. Carbon is sent to a commercial recycling facility where 9,000 pounds of the chemical are recovered for sale; the carbon is returned to the plant for reuse. 10,000 pounds per year of air emissions are lowered to 1,000 pounds per year.

WASTE REDUCTION

Internal floating roof installed on storage tank to reduce vapor formation by 90%. 1,000 pounds per year of vapors are generated, as opposed to 10,000 pounds per year under other waste management alternatives. 9,000 pounds per year reduction in air emissions saved for use as a raw material.

References

1. Colin L. Browne and Robert W. Work, "Man-Made Textile Fibers, <u>Riegel's Handbook of Industrial Chemistry</u>, James A. Kent, ed., 1983.

2. E.E. Schroeder, "Rubber," <u>ibid</u>.

3. Systems Application Inc., <u>Human Exposure to Atmospheric Concentrations of Selected Chemicals</u>, 1980.

4. JRB Associates, <u>Addendum to the Report Entitled: Assessments of the Impacts of Industrial Discharges on Publicly Owned Treatment Works</u>, McClean, VA, February 25, 1983.

5. Westat, Inc., "National Survey of Hazardous Waste Generators and Treatment, Storage and Disposal Facilities Regulated under RCRA in 1981," April 1984.

6. INFORM's calculations based on data in: Tom Lahre and Radian Corporation, "Characterization of Available Nationwide Air Toxics Emissions Data," prepared for the U.S. Environmental Protection Agency, Office of Air Quality Planning and Standards, Research Triangle Park, NC, June 13, 1984.

7. Office of Water Regulations and Standards, <u>Development Document for Proposed Effluent Limitations Guidelines and New Source Performance Standards for the Organic Chemicals and Plastics and Synthetic Fibers Industry</u>, EPA 440/1-83/009-b, U.S. Environmental Protection Agency, Washington, DC, February 1983.

8. The Conservation Foundation, <u>State of the Environment</u>, Washington, DC, 1984.

9. "Plants that Incinerate Poisonous Wastes Run into a Host of Problems," <u>The Wall Street Journal</u>, August 26, 1985, p. 1.

10. "Pollution Control and Hazardous Waste," <u>CMA News</u>, Vol. 10, No. 2, February 1982, p. 8.

Chapter 3

Summary of Findings

This study is the first ever to take a close detailed look at waste reduction practices in the organic chemical industry, examining the hazardous wastes it produces and ways in which it reduces these wastes. It breaks new ground by documenting in depth the actual practices of 29 individual organic chemical plants in three of the major hazardous waste producing states -- California, New Jersey and Ohio.

INFORM's findings from this documentation shed light on the problems and possibilities at these plants in reducing wastes. They also help clarify some broader issues related to the performance of the organic chemical industry as a whole. This industry is made up of over 1,000 plants and is the largest contributor of hazardous wastes in the country.

INFORM's findings address four critical questions:

1. How much and what types of hazardous wastes do the 29 organic chemical plants that INFORM studied generate?

2. To what extent have these plants chosen waste reduction at source, what practices have they implemented, how much waste reduction have they achieved, and what have been the economic costs and benefits? How does the use of hazardous waste reduction practices compare to the use of waste treatment and disposal methods?

3. What are the factors, regulatory, economic, and others, that have encouraged or inhibited these 29 plants, and, potentially other plants, in choosing waste reduction as a waste management option?

4. How complete a picture do regulatory agencies, and the plants themselves, really have of waste generation, waste disposal and waste reduction in this critical industry?

FINDINGS ON HAZARDOUS WASTE GENERATION

Despite the common assumption that solid waste generation far exceeds discharges to the air and water, hazardous chemical wastes were found to be generated in roughly equal amounts as air emissions, wastewater discharges, and as solid wastes. Based on INFORM's analysis of the quantities of wastes generated from the use of eight key chemicals at six plants in New Jersey (the only state that collected information that allowed a comparison between plants' air, water and solid wastes), wastewater discharges accounted for 42 percent; solid wastes, 35 percent; and air emissions, 23 percent.

The 29 plants generated hazardous chemical wastes in quantities ranging from only 55 pounds to millions of pounds per year. Du Pont's Deepwater, New Jersey, plant generates more hazardous waste overall than any other plant in this study (and perhaps more than any plant in the nation). Its solid wastes total 80 billion pounds a year. It is also the largest generator among INFORM's study plants of hazardous chemical air emissions -- 2.6 million pounds of methyl chloride and one million pounds of trichlorofluoromethane. USS Chemicals' Haverhill, Ohio, and Rhone-Poulenc's New Brunswick, New Jersey, plants dispose of the largest amount of hazardous chemicals in wastewater: each discharges 600,000 pounds per year of phenol. At the other end of the waste generation spectrum are some plants that produce little or no hazardous wastes. Scher's Clifton, New Jersey, plant loses a total of

55 pounds per year of hazardous chemicals in air and wastewater discharges. This plant, along with six others operated by Bonneau Dye (Avon, Ohio), Smith and Wesson (Rock Creek, Ohio), Halma (Lodi, New Jersey), Max Marx (Irvington, New Jersey), Fibrec (San Francisco, California) and Colloids (Richmond, California) had no reported solid wastes.

Plants using similar quantites of hazardous chemicals were found to generate greatly differing quantities of wastes. For example, plants operated by Exxon in Linden, New Jersey, and Rhone-Poulenc in New Brunswick, New Jersey, both handle large quantities of phenol. Exxon uses 5.7 million pounds per year and loses 4,430 pounds of the chemical (0.08 percent) as wastes, while Rhone-Poulenc, using 3.5 million pounds, generates a far greater quantity of wastes -- over 600,000 pounds (17 per cent). Similarly, Merck's Rahway, New Jersey, plant and Atlantic's plant in Nutley, New Jersey, each use 2,000 pounds per year of formaldehyde; Merck loses all of it as wastes while Atlantic loses only 11 pounds. Different processes and uses of chemicals at each plant account for some of the observed differences in quantities of waste generated.

FINDINGS ON WASTE REDUCTION PRACTICES

Twelve of the 29 (41 percent) organic chemical plants studied reported a total of 44 waste reduction practices in five categories. Despite a common assumption that waste generation stems chiefly from the manufacturing process itself, and that waste reduction must necessarily be directed at altering that process, INFORM found widespread potential for reducing wastes originating from all stages of the handling of hazardous chemicals. Wastes were reduced from storage, loading, transfer, and pollution control operations as well as in the manufacturing process itself. The five categories of waste reduction identified at the study plants are: process changes, operational changes, equipment changes, chemical substitutions and product reformulations, with some waste reduction practices belonging to more than one category.

Eighteen of the 44 (41 percent) waste reduction practices involved basic changes to the manufacturing process. Some representative examples are: new processes which eliminated the use of mercury and reduced chromium waste generation at CIBA-GEIGY's plant in Toms River, New Jersey; manufacturing processes at the USS Chemicals' plant in Haverhill, Ohio, producing phenol and aniline that both reduced hazardous wastes and upgraded a by-product (diphenylamine) to commercial grade so it could be sold rather than disposed of as waste. Since 1952 Monsanto's plant in Addyston, Ohio, has gradually changed its polystyrene process from one of batch reactions to a closed-system continuous reaction and achieved a 99 percent reduction in air emissions.

Fourteen waste reduction practices at six plants involved operational changes, often overlooked in considerations of ways to reduce waste, with some reductions of more than 90 percent. Simple changes in the ways materials are handled can achieve large reductions in waste generation. Borden, for example, saved rinsewaters from tank and filter washings at its Fremont, California, plant for reuse as raw materials, and educated employees on ways to minimize chemical losses. These initiatives resulted in a 93 percent reduction in solid waste generation from the plant's phenol operations and eliminated reliance on an evaporation pond as a means of treating and storing wastes.

Nine equipment changes were used at five plants to reduce wastes in both process and storage operations. To recover air emissions for use as a raw material, Perstorp installed a baghouse filter on the pentaerythritol process unit at its Toledo, Ohio, plant and USS Chemicals installed a cumene resin adsorption system and a condenser in its Haverhill, Ohio, plant. Both Exxon, in Linden, New Jersey, and USS Chemicals, in Haverhill, Ohio, installed floating roofs on their storage tanks for more than a 90 percent reduction in air emissions from the tanks. The Dow plant in Pittsburg, California, has replaced the use of nitrogen gas to push chemicals from their storage tanks to the reactor vessels by a pumping mechanism which eliminates the loss of the chemicals which became mixed with the gas while being transferred.

Only two examples of chemical substitutions and three of product reformulations were reported. Union Chemicals' La Mirada, California, plant was the only study plant to substitute a non-hazardous additive chemical for one containing the hazardous metal, mercury. CIBA-GEIGY, at Toms River, New Jersey, also achieved waste reduction by using a higher quality lime in its wastewater treatment plant to reduce sludge generation. The three instances of product reformulation were: Monsanto's Addyston, Ohio, plant and Stauffer's Richmond, California, plant each modified one of their products to minimize the generation of solid wastes. USS Chemicals (Haverhill, OH) was able to upgrade the quality of a hazardous waste by-product, diphenylamine, and market it as a commercial product instead.

The specific waste reduction practices reported were different at each plant. Only two of the study plants adopted the same waste reduction method: Exxon and USS Chemicals both use floating roofs. The diversity of the methods found within the five categories reflect the diversity of the organic chemical industry's operations, products and processes, and suggests the need to explore waste reduction opportunities carefully for each plant. For example, batch-processing plants more frequently reduced wastes through operational changes than continuous-processing plants, which found more often that wastes could be reduced through equipment changes. For some products, dyes at Atlantic and pharmaceuticals at Merck for example, the need for a high degree of purity limits the plants' ability to reuse solvents or rinsewaters and they explored other avenues of waste reduction.

FINDINGS ON IMPACTS OF WASTE REDUCTION PRACTICES

A total annual reduction in waste generation of seven million pounds was achieved at four plants reporting "pounds per year" savings. Five plants reporting economic benefits annually saved a total of $800,000. Plants operated by Merck in Rahway, New Jersey, and USS Chemicals in Haverhill, Ohio, reported the largest reductions in waste generation -- more than three million pounds per year at each plant. Three plants -- USS Chemicals (Haverhill, Ohio), Exxon (Linden, New Jersey) and Stauffer (Richmond, California) -- reported annual cost savings of more than $200,000.

An absence of detailed information made it impossible for INFORM to fully evaluate the impact of the 44 waste reduction practices. Companies reported material savings on a "pounds per year" basis for only eight of the 44 practices and annual cost savings for only nine. Actual material and cost savings were, therefore, considerably higher than the totals given here.

The millions of pounds of waste reduction reported by the study plants is only a minute fraction of the total wastes generated by them. Although total waste reduction can be measured in the millions of pounds, total hazardous waste generation at the 29 plants amounts to billions of pounds per year. The largest reported waste reduction figures of two to three million pounds per year are dwarfed by the largest reported wastestreams: 51.9 million pounds of carbon tetrachloride wastes at Du Pont's Deepwater, New Jersey plant; 17.5 million pounds of chromium wastes at International Flavors and Fragrances' plant in Union Beach, New Jersey; and 15.7 million pounds of phenol wastes at USS Chemicals' Haverhill, Ohio, plant.

WHY PLANTS DO OR DO NOT INITIATE WASTE REDUCTION

Only one regulation directly required waste reduction measures at any INFORM study plant. New Jersey's air pollution regulations for controlling organic chemical emissions required the Exxon plant in Linden to install floating roofs and conservation vents, both of which reduce wastes at their source. No other plants in INFORM's study were directly required by regulations to adopt waste reduction measures.

Cost factors, liability concerns, public scrutiny and the indirect impact of regulations, rather than plant characteristics such as size, age or location, were the dominant influences on whether or not a plant adopted waste reduction practices. Most regulations, rather than requiring waste reduction, simply restrict disposal alternatives or make them more costly. Limitations imposed by sewage treatment plants on Borden, in Fremont, California, and Sherwin-Williams in Cincinnati, Ohio, were instrumental in spurring those plants to adopt waste reduction.

Savings in material costs and in the costs of treatment and disposal were frequently cited as incentives for waste reduction. As the cost of cumene, a major raw material at USS Chemicals' Haverhill, Ohio, plant increased by six times, the plant found ways to reduce its cumene air emissions by more than a million pounds per year. By replacing leaking pump seals, Stauffer in Richmond, California, prevents the loss of $37,000 per year of raw material costs. As landfilling costs for Borden increased from $50 to $150 per cubic yard, the company's Fremont, California, plant found ways to reduce its solid waste generation from its resin operations from 350 to 25 cubic yards per year.

Regulatory exemptions and loopholes, the availability of low-cost disposal alternatives, and sporadic regulatory oversight all discourage the search for and adoption of waste reduction practices. Many waste management options are in use which are low-cost, convenient and not stringently regulated, providing attractive means for disposing of, rather than reducing, wastes. Some examples: Wastes can be burned as a supplemental fuel with fewer restrictions than if they were incinerated only to be disposed of as wastes. Deep well injection of hazardous waste is virtually unrestricted in terms of the types or quantities of chemicals that can be discharged, and is one of the cheapest means of waste disposal. Individual chemical plants can emit hundreds of thousands of pounds of hazardous organic chemicals to the air without violating federal air pollution regulations that regulate only six chemicals as hazardous. None of the 17 study plants (eight of which cooperated in this study) discharging to sewage treatment plants reported that fees charged by the sewage treatment plants were large enough to induce them to search for ways to reduce their discharges of wastewater. Sporadic regulatory oversight by federal and state agencies focuses attention on specific chemicals at select plants while virtually ignoring others.

Operating constraints were cited by six study plants as inhibiting waste reduction. Atlantic Industries' Nutley, New Jersey, plant reported that, because of its fixed costs for air pollution control and wastewater monitoring, potential savings from waste reduction are so small they inhibit the search for additional waste reduction measures. Several batch-manufacturing plants, including Atlantic, Merck (Rahway, New Jersey) and CIBA-GEIGY (Toms River, New Jersey), reported that while some waste reduction practices had been implemented, the highly variable nature of their operations, along with a need to maintain stringent quality control, minimized the opportunities for waste reduction.

Despite universal affirmation of waste reduction as the most desirable waste management option, it was implemented in nine practices at four study plants only after regulatory and/or operational pressures forced management to focus on waste reduction opportunities. For five of the nine cases, once waste reduction was adopted it was found to be the most economically advantageous choice for waste management -- an indication of how waste reduction can be the option of last, rather than first, resort. Exxon's plant in Linden, New Jersey, did not use floating roofs until they were required by regulations to do so, and Borden, in Fremont, California, did not adopt four waste reduction practices to minimize resin wastes until regulatory restrictions and problems with the plant's evaporation pond forced them to consider waste reduction alternatives. Both plants, however, found waste reduction to be cost-effective once in practice.

Four other practices -- two at Exxon's plant in Linden, New Jersey, and one each at CIBA-GEIGY's plant in Toms River, New Jersey and Sherwin-Williams' plant in Cincinnati, Ohio -- were spurred by regulatory and operational pressures; these study plants, however, provided no cost savings data on their impact.

ADEQUACY OF INFORMATION

Sixteen of the 29 study plants (55 percent) provided no information on hazardous waste reduction to INFORM. Despite a critical concern nationally for the problems of hazardous chemical wastes and the potential for waste reduction, the majority of the companies in this study would not provide information on their operations or waste reduction practices. Of the 20 large plants in the study, nine did not cooperate, including the largest chemical company in the country, Du Pont, as well as Chevron, Shell, and American Cyanamid. Of the nine small plants, cooperation was even more limited. Only two small plants -- Perstorp (Toledo, Ohio) and Scher (Clifton, New Jersey) -- provided information.

The 13 companies that cooperated in the study provided only limited information on waste reduction. Companies that cooperated were able for the most part to describe waste reduction practices and why they had been implemented, but the level of detail was limited either because the impacts of waste reduction had not been measured, or because the company considered such information confidential. None of the companies was able to supply anything like complete information either on the total wastes generated or on the impact of waste reduction on total plant-wide wastes, because most companies, spurred chiefly by regulatory reporting requirements, collect information on broad and ill-defined categories of waste from only limited aspects of their operations.

Government files contain information that is much too fragmented and incomplete to make possible any assessment of how much waste reduction was accomplished at the study plants. Government files had almost no detailed references to waste reduction and, indeed, provided no benchmark figures needed to assess progress in minimizing wastes, that is, up-to-date information on the amounts of specific hazardous chemicals used by plants and the amounts appearing as air, water or solid wastes.

One data file came close to providing the needed information: the "Industrial Survey" conducted by New Jersey in 1978. The survey collected information on the production and use of 155 hazardous chemicals and the generation of all wastes from the use of the chemicals. The data is chemical-specific and was collected from over 7,000 plants in the state so that comparisons of the fate of chemicals within and between plants was possible. But because this survey was conducted for one year only, this information base could not be used to document trends in either hazardous waste generation or in the extent and impact of hazardous waste reduction.

Chapter 4

The INFORM Study Plants

The 29 plants chosen by INFORM for study are listed in Table 4-1. The three states in which they are located -- California, New Jersey and Ohio -- were chosen both because they are among those having the largest concentration of organic chemical plants and because they differ in their types of industrial activity and the environmental regulatory frameworks of their state governments.

Differences in Three States

These three states contain 23 percent of all U.S. organic chemical plants and are ranked among the first five in the U.S. with respect to the number of such plants. The states vary in their overall manufacturing base and in the composition of their share of the organic chemical industry (see Table 4-2). In New Jersey, the manufacture of organic chemicals is a dominant industry; it accounts for three percent of total manufacturing activity in New Jersey, five times the relative contribution that the industry makes in Ohio and ten times that in California, both of which have more diversified manufacturing bases.

The three states differ in the types of chemicals produced. New Jersey's chemical manufacturing stresses end-use consumer chemicals -- the state is the leading U.S. producer of pharmaceuticals, detergents and toiletries -- as well as organic chemicals for U.S. industrial use. California is the leading producer

Table 4-1. INFORM's Study Plants

California

	PLANT NAME AND CITY	PARENT COMPANY (if different)
1.	Borden Chemical Company Fremont	Borden, Inc.
2.	Chevron Chemical Company Richmond	Standard Oil of California
3.	Colloids of California Richmond	Colloids, Inc.
4.	Dow Chemical U.S.A. Pittsburg	
5.	Fibrec, Inc. San Francisco	
6.	Polyvinyl Chemical Industries Vallejo	Beatrice Foods Company
7.	Shell Chemical Division Martinez	Shell Oil Company
8.	Stauffer Chemical Company Richmond	
9.	Union Chemicals Division La Mirada	Union Oil Company of California

Table 4-1 (con't). INFORM's Study Plants

New Jersey

PLANT NAME AND CITY	PARENT COMPANY (if different)
1. Atlantic Industries Nutley	
2. CIBA-GEIGY Corporation Toms River	
3. E.I. Du Pont de Nemours & Co., Inc. Deepwater	
4. Exxon Chemical Americas Linden	Exxon Corp.
5. Fisher Scientific Company Fair Lawn	Allied Corp.
6. J.E. Halma Company, Inc. Lodi	
7. International Flavors and Fragrances Inc. Union Beach	
8. Max Marx Color and Chemical Company Irvington	
9. Merck and Company Rahway	
10. Rhone-Poulenc Inc. New Brunswick	
11. Scher Chemicals, Inc. Clifton	

Table 4-1 (con't). INFORM's Study Plants

Ohio

	PLANT NAME AND CITY	PARENT COMPANY (if different)
1.	American Cyanamid Company Marietta	
2.	Bonneau Dye Corporation Avon	
3.	Carstab Division Cincinnati	Morton Thiokol, Inc.
4.	Frank Enterprises, Inc. Columbus	
5.	Monsanto Company Addyston	
6.	Perstorp Polyols, Inc. Toledo	
7.	The Sherwin-Williams Company Cincinnati	
8.	Smith and Wesson Chemical Company, Inc. Rock Creek	Lear Siegler, Inc.
9.	USS Chemicals Ironton	United States Steel Corporation

Table 4-2. Characteristics of the Organic Chemical Industry in INFORM's Study States and the U.S.

	United States	California	New Jersey	Ohio
Organic chemical industry as percent of total manufacturing activity *	1.5%	0.3%	3.1%	0.6%
No. of employees in organic chemical plants *	143,000	2,850	23,700	4,650
No. of employees per organic chemical plant *	148	48	217	91
Percent of large chemical plants built since 1970 **	15%	38%	19%	36%

* From "1982 Census of Manufacturers," Department of Commerce, Washington, DC, January 1985.

** From "Waste Disposal Site Survey," Subcommittee on Oversight and Investigations, Committee on Interstate and Foreign Commerce, House of Representatives, Washington, DC, October 1979.

of chemicals for paints, and is also a major producer of pharmaceuticals and agricultural chemicals. Ohio's organic chemical industry is diversified, contributing both basic industrial chemicals and consumer-related items.

California and Ohio have relatively newer chemical industries than New Jersey's. A Congressional survey of the largest chemical plants in the nation found that 38 percent of them in California and 36 percent in Ohio had been built since 1970, about twice the percentage of newer plants built in New Jersey or in the U.S. as a whole (see Table 4-2). The plants differ in the number of employees as well. New Jersey's organic chemical industry, whose older plants and consumer-related production tend to be labor-intensive, averages 217 employees per plant, as opposed to 91 per plant in Ohio and only 48 in California.

Regulatory Programs in the Three States

Of INFORM's three study states, and in the U.S. generally, California is commonly seen as taking a leadership role in environmental management. Its programs have served as models for national approaches to regulation, such as in the 1984 amendments to RCRA that incorporate the California definition of hazardous wastes which is broader and more stringent than the original federal definition.

New Jersey has launched vigorous programs of clean-up, regulation and enforcement that are significantly more stringent than federal requirements. New Jersey, for example, regulates eleven more hazardous air pollutants than does the federal EPA under the Clean Air Act. New Jersey and California were both among the states establishing a more stringent exemption for small quantity generators under RCRA, at less than 100 kilograms per month. Until the 1984 amendments to RCRA, the federal cut-off was 1,000 kilograms per month. In 1984, the federal law was changed to follow the example of these states.

Ohio has largely adopted the federal framework of environmental regulation without significant changes. However, state officials are able to (and often do) impose additional restrictions on hazardous waste generators in the process of permit writing and regulatory oversight.

Criteria for Choosing the Study Plants

The objective of INFORM's study was to assess the type and extent of hazardous waste reduction initiatives taken by the organic chemical industry as reflected in the practices of one group of facilities. These facilities were chosen to represent the diversity in size, products and ownership that is typical of the industry. Reputation with regard to environmental performance, good or bad, was not a factor.

The first criteria in selecting plants to be included in the study was that they must manufacture organic chemicals. INFORM's plant selection process was begun by compiling a master list of organic chemical plants in each study state. Given the absence of any single list for all organic chemical plants in any of the three states, INFORM's master list was compiled from two kinds of sources: environmental files maintained by federal and state government agencies, and commercial industrial directories (both state and national).

Included on INFORM's master list were plants (manufacturing facilities, as distinct from sales offices, company headquarters or warehouses) identified in at least one information source as belonging to Category SIC 286 of Industrial Organic Chemicals established by the U.S. Department of Commerce's "Standard Industrial Classification" code system. This category is one of the following eight main categories included under SIC 28 -- Chemicals and Allied Products:

28 Chemicals and Allied Products

  281 Industrial inorganic chemicals
  282 Plastic materials and synthetics
  283 Drugs
  284 Soaps, cleaners and toilet goods
  285 Paints and allied products
  <u>286 Industrial Organic Chemicals</u>
  287 Agricultural chemicals
  289 Miscellaneous chemical products

The plants whose operations were found to be mainly organic chemical manufacturing were next categorized by INFORM based on size of plant and types of products. Where plants were similar in these categories, ownership factors were checked. Table 4-3 summarizes the characteristics of the 29 study plants.

<u>Size of Plant</u>

In choosing plants for study, INFORM used the number of employees as the initial measurement of size. In general, figures on annual sales or volume of production for individual plants are not publicly available and the only data on size available to INFORM at the initial selection stage were the numbers of employees. Sixty percent of all U.S. organic chemical plants have fewer than 50 employees. INFORM chose 11 plants in this category. The smallest plant in the study is the Bonneau Dye Corporation in Avon, Ohio, which manufactures custom blended dyes and has two employees.

However, plants with 50 or more employees are responsible for 94 percent of annual sales in the U.S. industry as a whole. INFORM, therefore, chose a larger relative number of plants with employees of 50 or more -- a total of 18 -- to reflect the overall distribution of industry sales. The three largest plants were Du Pont's Chambers Works plant in Deepwater, New Jersey, by far the largest with 4,100 employees, CIBA-GEIGY's plant in Toms River, New Jersey, with 1,050 employees, and Monsanto's Port Plastics plant in Addyston, Ohio, with 800.

Table 4-3. Characteristics of INFORM's Study Plants

INFORM Study Plant	Year Plant Built (Bought)	Types of Products	Number of Employees	SIZE Sales ($)* T=thousands M=millions B=billions Division	Number of Plants in Division	Rank of Parent Company**	OWNERSHIP Parent Company (if not chemical)	Nationality	Public or Private
CALIFORNIA									
Borden	1959	resins, adhesives	50	680M (Div'82)	74	26	dairy		public
Chevron	?	agricultural chemicals	755	934M (Div'83)	16	34	petroleum		public
Colloids	1967	antifoam, dispersants	5	37T (plant)	4				private
Dow	1916	chlorinated organics	650	100M (plant)	32	2			public
Fibrec	1969	dyestuffs distributor	3	250T (plant)	1				private
Polyvinyl	1970(1982)	resins, coatings	18	1-5M (plant)	2				public
Shell	1931	metallic catalysts	59	3.2B (Div'83)	9	6	petroleum	Dutch/Eng	public
Stauffer	Early 1900s	pesticides	500	1.4B (Div'82)	66	22			public
Union	1949(1969)	polymers, solvents	100	1.1B (Div'83)	6	30	petroleum		public
NEW JERSEY									
Atlantic	1939	dyes	200	10M (Div.)	5				private
CIBA-GEIGY	1952(1981)	dyes, epoxy resins	1050	90M (plant'83)	?	20		Swiss	private
Du Pont	1917	additives, Freon, aromatics	4100	15.3B (Div'83)	41(major)	1			public
Exxon	1921	lubricants, solvents	550	2.1B (Div'79)	6	3	petroleum		public
Fisher	1955(1981)	reagent chemicals	325	1.7B (Div'83)	1	13			public
Halma	?	solvents, etchants, acids	7	1M (plant)	1				private
IFF	1952	perfume & fragrance chemicals	375	461M (Div'83)	5		food additives		public
Max Marx	1908	organic pigments	25	?	1				private
Merck	1903	pharmaceuticals, pesticides	500	3.6B (Div'84)	6		pharmaceuticals		public
Rhone-Poulenc	1949(1982)	aroma chemicals, pharmaceuticals	140	300M (Div'81)	11		national gov.	French	gov.owned
Scher	1956	additives, specialty chemicals	31	1-5M (plant)	1				private
OHIO									
Amer. Cyanamid	1915	dyes, additives, explosives	128	973M (Div'83)	36	17			public
Bonneau	?	custom blend dyes, candles	2	?	1				private
Carstab	1949(1982)	additives	250	478M (Div'83)	23				public
Frank	1970	specialty chemicals	5	500T (plant)	2				private
Monsanto	1952	plastics, adhesives, resins	800	1.8B (Div'83)	9(major)	4			public
Perstorp	1971(1977)	pentaerythritol, sodium formate	37	346M (Div'83)	4			Swedish	public
Sherwin-Williams	1966	saccharin, flavors, additives	200	94M (Div'83)	19		paint		public
Smith & Wesson	1968(1984)	chemical weapons	100	1-5M (plant)	?		firearms		private
USS Chemicals	1962(1965)	organic intermediates	224	715M (Div'83)	?	28	steel		public

? = information not available. * Sales figures given for plant when available or for Division to which plant belongs. If date is not given, then sales figure is an annual estimate. ** Rank among 50 largest chemical companies in terms of sales, for 1984 (from "Chemical and Engineering News," June 10, 1985, p. 34).

Types of Products

INFORM's 29 plants reflect the great diversity of products manufactured by the organic chemical industry. The products most frequently manufactured are dyes, resins and additives. The plants also produce intermediates, solvents, specialty chemicals, fragrances, agricultural chemicals, pharmaceuticals, adhesives, plastics, explosives, fibers, and metallic catalysts.

All of the plants produce more than one type of chemical: Some produce hundreds of different products each year. Du Pont's massive Chambers Works plant produces 700 chemical products, only slightly more than the number of dye products that can be manufactured at the much smaller Atlantic Industries plant. Perstorp's small plant produces two chemical products while the much larger USS Chemicals' facility produces only six.

All but one of the 29 plants manufacture the chemicals they sell. The Fibrec, Inc. plant in San Francisco, California, repackages bulk dyestuffs for distribution but does not manufacture them. One plant, Fisher Scientific in Fair Lawn, New Jersey, reformulates a wide variety of laboratory reagent chemicals, a few of which are manufactured on-site.

Plant Ownership

INFORM's study sample includes plants belonging to the four largest chemical companies in the U.S. (Du Pont, Dow, Exxon and Monsanto). While some of the plants are owned by small one-plant firms such as J.E. Halma or Bonneau Dye, others are owned by companies having 50 or more plants that make up a chemical division within a larger company. Borden Chemical, for example, has 75 plants throughout the U.S. and is a division of Borden, Inc., a large international corporation best known for its dairy products.

Twelve of the study plants are owned by companies that are subsidiaries of parent companies whose chief operations are outside of the organic chemical industry (petroleum, food, pharmaceuticals, paint, firearms and steel). Four belong to parent companies that are based abroad. Nine plants are privately-owned and one (Rhone-Poulenc) is owned by the French government. The remaining 19 are publicly-owned.

Table 4-4 displays the final distribution of INFORM's study plants according to the selection categories of size, product and ownership.

Problems in Obtaining Information for the Study

The four basic questions INFORM had on chemical use at the 29 study plants were:

1. What types and quantities of hazardous chemicals are in use at each facility?

2. Where, how and in what amounts are these chemicals released to the environment or otherwise disposed of?

3. What specific steps are being taken to reduce wastes and what actual impacts are they having on total plant wastes?

4. What factors are stimulating or impeding waste reduction efforts?

Master List

There was, however, a substantial degree of difficulty in collecting and organizing the information needed to answer them. (See Appendix B for a description of the primary data sources used.)

The first problem involved compilation of the master list of organic chemical plants. For any given plant several SIC codes may apply, depending on the variety

Table 4-4. Distribution of Study Plant Characteristics

GEOGRAPHIC LOCATION

	# of Plants in Study	# of Plants in SIC 286 Identified by INFORM	% of Plants in Study
California	9 plants	84	10.7
New Jersey	11 plants	181	6.1
Ohio	9 plants	95	9.5

SIZE OF PLANT

# of employees	% of INFORM Study Plants	% of U.S. Organic Chemical Plants*	% of Annual Sales of Organic Chemicals in U.S.*
less than 20	21	43	2
20-49	10	17	4
50-99	10	13	6
more than 100	59	27	88

TYPE OF PRODUCT

Products	# of INFORM Study Plants**	Products	# of INFORM Study Plants**
Dyes & pigments	6	Agricultural chemicals	2
Resins	5	Pharmaceuticals	2
Additives	5	Adhesives	2
Intermediates	3	Plastics	2
Solvents	3	Weapons, explosives	2
Specialty chemicals	3	Fibers	1
Fragrances	3	Metallic catalysts	1

OWNERSHIP

Parent Co. Is:	# of INFORM Study Plants	Parent Co. Is:	# of INFORM Study Plants**
Chemical	17	Foreign owned	4
Petroleum	4		
Food	3	Publicly owned	19
Pharmaceuticals	1	Privately owned	10
Paint	1		
Firearms	1		
Steel	1		
National government	1		

* From "1982 Census of Manufacturers," Department of Commerce, Washington, DC (annual sales based on value of shipments).

** These categories do not add up to 29 plants because one plant can produce more than one product and can represent more than one ownership category.

of products manufactured. Also, many information sources do not distinguish between manufacturing plants and headquarters offices, sales offices or warehouses. Since many references are updated infrequently, verification that facilities had not been sold or closed down required cross-checking of lists and, in many cases, calls directly to the facilities.

Government Files

The next type of information sought was data from government files on the use of hazardous chemicals and the types of hazardous wastes. The wide variety of regulatory files on hazardous chemicals and the specialized data they contain created several problems.

Government agencies at the federal, state and local levels all have individual systems for information collection and management, and their own language for referring to hazardous chemical wastes. As many as 30 files may exist for one plant. Some of these contain figures on the concentrations of a chemical in, for example, wastewater, while others reflect total quantities of the chemical discharged. Some data represent actual emissions, others are estimates and still others are maximum permissible amounts according to permit specifications. The information in some files detailed the hazardous component of wastes, in others, the overall quantity of waste material regardless of the amount of hazardous component, was given. Because of these various approaches to measurement there is no way to get an accurate picture of the total wastes generated by each plant. Nor can such a picture be drawn for specific chemicals handled at the plant and discharged as waste.

Government air permit files were the only files to relate waste generation (in the form of air emissions) specifically to the source of waste and to the steps taken (if any) to control emissions. Regulatory programs controlling solid wastes and wastewater discharges are not designed to collect information about the actual sources of waste within a plant -- the focus of these programs is to insure that the aggregate wastes from the individual sources within a plant are properly treated and disposed of as they leave the plant site.

Without specific information on individual sources, the impact of any waste reduction practices cannot be assessed. In addition, no government file makes any comparison between production levels and the quantities of waste a plant produces. It is impossible to know, therefore, whether increases or decreases in the quantities of wastes generated over a period of time at a plant stem from a change in waste management practices or simply from a production increase or decrease.

At a logistical level, the various agency files were so geographically and administratively scattered that a thorough review of them for a single plant took weeks of effort. Even then, access to information in government files was not always guaranteed. Federal EPA files generally classify information as confidential and non-confidential, and access to the latter was fairly easy. This distinction between confidential and non-confidential information was less clear, however, at the state and local agencies and access to files was sometimes delayed. For example, it took two months to gain permission to review air permit files in New Jersey. In California, the South Coast Air Quality Management District required INFORM to obtain permission from Union Chemicals for review of that company's files. When asked, the company did give blanket permission for the review of its air quality files.

Plant Interviews

Not all the study companies were so helpful, however. Only 13 out of the 29 study plants consented to interviews regarding their waste management practices (see Table 4-5). Du Pont, the largest chemical company in the U.S., generates more hazardous waste than any other plant in this study (and perhaps more than any other in the U.S.). It was, however, among those who did not give INFORM any information.

Table 4-5. Cooperation of Plants
in the Study

INFORM Study Plant	Cooperated Study	Did Not Cooperate in Study
CALIFORNIA		
Borden	X	
Chevron		X
Colloids		X
Dow	X	
Fibrec		X
Polyvinyl		X
Shell		X
Stauffer	X	
Union	X	
NEW JERSEY		
Atlantic	X	
CIBA-GEIGY	X	
Du Pont		X
Exxon	X	
Fisher	X	
Halma		X
IFF		X
Max Marx		X
Merck	X	
Rhone-Poulenc		X
Scher	X	
OHIO		
American Cyanamid		X
Bonneau		X
Carstab		X
Frank		X
Monsanto	X	
Perstorp	X	
Sherwin-Williams		X
Smith & Wesson		X
USS Chemicals	X	
TOTALS	13	16

The companies that did grant interviews were frequently unable to answer INFORM's questions. Plants do not routinely compute waste generation totals, nor are the consequent costs of waste management routinely recorded for individual processes or wastestreams. For example, Dow has a process accounting system at its Pittsburg, California, plant, but it does not fully document the costs of waste management for individual processes and cannot assess the economic impacts of specific waste reduction practices. Fisher Scientific and Merck told INFORM that more detailed accounting of their wastestreams was planned for their plants but had not yet been instituted. Generally, plants attributed their lack of detailed information about their wastestreams to the fact that government regulatory agencies did not require such information. Plants tend to collect only such information on their chemical wastes as is required by government forms.

Company concern about the confidentiality of product and process information was another factor that inhibited the plants' responses. Stauffer considered virtually all specifics relating to chemicals, processes, and procedures at its plant to be confidential. Perstorp readily agreed to meet with INFORM and provide information about the plant's history in Toledo and the overall impact of its waste reduction practices. It was considerably less willing, however, to provide details about how waste reduction actually came about at the plant. Dow cloaked the identity of a chemical when providing information on its reduction of air emissions, identifying the substance only by its legal classification in California as an "extremely hazardous waste."

A compounding factor throughout the study was the lack of understanding of what is meant by the term "waste reduction." A common response to INFORM's request for data on waste reduction was instead information on waste treatment systems.

Chevron's response to INFORM's request for information for this study is indicative of the problems encountered. Chevron plant officials declined to be interviewed.

When asked by INFORM to review their plant profile, which INFORM had developed from government sources and which stated that no indications of waste reduction had been found, Chevron provided some information on waste treatment at its plant including incineration and the use of evaporation ponds. Chevron did not provide any information on waste reduction, although it told INFORM that its reduction practices were documented in government files. When INFORM went back to the government agencies for further clarification, it was told again that they had no information regarding waste reduction practices at Chevron. One agency, the California Water Quality Control Board for the San Francisco Bay region, explained further that its files did contain descriptions of some waste reduction practices at Chevron's Richmond plant that Chevron had requested not be made available for public review, because its techniques were "trade secrets."

Chapter 5

Waste Generation, Disposal and Treatment at the Study Plants

While figures abound -- many were cited in Chapter 2 of this study -- which suggest the overall magnitude of hazardous wastes contributed by the organic chemical industry in the U.S., INFORM's research at the 29 plants explored the extent to which hazardous wastes for individual plants could actually be defined.

Such plant-specific waste information is clearly essential for forming an accurate picture of national or regional waste problems. However, in the context of this study, this information is a crucial tool to identify hazardous waste reduction opportunities for each plant, to evaluate the potential of waste reduction activities and to assess the results of individual waste reduction initiatives.

This chapter discusses what is and is not known about the type and amount of hazardous wastes generated at each of INFORM's 29 study plants. It also describes the waste treatment and disposal options employed by each plant. The emerging picture suggests the continuing overwhelming reliance by chemical plants on these alternatives as compared to their use of hazardous waste reduction practices.

All of the facts that INFORM amassed on the specific chemicals used at each of its 29 study plants, on wastes generated and on waste disposal and reduction methods implemented, are presented in detailed charts that follow the profiles of each plant in Part II.

Waste Generation

Enough data is available on plant wastes to conclude that the scale of waste materials ranged widely from millions of pounds of chemicals -- in individual and combined wastestreams -- down to a dozen pounds or less. However, for none of INFORM's 29 plants is data adequate to create a complete picture of the chemicals used or the amount and type of wastes disposed of to air, land, and water resources.

For nine of INFORM's study plants located in New Jersey, such comprehensive data was compiled for the year 1978 under provisions of the State's unique Industrial Survey. This survey collected production use and wastestream data on 155 chemicals at over 7,000 industrial plants in New Jersey and provides the clearest picture of the fate of specific chemicals within a plant.

Beyond this survey, the only data available for the 29 study plants in California, Ohio and New Jersey, came from government files that are oriented by regulatory programs for air, water, and solid wastes. The data, in general, cannot be compared between files because regulatory programs do not collect information uniformly. For instance, wastes reported as "volatile organic chemicals" under clean air programs cannot be compared with "solvent wastes" reported under solid waste programs because each category represents a composite of many different chemicals, some hazardous, some not.

While no overview of waste generation is possible for INFORM's study plants, a number of important points, that are relevant to considering waste reduction options, can be made about the nature of wastes generated at the plants.

Wastes Enter All Media

Existing data show clearly that hazardous wastes are being generated and discharged in large quantities to all three environmental media -- air, land and water. The figures on eight important chemicals, set out in Table 5-1, reveal the interesting contrast between plants. Overall, more than 10 percent of the total amount of the chemicals handled at the plants ends up as wastes. Of the total wastes for the eight common chemicals shown on this table, 23 percent are in the form of air emissions and 42 percent are wastewater discharges, with 35 percent as solid wastes.

The conventional use of the term "hazardous waste" is limited to solid wastes. Solid waste material generated by chemical plants is perhaps the most visible and tangible form of hazardous waste. Leaking drums at landfills across the country and government focus on Superfund sites have attracted the most public attention. However, as Table 5-1 shows, the wastes disposed of in the atmosphere and as wastewaters are of the same magnitude as the solid wastes for these plants. Thus, in this report, all types of hazardous wastes are considered equally when reviewing waste reduction opportunities at the study plants.

Wastes as a Percent of Chemicals Used at Plants Vary Greatly

The amount of wastes generated from the use of given chemicals in similar amounts varies quite dramatically, depending on the ways in which the chemicals are used at the plants. For example, two of the plants that use phenol reported losing less than 0.1 percent of the chemical as waste while two reported losing from seven to 17 percent. Exxon handles 5.7 million pounds per year of phenol, almost 60 percent more than Rhone-Poulenc, and yet Rhone-Poulenc generates more than 100 times as much phenol wastes. Also, the Du Pont plant disposes of most of its phenol wastes as solid wastes (over a million pounds per year) while

Table 5-1. Use and Waste Generation of Selected Hazardous Chemicals at INFORM Study Plants in New Jersey

Chemical/ INFORM Study Plant	Quantity Handled (lbs/yr)	Waste Generation (lbs/yr)				Total Wastes as Percentage of Total Handled
		Air	Water	Solid Waste	Total	
ANILINE						
Atlantic	224,000	110	240	0	350	0.16
Du Pont	confidential	0	550	865,200	865,750	?
Max Marx	42,000	0	0	0	0	0
Merck	850,000	0	0	0	0	0
CHROMIUM						
Atlantic	5,700	35	700	300	1,035	18.2
CIBA-GEIGY	35,461	0	2,946	12,460	15,406	43.4
Du Pont	confidential	0	2,000	47,580	49,580	?
COPPER						
Atlantic	12,600	0	1,000	0	1,000	7.9
CIBA-GEIGY	86,416	0	4,817	32,194	37,011	42.8
Du Pont	confidential	0	2,900	27,600	30,500	?
FORMALDEHYDE						
Atlantic	2,200	1	10	0	11	0.5
CIBA-GEIGY	757,100	161	0	0	161	0.02
Du Pont	confidential	19,350	0	0	19,350	?
IFF	confidential	0	0	0	0	0
Merck	2,000	75	1,925	0	2,000	100
Rhone-Poulenc	372,400	3	0	0	3	0.001
METHYLENE CHLORIDE						
Du Pont	confidential	23,170	9,100	16,320	48,590	?
IFF	confidential	0	0	0	0	0
Merck	136,000	5,075	450	130,475	136,000	100
PHENOL						
Atlantic	80,000	75	6,000	0	6,075	7.6
CIBA-GEIGY	709,100	0	150	0	150	0.02
Du Pont	confidential	2,210	670	1,047,600	1,050,480	?
Exxon	5,699,000	2,230	2,200	0	4,430	0.08
Rhone-Poulenc	3,553,400	81	598,230	4,290	602,601	17.0
TOLUENE						
CIBA-GEIGY	1,218,300	68,100	1,200	0	69,300	5.7
Du Pont	confidential	3,980	440	284,780	289,200	?
IFF	confidential	972	?	80,000	80,972	?
Merck	397,000	17,200	4,000	375,000	396,200	99.8
Rhone-Poulenc	304,900	270,951	28,980	0	299,931	98.4
ZINC						
Atlantic	1,300	2	75	0	77	5.9
CIBA-GEIGY	8,400	0	957	3,830	4,787	57.0
Du Pont	confidential	0	22,000	80,250	102,250	?
Merck	8,000	0	1,600	0	1,600	20.0
TOTALS		413,781	653,640	3,007,879	4,114,800	
TOTALS excluding Du Pont and IFF*	14,505,277	364,099	655,480	558,549	1,578,128	10.9
PERCENT OF TOTAL WASTES GENERATED excluding Du Pont and IFF*		23%	42%	35%	100%	

* Du Pont was excluded to avoid skewing the results. Wastes from this one plant exceed (by 30%) wastes from all the other plants combined, and over 96% of Du Pont's wastes of the listed chemicals are in the form of solid wastes. IFF was excluded because of the lack of complete data.

Source: 1978 New Jersey Industrial Survey.

Rhone-Poulenc's major phenol wastes are disposed of in wastewaters (over a half-million pounds per year).

For toluene, although CIBA-GEIGY, Merck and Rhone-Poulenc each handle large quantities of the chemical, CIBA-GEIGY's total waste generation of 69,300 pounds amounts to a loss of 5.7 percent of the chemical. Rhone-Poulenc and Merck, however, generate almost 300,000 pounds and 400,000 pounds of toluene waste respectively, representing losses of over 98 percent and 99 percent of total toluene use. Also, CIBA-GEIGY's and Rhone-Poulenc's toluene wastes are mostly in the form of air emissions while Merck discharges toluene mostly in the form of solid wastes.

Toluene is used by both Merck and Rhone-Poulenc during the intermediate stages of product manufacture. It is a component only of processing and not of the final product, so it must be completely isolated during the manufacturing process. As a result, virtually all of the toluene ends up as a waste material.

At CIBA-GEIGY's plant, on the other hand, toluene is a necessary component for the epoxies manufactured there; it is incorporated as part of the material leaving the plant as a commercial product. Because there is no need to separate the toluene from the product, CIBA-GEIGY does not generate the very high percentages of wastes of this particular chemical that are seen at the other study plants.

The organic chemical industry as a whole shows the same disparity in waste generation as INFORM's study plants in New Jersey. This was shown in a study to develop regulations for wastewater effluent, which was reported by the U.S. EPA in its 1983 <u>Development Document</u>.[1] The EPA measured the concentrations of 108 hazardous chemical pollutants in wastewater from 149 organic chemical and plastic plants. For over half of the 84 pollutants detected in the wastewaters the range of concentration values varied by over a thousand times.

Quantities of Wastes Differ Greatly

The information that is available on INFORM's 29 plants suggests the wide range of hazardous waste materials being generated by chemical facilities.

The six plants in Table 5-2 generated the largest solid wastestreams (as measured by pounds of waste per year) containing hazardous chemicals in the study. Most of these solid waste materials are burned or landfilled.

The largest solid waste producer in this study -- and perhaps the largest in the U.S. -- is the Du Pont Chambers Works plant in Deepwater, New Jersey. This plant's total annual discharge of 80 billion pounds of hazardous solid wastes[2] is equal to the total amount that the U.S. EPA originally estimated (according to the definition of hazardous waste under RCRA) to be generated per year by <u>all</u> plants in the U.S.[3] Even allowing for the fact that 99 percent of Du Pont's reported waste is water used for dilution, the remaining one percent -- 800 million pounds a year -- means that the Du Pont Chambers Works plant throws away more material than most chemical plants manufacture as product. The largest single chemical in a solid waste stream, phenol, was reported at over one million pounds per year from Merck's Rahway plant.

The largest air emissions of hazardous chemicals from INFORM's study plants are reported in Table 5-3. These differ from solid wastes in that air emissions are released directly into the environment, whereas solid wastes are often destroyed through incineration or contained in landfills.

Du Pont's Chambers Works plant again holds the lead among the plants in INFORM's sample. Du Pont's air emissions include more than 30 individual hazardous chemicals. Two of these -- trichlorofluoromethane

Table 5-2. The Largest Hazardous Solid Wastestreams from INFORM Study Plants

INFORM Study Plant	State	Solid Wastestreams containing Hazardous Chemicals
Du Pont	NJ	51,891,840 lbs/yr carbon tetrachloride manufacturing wastes
		16,365,888 lbs/yr ethyl chloride manufacturing wastes
		11,176,704 lbs/yr lead-containing wastes
IFF	NJ	17,537,400 lbs/yr chromium-containing wastes
USS Chemicals	OH	15,750,000 lbs/yr (2.1 million gallons) phenol process residues
CIBA-GEIGY	NJ	Over 6,000,000 lbs/yr (20,000 lbs/day) contaminated sludge
Dow	CA	6,000,000 lbs/yr still bottoms and other chlorinated organic wastes from carbon tetrachloride and perchloroethylene production
		3,000,000 lbs/yr chlorinated organic wastes from chlorine production
		Specific Chemicals in Solid Wastes
Merck	NJ	1,716,595 lbs/yr methylene chloride
Du Pont	NJ	1,047,600 lbs/yr phenol 899,660 lbs/yr nitrobenzene 865,200 lbs/yr aniline 658,600 lbs/yr chlorobenzene

Source: See individual plant profiles in Part II.

Table 5-3. The Largest Amounts of
Hazardous Air Emissions from INFORM Study Plants

INFORM Study Plant	State	Hazardous Air Emissions
Du Pont	NJ	2,606,000 lbs/yr methyl chloride 1,031,000 lbs/yr trichlorofluoro-methane 796,860 lbs/yr 1,2-dichlorobenzene
Perstorp	OH	602,000 lbs/yr formaldehyde*
USS Chemicals	OH	581,400 lbs/yr phenol*
Rhone-Poulenc	NJ	270,951 lbs/yr toluene
Monsanto	OH	185,200 lbs/yr acrylonitrile
Borden	CA	170,400 lbs/yr formaldehyde*

* The U.S. EPA considers these figures unreliable because of the estimation techniques used.

Source: See individual plant profiles in Part II.

and methyl chloride -- have been reported as emitted in excess of one million pounds per year. Air emissions of phenol from the USS Chemicals plant in Haverhill, Ohio, amount to 581,400 pounds per year. USS Chemicals' cumene emissions total 290,000 pounds per year and that is after emissions of this hazardous chemical were reduced by 715,000 pounds in 1979, using an adsorption unit, and reduced an additional 400,000 pounds in 1983 with a condenser. Other plants not included in Table 5-3 may also have sizable emissions of hazardous chemicals. Dow Chemical's Pittsburg, California plant, for instance, emits 598,000 pounds per year of organic chemical vapors, but government files do not specify if the chemicals released are hazardous or not.

Wastewater discharges can also contain strikingly large amounts of chemicals (see Table 5-4) which are either treated at sewage treatment plants or directly discharged to surface waters or into underground wells. The USS Chemicals plant in Ohio disposes of more than 600,000 pounds per year of phenol in its wastewater discharge to underground injection wells. The Rhone-Poulenc plant in New Brunswick, New Jersey, discharges almost 600,000 pounds per year of phenol and 30,000 pounds per year of toluene to the Middlesex County sewage treatment plant.

Other plants in INFORM's study with the largest discharges of wastewaters treat the wastes before discharging them, but still discharge thousands of pounds of hazardous chemicals each year. The Du Pont plant treats 36 million gallons of wastewaters daily and discharges 28,400 pounds of arsenic and 22,000 pounds of zinc per year into the Delaware River.

Some of the plants in INFORM's sample lie at the opposite end of the waste-generating spectrum. For example, the Atlantic Industries plant in Nutley, New Jersey, generates 40,000 pounds a year of solid waste sludge containing relatively small quantities (300 pounds or less) of individual hazardous chemicals. Seven study plants -- Bonneau Dye, Colloids, Fibrec,

Table 5-4. The Largest Amounts of Hazardous Chemicals
In Wastewater from INFORM Study Plants

INFORM Study Plant	State	Hazardous Components in Wastewater
USS Chemicals	OH	600,000 to 3,600,000 lbs/yr (2,000 to 12,000 lbs/day) phenol
Rhone-Poulenc	NJ	598,230 lbs/yr phenol 28,980 lbs/yr toluene
Du Pont	NJ	28,400 lbs/yr arsenic 22,000 lbs/yr zinc 19,000 lbs/yr 2,6-dinitrotoluene 16,000 lbs/yr 2,4-dinitrotoluene 12,000 lbs/yr 1,4-dichlorobenzene 9,100 lbs/yr methyl chloride 7,500 lbs/yr cyanide 6,200 lbs/yr lead
American Cyanamid	OH	41,530 lbs/yr zinc chloride Over 27,000 lbs/yr (900 lbs/day) chlorobenzene
Merck	NJ	22,000 lbs/yr benzene
Atlantic	NJ	6,000 lbs/yr phenol 5,000 lbs/yr formic acid

Source: See individual plant profiles in Part II.

J.E. Halma, Max Marx Color and Chemical, Scher Chemicals, and Smith and Wesson -- have no reported solid wastes whatsoever from their operations.

Air emissions and wastewater discharges are similarly small at the Scher plant. Total combined air emissions and wastewater discharges of tetrachloroethylene were reported as less than 10 pounds per year, roughly equal to an annual loss of a gallon of this chemical. Equally small quantities of acrylonitrile (40 pounds) and maleic anhydride (five pounds) were also disposed of.

At first glance, the figures in Table 5-2 for solid waste generation, which range up to 50 million pounds per year, may appear to overwhelm air emissions or wastewater discharge data, which range from several thousand to three million pounds per year. These figures require careful interpretation, however. The solid waste data usually represent chemical conglomerates of hazardous and non-hazardous materials, and often include large quantities of water or soil. IFF's disposal of 17.5 million pounds of chromium waste, for instance, contains an unknown quantity of the hazardous metal chromium. Only one percent of Du Pont's 80 billion pounds of solid wastes generated annually contains hazardous chemicals; the rest is water.

The solid wastestreams where the specific quantities of hazardous chemicals are known (listed for Merck and Du Pont at the bottom of Table 5-2) are the same scale as the chemical-specific disposal data for air emissions and wastewater discharges. Thus, the information in Tables 5-2 to 5-4 supports INFORM's earlier conclusions from the data in Table 5-1: air emissions, wastewater generation and solid waste disposal are all of the same order of magnitude and equally in need of attention regarding the potential for waste reduction.

Methods of Waste Treatment and Disposal

Methods of waste management, aside from waste reduction, used by the 29 INFORM study plants include all the methods of treatment and disposal available to organic chemical plants generally. These are summarized in Table 5-5. Briefly, chemical wastes that are not removed by air pollution control equipment are emitted into the atmosphere by being vented from tanks, process equipment, burners and other sources. Wastewaters are either 1) treated and discharged to surface waters (oceans, lakes, rivers, streams); 2) sent to municipal sewage treatment plants, some being "pretreated" first; 3) injected into deep underground wells; or 4) discharged to surface lagoons or evaporation ponds. Solid wastes are 1) disposed of in landfills, 2) incinerated or 3) treated to recover chemical components for reuse.

Air Emissions

Many sources of air pollutants can exist at any given plant: smokestacks, vents for gases and dust, incinerators, evaporation from storage and process tanks, equipment leaks and other unintentional discharges. INFORM found evidence in various environmental files that 23 of the study plants dispose of wastes as air emissions. No evidence of air emissions was found for the other six.

Because three types of air emissions sources are not always monitored or regulated by government agencies, it is close to impossible to identify the actual number of emissions sources even when a plant is known to have air emissions. The first type — fugitive emission sources such as leaks and evaporative losses from ponds — are not generally counted, either because they are not recognized as sources or because there is no regulatory requirement to do so. The second type includes sources that are excluded from regulatory scrutiny because they existed before the federal air regulations went into effect. These are called "grandfathered" sources. Finally, other air emissions not formally controlled are those that emit chemical quantities below minimum discharge cut-off levels established by state or federal regulations.

Table 5-5. Waste Treatment and Disposal Methods at INFORM Study Plants*

INFORM Study Plant	AIR		WASTEWATER				SOLID WASTE			
	On-site Incineration	Other** Air Sources	On-site Wastewater Treatment	Discharge to Sewage Treatment Plant	Surface Discharge	On-site Discharge to Ponds, Wells	On-site Land-filling	On-site Incineration	Off-site Land-filling	Off-site Incineration	Other*** Off-site Disposal
CALIFORNIA											
Borden		X		X		O					X
Chevron		X	X	X	X	X					X
Colloids				X							
Dow		X	X		X,O	X	X	X		X	X
Fibrec											
Polyvinyl		X									
Shell		a	X		X	X,O					
Stauffer	X	X	X	X	X	X			X		X
Union	O	X	X	X							X
NEW JERSEY											
Atlantic		X	X	X							X
CIBA-GEIGY	X	X	X		X	X	X,O		X	X	X
Du Pont		X	X		X	X	X	X	X	X	X
Exxon		X	X		X	X			X	X	X
Fisher		X	X						X	X	X
Halma				X							
IFF		X	X	X		X		X			X
Max Marx				X							
Merck		X	X	X	X,O			O	X	X	X
Rhone-Poulenc		X	X	X	O				X		X
Scher		X		X							
OHIO											
American Cyanamid		X	X	X	X	X	O		X	X	X
Bonneau		X	O	X						X	
Carstab		X		X							X
Frank		X	X	X							
Monsanto	X	X	X		X	X				X	X
Perstorp		X		X					X		
Sherwin-Williams		X	X	X						X	X
Smith & Wesson		X									
USS Chemicals		X	X	X	X	X,O		X,O	X	X,O	X
TOTALS (active)	3-X	23-X	17-X	17-X	11-X	11-X	3-X	5-X	9-X	11-X	19-X
(discontinued)	1-0		1-0		3-0	3-0	2-0	2-0		1-0	

X-active facilities   O-discontinued facilities   a-chemical plant emissions cannot be distinguished from refinery emissions at the same industrial plant site
Footnotes are on the following page.

Footnotes for Table 5-5.

* Evidence of treatment and disposal methods at the study plants was primarily found in government files. It is this information that is included in this table. Some plants, particularly those that did not grant interviews for this study, might use methods not indicated here for which there was no reference in any of the files searched by INFORM.

** Includes chemical emissions that are vented to the atmosphere as well as those vented to pollution control equipment such as scrubbers or filters. Emissions may stem from vents, stacks, evaporation from tanks, or equipment leaks and other unintended sources, but do not include routine emissions from conventional fuel-burning equipment.

*** Unidentified disposal practices likely to include both landfilling and incineration.

Those study plants for which INFORM found evidence of the discharge of any type of air emissions are listed in Table 5-5. The information on specific types of treatment is not included because a plant can have hundreds or even thousands of sources of air emissions and there exist many types of air pollution control technologies (e.g., filters, scrubbers, adsorbers, condensers).

However, cases have been noted where numerous sources were vented together and sent to a burner for incineration. Three INFORM study plants (Stauffer, Du Pont and Monsanto) incinerate their chemical gases and two (Du Pont and Monsanto) burn their waste gases as fuel for energy recovery.

Wastewater Treatment and Discharge

An industrial plant has several choices for treating and disposing of its wastewaters. It can treat them on-site and then discharge them to surface waters; it can discharge them to municipal sewage treatment plants or it can inject them underground into wells, with or without treating them first.

On-Site Wastewater Treatment: Wastewater treatment can include physical treatment, such as letting solids settle out of the wastewater before discharge, or biological or chemical treatements designed to change hazardous chemicals into less hazardous or non-hazardous substances. Treatment systems can be as simple as the holding tank used at Atlantic Industries to settle solids, or as complex as the multi-million dollar state-of-the-art treatment system at Du Pont which uses activated carbon and biological treatment to either destroy hazardous chemicals or concentrate them for disposal as solid wastes.

INFORM found evidence in various government files that 24 of the 29 study plants generated wastewaters that must be treated and/or disposed of. However, only 17 of the study plants treat their wastewaters on-site before discharging them. Three of the study plants have discontinued use of some of their on-site treatment facilities in recent years: Borden and Shell use on-site treatment at a reduced level and Carstab now relies solely on discharging to a municipal sewage treatment plant.

Discharge to Sewage Treatment Plants: Seventeen INFORM study plants reported discharges to municipal sewage treatment plants -- five small ones and 12 large ones. None of the five small plants reported on-site treatment of their wastewaters prior to discharge into the municipal system. On the other hand, of the 12 large INFORM study plants discharging to municipal sewage treatment plants, 10 provide some form of on-site treatment first.

According to a 1983 U.S. EPA report, the U.S. organic chemical industry (including plastics manufacturing) accounts for 80 percent of all industrial discharges of toxic organic chemicals to sewage treatment plants.[4]

Discharge to Surface Waters: Eleven study plants reported discharging to surface waters. Ten of these have on-site treatment capability. As the federal-state permit programs regulating surface water discharges are implemented, some plants have found the best option is to discharge to municipal sewage treatment plants. Merck, for example, diverted two discharge streams in order to eliminate the need to apply for permits for surface discharges.

Permit restrictions ordinarily require monthly monitoring reports measuring concentration levels of pollutants in the discharges, although additional tests may also be imposed. Both the Stauffer and CIBA-GEIGY study plants, for example, are required to conduct fish-survival tests, placing fish in tanks of wastewater to see how many survive, as a test of the wastewater's toxicity.

Only one of INFORM's study plants (CIBA-GEIGY) uses ocean disposal for some of its wastewaters. According to a 1983 Chemical Manufacturers Association Survey, less than one percent of the large U.S. chemical companies discharge wastewaters directly to the ocean.[5]

Deep Well Injection: USS Chemicals in Haverhill, Ohio, is the only INFORM study plant to use deep well injection to dispose of liquid wastes. Into two on-site deep wells, USS Chemicals injects several hundred thousand gallons of pretreated process wastewaters a day. Each day's wastes contain an estimated 2,000 to 12,000 pounds of phenol, 500 pounds of aniline, and other waste chemicals. Eight of the other companies represented in INFORM's study use deep well injection to dispose of wastes at 23 of their other plant sites (see Table 5-6).

Deep well injection of hazardous wastes is a controversial disposal method. However, USS Chemicals considers it a secure disposal method and as Table 5-6 shows, Du Pont, Monsanto and Shell Oil are all large users of wells at plants in states not studied by INFORM. But two of the companies in this study (Monsanto and Dow) reported to INFORM that they have corporate waste management policies designed to discourage continued use of this disposal method. In effect, Dow's policy says that if hazardous wastes cannot be recycled or incinerated, then land disposal may be considered, but that deep wells are no longer an option for hazardous waste disposal. The company has eliminated use of all of its deep wells. Monsanto's policy is to reduce dependency on deep well injection in anticipation of future regulatory restrictions on this option. The 1984 RCRA amendments have banned disposal of hazardous wastes in deep wells located near potential underground sources of drinking water.

In the U.S. organic chemical industry as a whole, deep well injection of hazardous wastes remains a widely used disposal practice.[6] According to a 1983 Chemical Manufacturers Association survey of over 500 large chemical companies, it is the most widely used practice. These companies dispose of 63 percent of their 5.7 million tons of hazardous solid wastes and

Table 5-6. Active Class I Underground Injection Wells Used by Companies Represented in INFORM's Study

Company	States	# of Facilities	# of Wells
Allied	Illinois	1	1
American Cyanamid	Louisiana	1	5
Borden	Louisiana	1	3
Chevron	Louisiana	1	2
Du Pont	Kentucky	1	2
	Louisiana	1	7
	Texas	5	25
Monsanto	Florida	1	3
	Louisiana	1	2
	Texas	2	4
Shell Oil	Louisiana	3	12
	Texas	1	2
Stauffer	Alabama	1	2
	Louisiana	1	2
USS Chemicals	Indiana	1	1
	Ohio*	1	2
TOTALS	8 states	23	75

* Only the two injection wells at USS Chemicals' plant in Haverhill, Ohio are included in INFORM's study.

Source: U.S. Environmental Protection Agency, "Report to Congress on Injection of Hazardous Waste," EPA 570/9-85-003, Washington, DC, May 1985.

another three percent of their 700 million tons of hazardous wastewaters in underground injection wells.

In the U.S. there are 195 active Class I wells (those wells permitted by the EPA to receive hazardous wastes). A 1985 report by the U.S. EPA's Office of Drinking Water found that manufacturers of organic chemicals accounted for 51 percent of the volume of hazardous wastes being disposed of in these wells.[7]

Most Class I wells are located in the Gulf Coast states and near the Great Lakes where they were originally dug by the oil and gas industry. Only 15 states in all have Class I wells and all but 18 of the 195 Class I wells in the U.S. are located on the sites of the plants using them. Among the states represented in this study, Ohio has 14 active Class I wells, California has two and New Jersey has none.

Discharge to Evaporation Ponds and Surface Lagoons:
Another method for treatment of wastewaters is to discharge them into surface impoundments (ponds or lagoons, for instance), which allow solids in the water to settle to the bottom of the impoundment and other pollutants to evaporate. In some cases, the impoundments are also designed to allow wastewater to percolate through the soil into groundwater.

Evaporation ponds as a means of wastewater disposal are used by four INFORM study plants, all in California: the Chevron, Dow and Stauffer plants in Richmond and Shell's plant in Martinez. Such ponds can only be used effectively in Western states having low rainfall, a sunny climate and a high annual evaporation rate.

Reliance on such surface impoundments by INFORM study plants is decreasing because of the difficulties of preventing leaks over the long-term. The risks of groundwater contamination have led to state restrictions. The Dow plant in Richmond, for example, is now required by the state of California to close its evaporation ponds within 12 years. The Borden plant in Fremont, California, closed its pond in 1982 because of state limits on effluent concentrations of phenol that would have required a much larger pond. The company replaced it by combined waste reduction and waste treatment programs. The Shell plant in Martinez is having acute problems with its pond, which is now listed by the U.S. EPA as one of 786 Superfund sites that must be excavated and cleaned up because they pose a threat to human health. The ponds used by Chevron and Stauffer are still operating.

Two of the INFORM study plants (CIBA-GEIGY and International Flavors and Fragrances) have been issued permits by the state of New Jersey which require monitoring of the groundwater near these facilities. In all, 40 percent of the INFORM study plants use surface lagoons. Three of them (Borden, Shell, and Carstab) have closed one or more of their surface impoundments in recent years. Their reasons vary, including state restrictions and risks of potential groundwater contamination.

Hazardous Solid Wastes Disposal

Twenty out of the 29 INFORM study plants reported that they generate hazardous solid wastes that must be disposed of. The methods of treatment and disposal include incineration, landfills, and processing through which chemicals are recovered for reuse. These forms of waste management can take place either on the plant site or at off-site disposal facilities.

Five of the INFORM study plants incinerate their hazardous solid wastes on-site. Two of these plants (International Flavors and Fragrances, and USS Chemicals) burn these chemical wastes in boilers as fuel for energy recovery. The Merck study plant ceased burning its waste solvents as supplemental fuel in 1983.

Three of the study plants -- Dow, CIBA-GEIGY and Du Pont -- have on-site landfills. The CIBA-GEIGY plant at Toms River in New Jersey has had to close part of its landfill operations because of groundwater contamination. The site has been listed as a Superfund site and is being evaluated for cleanup.

Twenty of the study plants transport their solid wastes off-site for disposal by incineration, landfilling and, in at least eight cases, for recovery and recycling. Many of the off-site facilities receiving these solid wastes are located in other states, including Alabama, Louisiana, Texas, Maryland, Illinois and Indiana.

A 1984 report by the U.S. EPA[8] compared the methods of treating and disposing of the 580 billion pounds of hazardous solid waste generated in 1981, 68 percent of which was generated by the chemical industry (separate figures for the organic chemical industry were not reported). Twenty percent of this hazardous waste material was disposed of in landfills or other surface impoundments (approximately 50 billion pounds) or by injection into deep wells (70 billion pounds). Less than one percent (3.7 billion pounds) was incinerated. Hazardous wastes not directly treated or disposed of were stored in tanks and ponds.

References

1. Office of Water Regulations and Standards, Development Document for Proposed Effluent Limitations Guidelines and New Source Performance Standards for the Organic Chemicals and Plastics and Synthetic Fibers Industry, EPA 440/1-83/009-b, U.S. Environmental Protection Agency, Washington, DC, February 1983, p. VI-5.

2. Philip Shabecoff, "Hazardous Waste Exceeds Estimates," New York Times, p. A1.

3. Office of Technology Assessment, Technologies and Management Strategies for Hazardous Waste Control, Congress of the United States, Washington, DC, 1983, p. 121.

4. JRB Associates, Addendum to the Report Entitled: Assessments of the Impacts of Industrial Discharges on Publicly Owned Treatment Works, McClean, VA, February 25, 1983, p. 2-7.

5. Chemical Manufacturers Association, The CMA Hazardous Waste Survey for 1981 and 1982, prepared by Environmental Resources Management, Inc., West Chester, PA, Washington, DC, September 1983, p. 4-25.

6. Ibid., p. 4-26 and 4-31.

7. Office of Drinking Water, Report to Congress on Injection of Hazardous Waste, U.S. Environmental Protection Agency, Washington, DC, March 1985, p. II-12.

8. Westat, Inc., "National Survey of Hazardous Waste Generators and Treatment, Storage and Disposal Facilities Regulated under RCRA in 1981," April 1984.

Chapter 6

Waste Reduction at the Study Plants

After three years of intensive research into the hazardous waste reduction practices of 29 organic chemical plants, what has INFORM concluded about the types of waste reduction being practiced and the extent of hazardous waste reduction being achieved? The encouraging news is that INFORM found reports of 44 interesting and innovative waste reduction practices. They involved a variety of process, product, equipment and operational changes that substantially reduced or eliminated individual chemicals in wastestreams at the plants. To the extent that INFORM was able to document the actual impact of the practices, it was found they prevented the generation of at least seven million pounds of hazardous chemical wastes and saved companies at least $800,000 in reduced raw material and waste disposal costs. These 44 practices taken together suggest the range of possiblities that exist for the more than 1,000 U.S. organic chemical plants to reduce wastes at the sources.

The discouraging news is what we do not know about the extent of waste reduction and its impact on overall wastes at any individual plant or in this critical industry generally. Despite industry claims of support for waste reduction, INFORM was only able to document waste reduction at 13 of the 29 study plants. And for none of these plants was the available information sufficient to relate the achievements of the waste reduction practices to the plant's total wastestreams. If waste reduction is to be a focus of public policy, the extent and impact of waste reduction on all wastes at a plant must be known. This chapter serves to show what kind of information is currently available, what kinds of assessments can be made, and what further information is needed.

Data on Hazardous Waste Reduction: An Inadequate Picture

One of the most important findings of this study is how little information is available to assess the extent of hazardous waste reduction being accomplished. While hazardous waste reduction is nearly universally supported by many chemical companies and their trade associations (see Chapter 1), information is not readily available to document progress in source reduction.

<u>Limited government data on hazardous waste reduction</u>

The voluminous public files (as many as 30 separate files for some plants) reviewed with great effort and expense by INFORM yielded no up-to-date information to document the amount of hazardous chemicals used by each plant, or the amounts of discharges of the chemicals to the air, water and land -- the benchmark figures needed to measure reductions in wastes that are achieved over time. The one data file that came closest to providing such information is the remarkably straightforward "New Jersey Industrial Survey" conducted by the New Jersey Department of Environmental Protection (described in Chapter 5). But because it was used only once in 1978, even this information base cannot be used to document the trends in hazardous waste generation, without which any assessment of hazardous waste reduction is impossible.

Beyond the baseline data, INFORM searched government files for information on waste reduction practices at the individual plants. Such information was found for only six waste reduction practices at five study plants (Exxon, Perstorp, Sherwin-Williams, Union and USS Chemicals).

Limited corporate cooperation

The only way to find out about hazardous waste reduction and the impact it is having on total wastes at an individual plant is for the plant management to supply the information. Obtaining such information is often not possible. Only 13 of the 29 companies included in our study were willing to provide any information (see Table 6-1). Among those refusing to cooperate were some of the largest in the country, including Du Pont, Shell, Chevron and American Cyanamid. Generally, however, the large plants (those with 50 or more employees) cooperated more frequently than the small ones -- 11 of the 20 large plants in the study provided information to INFORM while only two of the nine small ones did, and only one of these (Perstorp) reported waste reduction practices. Therefore, most of the conclusions reported here are based upon the reported experience of the larger firms.

At none of the plants cooperating in this study was baseline data and the full impact of waste reduction practices on total plant wastes measured. From the plant manager's perspective it may not be necessary to keep track of the amount of wastes not generated (that is, reduced). Instead, the major concern is knowing whether regulations are being met and, as explained above, the regulations do not require such data. This lack of information has implications for whether or not the search for waste reduction is undertaken. Justifying waste reduction practices is more difficult if the overall savings in terms of materials and costs is not known.

Two study plants have achieved waste reduction through monitoring their wastestream.

The importance of the baseline data for identifying waste reduction practices was seen at two of INFORM's study plants. Both Borden and Exxon have instituted wastewater monitoring systems that track the level of organics in the wastestream and can quickly spot any unusual changes in their concentrations and trace

Table 6-1. Cooperation and Waste Reduction Practices Identified At Large and Small Plants in INFORM's Study

INFORM Study Plant	LARGE PLANTS (50 or more employees)			INFORM Study Plant	SMALL PLANTS (fewer than 50 employees)		
	Cooperated in the Study		Waste Reduction Practices Identified By INFORM		Cooperated in the Study		Waste Reduction Practices Identified By INFORM
	Yes	No			Yes	No	
CALIFORNIA							
Borden	X		X	Colloids		X	
Chevron		X		Fibrec		X	
Dow	X		X	Polyvinyl		X	
Shell		X					
Stauffer	X		X				
Union	X		X				
NEW JERSEY							
Atlantic	X		X	Halma		X	
CIBA-GEIGY	X		X	Max Marx		X	
Du Pont		X		Scher	X		
Exxon	X		X				
Fisher	X						
IFF		X					
Merck	X		X				
Rhone-Poulenc		X					
OHIO							
American Cyanamid		X		Bonneau		X	
Carstab		X		Frank		X	
Monsanto	X		X	Perstorp	X		X
Sherwin-Williams		X	X				
Smith & Wesson		X					
USS Chemicals	X		X				
TOTALS	11	9	11		2	7	1

them back to their source. Borden began monitoring ways in which phenol wastes from its resin operations were entering the plant's wastewater streams when regulations limited the amount of phenol that could be discharged. Sources of the wastes and waste reduction opportunities were identified through the monitoring and a 93 percent reduction in the level of organics in the wastewater stream resulted.

Exxon began monitoring in order to reduce the large fluctuations in the composition of its wastes which were interfering with the operation of its on-site wastewater treatment plant. Exxon has used its wastewater monitoring system to set waste reduction goals for each of its processes and has achieved an overall reduction of 75 percent in the level of organic chemicals in the plant's wastewater.

Extent of Hazardous Waste Reduction Practices

Despite the difficulties in obtaining and interpreting data concerning source reduction at individual organic chemical plants, this study has provided insights into the extent of hazardous waste reduction being employed at INFORM's 29 study plants.

<u>Hazardous waste reduction is being practiced to some degree at at least 12 of the study plants.</u>

This study identified a total of 44 distinct waste reduction practices at 12 different plants. Of the 13 plants cooperating in this study, 11 reported one or more waste reduction practices. Of the 16 non-cooperating plants, it was possible to document hazardous waste reduction practices at only one (Sherwin-Williams). It is not possible to determine whether there is any correlation between the lack of cooperation and the lack of hazardous waste reduction.

About 35 percent of the reported hazardous waste reduction practices have substantially or entirely reduced wastes from the sources affected by them.

Five of the 44 reported waste reduction practices totally eliminated a chemical from a wastestream. Another 10 achieved from an 80 to 100 percent reduction. The degree of reduction for the reported practices varied from 10 percent to 100 percent, with the highest reductions occurring through operational and equipment changes. Those practices that reduced air emissions (primarily equipment changes) were found to achieve some of the largest reductions. Solid waste reduction was hardest to achieve and its impact least often measured.

About 75 percent of the reported hazardous waste reduction practices impact only single wastestreams within the plant.

Thirty-four out of the 44 practices reduced wastes in one type of wastestream only -- to air, water or as solid waste. Only waste reduction measures that included process or operational changes affected more than one type of wastestream. The most frequently encountered type of this kind of waste reduction was the reduction of organic materials in wastewater which in turn reduces the quantity of sludge produced during treatment of the wastewaters. CIBA-GEIGY reported the one example of reduction of wastes in all three media. In that case a process change made possible the complete elimination of the use of mercury at the plant and the consequent elimination of mercury air emissions, wastewater discharges and solid waste disposal.

The 34 reported practices that reduced wastes in a single type of wastestream only are approximately evenly divided among the types of wastes -- 12 reduced air emissions, eight reduced wastewater discharges and 14 reduced solid wastes. The practices reducing air emissions were, for the most part, additions of equipment designed to prevent the vaporization and loss of chemicals from storage tanks and process operations. The practices reducing wastewaters and solid wastes were more varied and included examples from all of the five waste reduction categories, discussed below.

It was not possible to document many instances in which hazardous waste reduction practices resulted in substantially diminished amounts of wastes on a plant-wide basis.

The percentage of waste reduced was reported to INFORM for less than half (20 of the 44) of the examples in the study. For another 20 percent (nine of the 44) the actual but not relative amounts of waste reduced were available. Most often this information was unavailable for process and operational changes and for practices reported as reducing solid wastes.

Waste Reduction: Five Categories and 44 Practices Identified by INFORM

Although waste reduction is often thought of as changes in the chemical manufacturing process, INFORM identified five major categories of waste reduction practices -- process changes, product reformulations, chemical substitutions, equipment changes and operational changes. These reduced wastes not only from manufacturing processes but also from non-process sources of waste, such as storage and loading activities or from plant maintenance operations. Process changes were indeed the most common category of waste reduction practices at the 29 plants, but operational changes and equipment changes were also widely used:

Category of Waste Reduction	No. of Practices
PROCESS CHANGES	18
OPERATIONAL CHANGES	14
EQUIPMENT CHANGES	9
PRODUCT REFORMULATIONS	3
CHEMICAL SUBSTITUTIONS	2

Note: Some practices fall into more than one category and four were identified by the study plants but not described sufficiently to allow categorization.

The extent to which these five categories of waste reduction are in use, and the reasons for their adoption, have been the focus of INFORM's study. Examples of all five categories were found in two or more of the study plants. The specific practices were quite varied and different, however, and are briefly described below. Full descriptions of the waste reduction practices, along with a description of plant operations and waste generation at each study plant, are presented in the 29 plant profiles in Part II of this study.

Process changes that reduce wastes can range from simple alterations of process conditions such as temperature or pressure, to discovering new chemical pathways and production technologies that advance the state-of-the-art of manufacturing as they help to achieve waste reduction.

Nine INFORM study plants reported making a total of 18 process changes that reduced air, water or solid wastes (see Table 6-2). Process changes, while the most common type of waste reduction practice found in this study, represent one category of practice that is particularly difficult to learn about because they lie at the heart of chemical company concerns about trade secrets.

The amount of information that we have describing process changes at Sherwin-Williams is typical of the

Table 6-2. Waste Reduction Practices at Study Plants: Process Changes

Waste Reduction Practice	Type of Waste Reduced	Percent-age Reduced	Amount Reduced	Year Practice Implemented
ATLANTIC				
—Changes in diazo dye manufacturing process	Water	?	55,000 lbs/yr organics 250,000 lbs/yr inorganics 500,000 gallons/yr water use	?
CIBA-GEIGY				
—Use of a new process chemistry for manufacturing dyes to eliminate mercury as a catalyst	Air Water Solid waste	100%	?	1983
—New solvent-based process increases retention of chromium in product	Water Solid waste	25%	?	early 1980s
—Optimized acid additions to process through research and development tests	Water Solid waste	10–40%	?	1983
DOW				
—Isolation of hydrochloric acid for reuse as a process raw material or for sale as a commercial product	Water	?	?	?

Table 6-2 (cont.). Waste Reduction Practices at Study Plants: Process Changes

Waste Reduction Practice	Type of Waste Reduced	Percentage Reduced	Amount Reduced	Year Practice Implemented
MERCK				
-Internal solvent recovery of methylene chloride in new Primaxin process	Solid waste	?	2.6 million lbs/yr	?
-Internal solvent recovery of isoamyl alcohol in new Primaxin process	Solid waste	?	423,000 lbs/yr	?
-Internal solvent recovery of acetone in new Primaxin process	Water	?	230,000 lbs/yr	?
-Recovery and reuse of scrubbed hydrogen chloride gas emissions from TBZ process	Air	?	?	?
MONSANTO				
-Transition from batch to continous polystyrene manufacturing process	Air	99.7%	From 50 to 0.17 lbs per 1,000 lbs of product	1952 to present
PERSTORP				
-Improvements in product quality eliminating need to mix high- and low-grade PE	Air	100%	?	?

Table 6-2 (cont.). Waste Reduction Practices at Study Plants: Process Changes

Waste Reduction Practice	Type of Waste Reduced	Percentage Reduced	Amount Reduced	Year Practice Implemented
SHERWIN-WILLIAMS				
-"Improved practices program" to capture and recycle TCB	Water	?	?	1982
-Process change to eliminate discharge of sodium sulfide solution to sewer	Water	100%	?	?
STAUFFER				
-Change in DEVRINOL manufacturing process	Water	?	?	1976
-Process and product reformulation changes in VAPAM manufacturing	Solid waste	?	535 drums per year	?
USS CHEMICALS				
-Use of non-conventional aniline manufacturing process	Solid waste	?	?	1982
-Use of newer, more efficient phenol manufacturing unit	Solid waste	?	?	1981
-Product and process changes to improve quality of DPA to commercial grade	Solid waste	?	2 million lbs/yr	1982

detail obtained for this category of practice. In the early 1980s, the Metropolitan Sewer District of Greater Cincinnati became concerned about risks to its workers' health and safety because of the high levels of trichlorobenzene (TCB) that Sherwin-Williams was discharging to the city's sewage treatment plant. In reports in government files, Sherwin-Williams indicated that, through process changes as well as other waste handling techniques, which the company would not describe, it reduced TCB levels by 25 to 30 percent in 1982. In that same year the company also reported that its discharges of sodium sulfide in its wastewaters were completely eliminated through an improvement in process chemistry.

Atlantic undertook changes in its diazo dye manufacturing process to improve product yields in order to stay competitive. The changes also reduced the plant's waste load and water use. The process changes included increased process chemical concentrations, a lower reaction temperature and a new method for combining dye components. These measures not only increased the product yield by eight percent but also reduced the volume of water used by 500,000 gallons per year, the amount of organics in the wastewater by 55,000 pounds per year, and the amount of inorganics by 250,000 pounds per year.

Monsanto greatly reduced air emissions at its plant, which began operation in 1952, through gradual changes in its polystyrene manufacturing from a batch to a continuous process. The plant's original production level was 90,000 pounds of polystyrene per day with air emissions totalling five percent of total production. The new closed-system process is capable of producing 1.2 million pounds per day with air emissions of less than 0.2 percent of total production. The process changes have increased production levels 13-fold while reducing air emissions from 4,500 to 200 pounds per day.

Perstorp's process and equipment changes improved the quality of its major product, pentaerythritol (PE --

an intermediate chemical used in paints, printing inks and synthetic lubricants) and reduced several waste sources. Before the changes, Perstorp produced a low grade of PE and blended it with a higher-grade, imported PE. The waste reduction changes eliminated the mixing step, thereby eliminating PE dust emissions from mixing.

CIBA-GEIGY reported three examples of process changes, one of which completely eliminated the use of mercury. In 1978, the company used 2,280 pounds of mercury at its Toms River, New Jersey, plant, almost all of which ended up as air, water and solid wastes. Evidence of environmental contamination and serious human illness resulting from mercury poisoning led CIBA-GEIGY's corporate research group in Switzerland to develop a new dye manufacturing process that circumvented the sulfonation step in making anthraquinone dyes, materials widely used for dyeing cotton, and required no mercury.

CIBA-GEIGY also developed a new solvent-based dye manufacturing process that uses chromium more efficiently than its old water-based process. The company estimated that the new process reduces the loss of chromium in its wastewaters and solid wastes by 25 percent, but indicated that the cost of producing the dyes has risen, primarily because of the added expense of using solvent instead of water.

Also at CIBA-GEIGY, laboratory and plant-scale experiments helped determine the minimum amounts of sulfuric acid needed to complete the major sulfonation reactions in dye manufacturing, reducing the quantity of excess acid discharged. Each dye manufacturing operation is being independently evaluated for the minimum acid needed. So far, CIBA-GEIGY estimates that total sulfuric acid waste discharges have been reduced by 10 to 40 percent.

USS Chemicals is the only U.S. plant that produces aniline (used in the manufacture of polyurethane) from phenol and ammonia rather than by the conventional process using nitrobenzene and hydrogen as raw materials. This process and the new phenol unit generate less waste. There are, however, no figures available on the amounts reduced.

Dow was able to identify a market for hydrochloric acid and, thus, beginning in the late 1960s, installed incineration units, additional storage capacity and equipment to concentrate hydrochloric acid and make it into a saleable product. Hydrochloric acid formerly was a by-product of the processes and air pollution control devices producing perchloroethylene, carbon tetrachloride, pesticides and other products. The wastewaters containing the acid had been discharged into the New York Slough, an estuary leading into San Francisco Bay.

Merck reported three waste reduction practices that were introduced during the development of a new process to produce Primaxin (an antibiotic) at its Rahway, New Jersey, plant. By designing internal solvent recovery steps in this process, Merck estimates that it prevents the annual loss of 2,600,000 pounds of methylene chloride, 423,000 pounds of isoamyl alcohol and 230,000 pounds of acetone.

Merck also reported that the hydrogen chloride gas from its thiabendazole (TBZ) manufacturing process, is now partially captured and reintroduced into another stage of the TBZ manufacturing process. The gas previously was removed from air emissions and disposed of.

In 1976 Stauffer combined a process change and a product reformulation in a single waste reduction step that produced a cleaner product (VAPAM, a soil fumigant) and reduced solid wastes by minimizing the need to filter its product (and dispose of the filters). It also modified the process by which its selective weedkiller, DEVRINOL, is made, to improve raw material yields. The company estimates that the DEVRINOL process change will save $200,000 at 1985 production levels and the VAPAM changes will save $28,000 in disposal costs.

Product reformulations can be achieved without changing the fundamental manufacturing process. For example, creating a chemical product in the form of pellets rather than as a powder can reduce the generation of waste dusts as the material is transferred during final packaging operations. Materials that become wastes when they are removed from the final product stream (e.g., by filtration or distillation) can also sometimes be left in the product without impairing product quality.

Three INFORM study plants reported that they had reduced wastes by reformulating a product (see Table 6-3). USS Chemicals was able to formulate a higher-grade of DPA (diphenylamine), a by-product of aniline manufacture. The research center of U.S. Steel, parent company of USS Chemicals, identified modifications that resulted in a grade of DPA meeting commercial specifications, and now, instead of being burned as waste, as was originally planned, DPA is sold as a product.

On the other hand, the experience of Monsanto shows how difficult product reformulation can be to achieve. Monsanto reformulated a specialized industrial adhesive to avoid filtering it, hence eliminating the need to use and dispose of filters. However, Monsanto had to convince its customer that the particulate matter formerly removed by the filters could remain in the product without affecting its adhesive qualities. From the time the company researchers came up with the idea of reformulating the product, two years of effort involving both Monsanto's Research and Marketing Divisions was needed before the reluctance of the purchaser to accept a 'different' product was overcome and the change could be made.

Table 6-3. Waste Reduction Practices at Study Plants: Product Reformulations

Waste Reduction Practice	Type of Waste Reduced	Percent-age Reduced	Amount Reduced	Year Practice Implemented
MONSANTO				
-Hazardous particulate matter retained in adhesive product instead of being filtered out	Solid waste	100%	?	?
STAUFFER				
-Process and product reformulation changes in VAPAM manufacturing	Solid waste	?	535 drums per year	?
USS CHEMICALS				
-Product and process changes to improve quality of DPA to commercial grade	Solid waste	?	2 million lbs/yr	1982

Chemical substitutions of non-hazardous chemicals for hazardous ones can eliminate the potential for creating hazardous wastes. A large chemical manufacturing facility uses hundreds of materials for a wide variety of essential operations outside of the manufacturing process itself, such as cleaning and maintenance, pollution control, and corrosion inhibition. If non-hazardous chemicals can be used for these purposes, hazardous wastes can be reduced.

Instances of chemical substitutions to reduce specific chemicals in waste discharges were reported by two study plants (see Table 6-4). Union, concerned (as was CIBA-GEIGY) with potential environmental and toxicity problems related to mercury, substituted other chemicals and, in 1971, eliminated the use of mercury as a biocide in the latex it was selling to paint formulators.

CIBA-GEIGY has substituted a higher quality lime in its treatment of acidic wastewaters. In this case, the source of the waste being reduced is CIBA-GEIGY's on-site wastewater treatment plant. The acidity of the wastewaters must be reduced before they can be discharged into the Atlantic Ocean. The neutralization treatment generates large quantities of sludge which must then be disposed of in an on-site landfill. By substituting a high-quality lime for a lower quality one, more acid can be neutralized and less sludge generated per pound of lime added to the wastewater. The benefits of reduced sludge generation outweigh the added cost of the lime. The total quantity of lime needed at the plant is also reduced by mixing in caustic wastewaters from other nearby industrial operations, reducing the quantity of sludge generated even further.

Table 6-4. Waste Reduction Practices at Study Plants: Chemical Substitutions

Waste Reduction Practice	Type of Waste Reduced	Percentage Reduced	Amount Reduced	Year Practice Implemented
CIBA-GEIGY				
−Use of high-quality lime to minimize sludge generation	Solid waste	?	?	early 1980s
UNION				
−Use of biocides containing mercury discontinued	Water	100%	?	1971

Equipment changes, modifications and additions can occur in every stage of the manufacturing process for storing, moving, mixing, and reacting chemicals. Because equipment is used in all aspects of a plant's operations, there are numerous opportunities for waste generation and as many chances to implement waste reduction through equipment changes.

Nine instances of equipment changes that reduced wastes were reported at five INFORM study plants, with three plants providing two or more examples (see Table 6-5). The vast majority of changes involved rather simple modifications, most of which had their impact primarily on the amount of air emissions, rather than any other wastestream.

Dow found that it could significantly reduce its generation of hazardous chemical gases by replacing the pressurized gas used to transfer materials from storage tanks into reactor vessels with a pumping mechanism. Nitrogen gas had been used to push the raw materials from the storage tank into the reactor vessel, but some of the raw material was lost when it became mixed with the gas. By replacing the gas-pressurizing technique with a pumping mechanism, raw material losses were eliminated. Dow also has plans to add the pump to the first material transfer step (tank car to storage tank).

An even simpler measure was undertaken at the USS Chemicals plant. In 1983, after an operator in the plant reported cumene odors (discovered to be the result of uncontrolled vapors escaping from a pressure control vent), a condenser was added to return these hazardous emissions directly to the phenol process unit. This equipment recovered 400,000 pounds of cumene, one of the plant's major raw materials, in the first year it was installed, producing a savings of $100,000 compared with the $5,000 cost of installation. This employee-generated idea arose through a company-wide program called Suggestions for Cost Reduction (SCORE).

Table 6-5. Waste Reduction Practices at Study Plants: Equipment Changes

Waste Reduction Practice	Type of Waste Reduced	Percent-age Reduced	Amount Reduced	Year Practice Implemented
DOW				
-Pressurizing technique transferring raw material from storage tank to reactor vessel replaced by pumping mechanism	Air	100%	?	?
EXXON				
-Addition of floating roofs to 16 storage tanks	Air	90%	681,810 lbs/yr	1975-1983
-Addition of conservation vents to storage tanks and process sources	Air	30-75%	?	1976-1983
PERSTORP				
-Improvements in product quality eliminating need to mix high- and low-grade PE	Air	100%	?	?
-Addition of baghouse filter to process source	Air	99%	?	1978

Table 6-5 (cont.). Waste Reduction Practices at Study Plants: Equipment Changes

Waste Reduction Practice	Type of Waste Reduced	Percentage Reduced	Amount Reduced	Year Practice Implemented
STAUFFER				
-Packed seals on transfer pumps replaced with mechanical seals	Solid waste	?	2,600 gallons per year	?
USS CHEMICALS				
-Installation of resin adsorption systems to prevent loss of cumene vapors	Air	80%	715,000 lbs/yr	1981
-Installation of condenser to prevent loss of cumene vapors	Air	?	400,000 lbs/yr	1983
-Installation of floating roofs on both phenol units	Air	?	?	1969, 1981

USS Chemicals has also reduced cumene air emissions through the use of a resin adsorption system, suggested by the research center of its parent company (U.S. Steel). In 1979, during planning for a new phenol production unit, the use of resin adsorbers was proposed to soak up cumene vapors, preventing their loss to the atmosphere and allowing them to be recovered. Emissions from USS Chemicals' original phenol unit were reduced 80 percent, a reduction of 700,000 pounds per year.

Perstorp used a baghouse filter to substantially reduce air emissions of pentaerythritol (PE) dust. The baghouse filter captures particles from the PE manufacturing process for reuse, allowing 99 percent of the dust to be recaptured.

Impacts of equipment changes can be both large and small. Stauffer reported an equipment change which resulted in a small, but significant, reduction in its hazardous wastes. On its transfer pumps Stauffer has replaced "packed seals" (seals with densely packed material such as graphite and synthetic fibers which are prone to leak) with "mechanical seals" (machined rotating elements) virtually eliminating leaks in the equipment. Stauffer estimates that this change has reduced product and raw material losses by 2,600 gallons per year and results in an annual savings of $37,000.

A simple equipment change, which has had one of the greatest impacts on hazardous waste reduction found in INFORM's study, was the installation of floating roofs on liquid storage tanks at Exxon. A floating roof, whose purpose is to minimize the evaporation losses from large tanks, is a cover that rests on the surface of the liquid being stored in the tank. The roof rises and falls as the level of the liquid changes. Since 1975 Exxon has installed 16 floating roofs and seven conservation vents (another waste reduction practice). Exxon's floating roofs reduce air emissions by an average of 90 percent and currently save over 680,000 pounds of chemicals per year. They have saved over five million pounds of raw materials since they were installed.

Floating roofs are also used at the USS Chemicals plant where they were installed as a waste reduction measure for acetone storage as far back as 1969.

Operational changes are often overlooked in considerations of ways to reduce waste. These are changes in the way hazardous materials are handled at a plant and can include careful observation and control of material and process conditions and employees' habits so as to minimize spills, process upsets, the use of excessive chemicals or other problems that can generate waste.

The 14 operational changes reported by six study plants (see Table 6-6) more often reduced wastewater and solid wastes discharges, in contrast to the equipment changes which primarily reduced air emissions.

Borden and Exxon provide perhaps the most noteworthy examples of what can be achieved by simple operational changes. Strict limits on phenol in wastewater discharges imposed by the municipal sewage treatment plant, and the problems encountered with Borden's on-site evaporation pond, led to a search for operational changes that would reduce phenol wastes associated with its resin manufacturing operations. One resulting change involves the rinsewater used to clean filters, which remove large particles of resinous material as the product is loaded into tank cars. The filters are rinsed and the rinsewater is collected in a 250-gallon recovery tank instead of being sent to floor drains. Since 1982, the rinsewater has been reused when a new batch of phenolic resin is produced.

Learning from these changes, Borden also modified its urea resin loading filter system. The urea resin loading pump is now reversed, and any resin on the filters is sucked back into the product storage tank. No special recovery tank is needed. Borden considers

Table 6-6. Waste Reduction Practices at Study Plants: Operational Changes

Waste Reduction Practice	Type of Waste Reduced	Percentage Reduced	Amount Reduced	Year Practice Implemented
BORDEN				
-Phenol resins drained from filters for reuse as raw material	Water Solid waste	Total of 93% reduction of organics in wastewater for all these practices	25 gallons per filter rinse	1982
-Phenol rinsewater from reactor vessel reused as raw material	Water Solid waste		?	1982
-Urea resins drained from filters, pumped back into product tanks	Water Solid waste		?	1983
-Changes in employees' habits to reduce spills	Water Solid waste		?	1980-1983
-Storage and reuse of adhesives samples taken for quality control	Solid waste	?	?	1983
-Reuse of adhesive mixer rinses as raw material	Solid waste	?	?	1983

Table 6-6 (cont.). Waste Reduction Practices at Study Plants: Operational Changes

Waste Reduction Practice	Type of Waste Reduced	Percentage Reduced	Amount Reduced	Year Practice Implemented
DOW				
-Pressurizing technique transferring raw material from storage tank to reactor vessel replaced by pumping mechanism	Air	100%	?	?
-Isolation of hydrochloric acid for reuse as a process raw material or for sale as a commercial product	Water	?	?	?
EXXON				
-Plant-wide stewardship program to monitor wastewater streams for organics	Water	75%	?	early 1970s
MONSANTO				
-Company-wide program to reduce accidents and spills	?	?	?	?
-Employment of trained specialists to segregate hazardous from non-hazardous wastes	Solid waste	?	?	?
-Plant-wide program to identify and minimize fugitive emission sources of acrylonitrile	Air	?	?	1978

Table 6-6 (cont.). Waste Reduction Practices at Study Plants: Operational Changes

Waste Reduction Practice	Type of Waste Reduced	Percentage Reduced	Amount Reduced	Year Practice Implemented
SHERWIN-WILLIAMS				
-"Improved practices program" to capture and recycle TCB	Water	?	?	1982
STAUFFER				
-Kerosene reused several times before being discarded	Solid waste	?	800 gallons per year	1980

this the better technique, but explains it is possible only because of the high volume of urea resin handled at the plant.

A third change in Borden's resin manufacturing operations is a revised procedure for rinsing the reactor vessels between batches. Previously these closed, pressurized 11,000 to 15,000-gallon chambers had been cleaned by filling them with water, heating and stirring the water to remove residues and then draining the rinsewater into the plant's wastewater. During 1982, trial-and-error tests showed that a two-rinse method could reduce the amount of organic materials going into the wastewater as well as recover product material for reuse. The current procedure removes most of the residual material in the reactor vessel with a small first rinse of 100 gallons of water. This rinsewater is stored for reuse as material in a later batch. A second, full-volume rinse is required to fully clean the vessel. The second rinse is discharged as wastewater but has reduced phenol concentrations compared to the original one-rinse procedure.

In a fourth area of Borden's resin manufacturing operations, waste reduction was also possible. As part of its program to review employees' habits for ways to reduce wastes, the operation of transferring phenol from tank cars to storage tanks was changed. When truck deliveries of phenol are transferred to storage tanks, the transfer is made through a hose. Formerly, when the hose was disconnected the small amount of phenol remaining in the hose dripped through a floor drain into the plant's wastewater. Even small amounts of phenol create problems for Borden in trying to meet the strict regulatory limitations for phenol (no more than one to five mg/l) in its wastewater. Now the hose is flushed with a few gallons of water to rinse the last bit of phenol into the storage tank.

Borden estimates that all four of these operational changes -- filters, reactor vessel and hose rinsing procedures -- and other changes in employees' habits have resulted in a 93 percent reduction of organic materials in its wastewater. In turn, most of the hazardous solid wastes generated from the resin manufacturing processes have been eliminated because Borden was able to discontinue the use of the on-site evaporation pond to treat its wastewaters.

Borden also reported two changes in their adhesives manufacturing operations. In one, samples of adhesives from each batch, which had previously all been thrown out, are now stored and returned to the process. The other involved a new rinsing procedure for the adhesive mixers begun in 1983. It is similar to the one initiated for the resin reactor vessels: residual materials are recovered from the rinsewater when possible for use as a raw material.

Exxon, like Borden, reviewed employees' practices and monitored its wastewater for ways to reduce wastes. Exxon's Stewardship Program, begun in the 1970s, was instituted at this study plant when it was experiencing problems with large fluctuations in the quantity and quality of the wastewaters arriving at its on-site treatment facility. Under this program Exxon set up a series of in-plant wastewater sampling stations to monitor concentrations of organic chemicals. Plant operators can, as a result, trace unusual changes in these concentrations back to their source and quickly remedy the problem.

Exxon also uses the monitoring information to charge each manufacturing process its proportional share of the cost of operating its wastewater treatment system. Plant officials reported that a combination of cost charges and targets set for waste reduction under the Stewardship Program has resulted in an estimated 75 percent reduction in the level of organic wastes entering its wastewater treatment plant.

Monsanto relies on a plant-wide approach to identify and undertake operational changes to reduce hazardous wastes. One goal of Monsanto's approach is to reduce the number of accidents and spills. Operational changes under this program include revised inspection procedures to ensure proper functioning of automatic cut-off devices and overflow pipes, installation of an automatic alarm that alerts personnel to potential spills, and observation of tanks while they are being filled with hazardous materials.

Monsanto also reported that operational and maintenance changes had been made to reduce fugitive emissions of acrylonitrile from such sources as leaking equipment and the open trenches used to convey wastewaters to the plant sewers.

Another operational change that reduces hazardous wastes at Monsanto is the segregation of solid wastes to avoid contaminating non-hazardous wastes with hazardous materials. Non-hazardous trash is routinely generated at the plant. In the past, small quantities of hazardous materials, such as contaminated packaging, were sometimes included in the trash, requiring that entire loads be handled and disposed of as hazardous wastes. Now, however, three specially-trained trash specialists keep hazardous and non-hazardous trash separate, preventing the unnecessary contamination of entire waste loads.

Stauffer reported one operational change involving the use of kerosene as a solvent to rinse tanks and equipment. The company began in 1980 to save and reuse kerosene, formerly used only once, several times before discarding it. The company estimates that this practice has reduced kerosene wastes by 800 gallons per year and saves $1,000 annually.

Unexplained Waste Reduction Practices

Four of the INFORM study plants reported some wastes reduced at their plants but did not supply any details of the practices that brought about such reductions (see Table 6-7). In some cases the amount of reported hazardous waste reduction was substantial. For example, CIBA-GEIGY estimates that nitrobenzene in all wastestreams has been reduced by 90 percent since 1972,

Table 6-7. Waste Reduction Practices at Study Plants: Unexplained Changes

Waste Reduction Practice	Type of Waste Reduced	Percentage Reduced	Amount Reduced	Year Practice Implemented
ATLANTIC				
-Less dust is generated by drying, grinding and blending operations	Air	?	?	?
CIBA-GEIGY				
-Improvements to reduce nitrobenzene wastes	Air Water Solid waste	90%	?	?
EXXON				
-General program to reduce solid wastes	Solid waste	55%	?	?
PERSTORP				
-General improvements to reduce organics in wastewater	Water	50-70%	?	?

and Exxon provided aggregate data for both of its New Jersey chemical operations -- the Bayway and Bayonne plants -- that show a 55 percent reduction in hazardous waste generated per ton of product manufactured. Perstorp reported that general improvements have resulted in a 50 to 70 percent reduction in the level of organics in the plant's wastewater. Atlantic reported significant progress in reducing product dust from its drying, grinding and blending operations.

Chapter 7

Factors Influencing Waste Reduction Practices at the Study Plants

The 44 examples of waste reduction practices identified in INFORM's research indicate some of the potential this option offers. One key to tapping that potential is to identify the factors that encourage waste reduction or deter it. To do this effectively, much more information -- detailed records of actual chemical use and discharges in particular -- is needed. An encouraging finding of INFORM's study, however, is that the organic chemical industry has a remarkable capacity -- economic, technological and imaginative -- to meet the challenge.

In analyzing data from its 29 case study plants, INFORM found that the factors that influence waste reduction fell into four broad categories: plant characteristics, regulations, costs, and liability risks. All of the factors examined had the potential to influence company practices at our study plants or at other facilities. Hence, they were worth examining and are discussed here.

Only two of the plant factors -- product type and process type -- were identified as influencing hazardous waste reduction at INFORM's 29 study plants. Regulatory, cost and liability factors varied in their impact, often involving trade-offs in waste management practices. The primary impact of regulations and costs on hazardous waste reduction was found to be indirect; they did not, as might be presumed, always encourage hazardous waste reduction at the study plants, but sometimes actually deterred it. The impacts of all the various factors emerge more clearly in the detailed examination that follows.

Plant Characteristics

INFORM investigated seven characteristics inherent in companies and their operations:

>Product type (e.g. dye, drug, bulk chemicals)
>Process type (batch or continuous)
>Plant size (small, large)
>Age of plant (new, old)
>Geographical location (California,
>   New Jersey, Ohio)
>Company type (e.g. steel, dye, oil)
>Ownership structure (private, public)

<u>Only two of these factors, product type and process type, were found to significantly influence hazardous waste reduction.</u>

<u>Product Type</u>: The type of product manufactured can exercise a large but very case-specific influence on the ability of a company to achieve hazardous waste reduction. For example, Borden was able to achieve dramatic waste reduction in its resin operations in part because its resin products are self-compatible; that is, the residue from one product can be mixed in as a raw material for a different product without affecting the quality of its resins. Borden's adhesive-manufacturing operations, on the other hand, are considerably more constrained. The added diversity and absence of self-compatible wastes in Borden's adhesives operations limited the plant in its ability to achieve the same degree of waste reduction. Thus, the nature of the two types of products manufactured at Borden -- resins and adhesives -- has a large influence on waste reduction possibilities at the plant.

The need for a high degree of purity in certain types of products can also limit waste reduction initiatives. Atlantic Industries reported that thorough and complete rinsing of equipment when switching production from one dye product to another was essential to maintain product quality: dyes must be absolutely identical from batch to batch. Even minor amounts of contamination from a blue dye in one batch, for example, could ruin a subsequent batch of yellow dye. With 300 to 400 dyes manufactured annually at this plant in 50 batch-reaction vessels, frequent rinsing of equipment is required. As a result, Atlantic generates up to 400,000 gallons per day of wastewater contaminated with process and product chemicals, and the plant management sees little opportunity for reducing these wastes at the source. Similarly, Merck reported that its Rahway, New Jersey, plant must maintain such an extremely high degree of purity for its pharmaceutical product that waste reduction steps involving reuse of chemicals from other processes are not usually considered, due to the potential for contamination.

Process Type: INFORM found examples of waste reduction practices at plants using both batch and continuous processes, but found that the type of waste reduction practice adopted varied for the two types of plant processes. The majority of INFORM's 29 study plants use batch processing (see Table 7-1). Of these, eight reported waste reduction practices in their batch operations. Thirty-six percent were operational changes and chemical substitutions. Five plants reported waste reduction practices in their continuous operations, 30 percent of which were equipment changes, and none involved chemical substitutions.

There are fundamental distinctions between batch and continuous processes in terms of waste generation; continuous operations generate less waste per unit of product produced than batch operations during the manufacturing process (see Table 7-2). Most plants constructed to use batch or continuous processes do not alter process type. However, the Monsanto plant that INFORM studied, an unusual case of a plant that converted from a batch to a continuous process for its polystyrene production, affords a good example of the overall differences that can occur. In 1952,

Table 7-1. Product and Process Characteristics of INFORM Study Plants

INFORM Study Plant	Type of Products	Type of Process
CALIFORNIA		
Borden	resins, adhesives	batch/continuous
Chevron	agricultural chemicals	unknown
Colloids	antifoam, dispersants	batch
Dow	chlorinated organics	continuous
Fibrec	dyestuffs distributor	batch
Polyvinyl	resins, coatings	batch
Shell	metallic catalysts	unknown
Stauffer	pesticides	batch/continuous
Union	polymers, solvents	batch
NEW JERSEY		
Atlantic	dyes	batch
CIBA-GEIGY	dyes, epoxy resins	batch
Du Pont	additives, Freon, aromatics	batch
Exxon	lubricants, solvents	batch
Fisher	reagent chemicals	batch/reformulator
Halma	solvents, etchants, acids	batch
IFF	perfume & fragrance chemicals	batch/continuous
Max Marx	organic pigments	batch
Merck	pharmaceuticals, pesticides	batch
Rhone-Poulenc	aroma chemicals, pharmaceuticals	unknown
Scher	additives, specialty chemicals	batch
OHIO		
American Cyanamid	dyes, additives, explosives	batch
Bonneau	custom blend dyes, candles	batch
Carstab	additives	batch/continuous
Frank	specialty chemicals	batch
Monsanto	plastics, adhesives, resins	continuous
Perstorp	pentaerythritol, sodium formate	continuous
Sherwin-Williams	saccharin, flavors, additives	batch(70%)/cont(30%)
Smith & Wesson	chemical weapons	unknown
USS Chemicals	organic intermediates	continuous

Table 7-2. Organic Chemicals in Wastewaters from Organic Chemical Industry Operations

Type of Process	Pounds of Organic Chemicals* per 1,000 Pounds of Wastewater
Continuous -- Non-Aqueous Processes	0.3 - 3.7
Continuous -- Water Use Limited	0.47 - 21.5
Continuous -- Aqueous Processes	1.9 - 385
Batch -- Aqueous Processes	180 - 4,800

* Based on measurements of chemical oxygen demand (COD), a commonly used index of organic chemical contamination.

Source: U.S. Environmental Protection Agency, "Development document for effluent limitations guidelines and new source performance standards for the major organic products segment of the organic chemicals manufacturing point source category," Washington, DC, April 1974.

the plant's original process included a large number of parallel batch reactors, with air emissions amounting to about five percent of total production. Now, however, the process has been changed to a closed-system continuous process and air emissions have dropped to less than 0.02 percent of production.

A characteristically large source of wastes in batch process facilities — the largest source for the Fisher study plant in Fair Lawn, New Jersey — is the cleaning operations. Because the same reaction vessels are used over and over again to produce a wide array of materials in different batches, residues from one batch must often be cleaned out of the vessels before the next batch is begun. Cleanup may require not only the rinsing of reactors but the complete purging of all associated piping, pumps, storage units and other equipment.

The Borden plant found ways to reuse rinsewaters resulting from cleaning its reaction vessels between batches. Another operational change occurred at the Exxon plant. Exxon has instituted monitoring of its production processes to provide advance warning of process upsets so that operating staff can respond quickly to potential problems that would otherwise result in excessive waste generation.

Continuous process operations are rarely interrupted for cleaning as the equipment is dedicated to the production of one or a few products. INFORM's plants with continuous operations were found to have had more success in reducing wastes with equipment changes, particularly process-related add-on equipment such as filters, condensers and adsorbers that are used to capture materials for reuse in the production process. Perstorp, for example, installed an air pollution control device known as a baghouse filter that allowed the company to capture emissions of chemical dust and reuse the material in its continuous manufacturing process.

Such equipment as baghouse filters are certainly not exclusive to continuous process plants. Atlantic Industries' batch plant makes extensive use of filters for air pollution control, but not waste reduction. The large variety of materials used and trapped in the filters makes it infeasible to reuse the particulate materials and they are discarded as wastes.

It is important to note that, because the volume of product is generally higher at plants with continuous operations, the total volume of waste can be larger than at a plant with batch operations, even though the former produce less waste for each pound of product manufactured. In addition, the differences discussed here regarding batch and continuous plants apply to waste from the manufacturing process, but do not apply to wastes from other aspects of a plant's operations such as loading or storage.

<u>Five of the seven plant characteristics could not be clearly identified as having any significant influence on hazardous waste reduction.</u>

<u>Plant size</u>: It was impossible to obtain enough data to reach any conclusions about the influence of plant size on waste reduction practices due to the limited cooperation of smaller plants (those with fewer than 50 employees). Of the 13 plants that cooperated in INFORM's study (out of a total of 29) only two were small facilities.

There are many possible influences that can be associated with size. Large and small plants are likely to differ in their access to technical expertise, availability of capital, or the impact of regulations and liability (for instance, small businesses are exempt from some regulations). The impact of these influences on waste reduction practices merits further investigation. As evidenced by the sketchy profiles on small companies in Part II of this study, few details are available on the types and extent of wastes generated by small plants, or the impact that waste reduction is or could be having on these facilities.

Age of plant: There was no apparent relationship between the age of a plant and waste reduction. Relatively new facilities like the Perstorp plant (built in 1971) and USS Chemicals (1962) did not uniformly differ in the type or impact of waste reduction practices from older facilities such as Exxon (1921) or Stauffer (early 1900s). The date a plant was built may not bear a direct relationship to the age of physical facilities or processes at the plant. Even though Merck's plant in Rahway began operations in 1927, most of the processes at the plant are relatively new, as is the equipment in use.

Geographical location: There was no indication that plants in any one of INFORM's three study states were achieving significantly more hazardous waste reduction than in the other states, despite differences in the types and prevalence of waste management and disposal practices. These latter differences stem from both the differing regulatory programs in each state and geographical differences in available waste management options. Regulatory controls on specific toxic chemical air emissions in New Jersey, for example, are far more stringent than in California or Ohio, leading to different pollution control practices in New Jersey plants. Geographical variation makes evaporation ponds a viable waste management option only in California; the absence of appropriate geological sites for construction of deep wells rules out this waste management option in New Jersey. While such differences may influence plants' waste treatment and disposal choices, INFORM found they had no overall impact on whether waste reduction was chosen over waste treatment and disposal.

Company type or ownership: Neither the nature of a company's major line of business nor the type of ownership could be identified as exercising an influence over waste reduction practices in INFORM's group of study plants. Among these were plants belonging to privately-owned companies and those belonging to publicly-owned (and in one case, Rhone-Poulenc, owned by the French government). The primary business of some of the companies was chemical manufacturing while for others it was in such other areas as steel, food, and petroleum.

The Impact of Environmental Regulations on Hazardous Waste Reduction

More than 20 federal laws, along with hundreds of legislative counterparts at the state and local level, govern the manufacture, use, transportation, management and ultimate disposal of hazardous chemicals. INFORM found that these laws, implemented in quite different ways by the many different regulatory agencies, could in some cases encourage and in other cases inhibit the search for and adoption of waste reduction practices.

Only one regulation directly mandates waste reduction.

Despite the great variety of regulations that exist addressing hazardous waste problems, only one was found by INFORM that directly required waste reduction by organic chemical plants. It affected practices at just one of the 29 plants in this study. The state of New Jersey's regulations for the control of emissions of volatile organic substances (VOS) required Exxon Chemicals to install floating roofs and conservation vents that prevent organic vapors from forming and being released. Both conservation vents and floating roofs are waste-reduction devices, that is, equipment added to an emissions source that prevents the generation of excess air emissions (as distinct from air pollution control devices such as scrubbers, which merely transfer pollutants from the air into wastewater).

New Jersey's regulations for the control of VOS states that:

> No person shall cause, suffer, allow or permit the storage of a VOS in any stationary storage tank having a maximum capacity of 10,000 gallons (37,850 liters)

or greater unless such stationary storage tank is equipped with control apparatus as determined in accordance with the procedure (detailed elsewhere in the regulations).

The procedure referred to categorizes each tank as belonging to one of three "Ranges" according to the size of the tank and the vapor pressure (an indication of the ease of evaporation) of the material in storage. The regulations call for the following controls:

Range I: No control apparatus required.
Range II: Conservation vent required.
Range III: Floating roof required.

Exxon's chemical plant in Bayway, New Jersey, had numerous storage tanks in the Range III category and installed the required floating roofs, reducing emissions from each tank by 85 percent or more. These roofs proved to be so cost-effective -- that is, by preventing the loss of valuable materials, they saved more money than they cost in terms of installation and operating expenses — that Exxon installed floating roofs elsewhere that were not mandated by regulations.

Three ways in which regulations indirectly stimulate hazardous waste reduction.

Despite an almost complete absence of direct regulations requiring waste reduction, environmental regulations, individually and collectively, are stimulating hazardous waste reduction in at least three different ways: a) by restricting hazardous waste handling alternatives; b) by increasing regulatory requirements for current waste generation and handling practices; and c) by drawing attention to poor company practices.

By restricting hazardous wastes handling alternatives:
The impact of increasing regulatory restrictions is well-illustrated by the actions of two study plants, Borden and Sherwin-Williams, in response to restrictions

imposed by the sewage treatment plants that receive their wastewaters.

Sherwin-Williams implemented a variety of operational changes and process improvements to reduce trichlorobenzene in its wastewaters in response to a cease-and-desist order issued by the Metropolitan Sewer District (MSD) of Greater Cincinnati. The MSD receives wastewaters from numerous industrial facilities, including two plants -- Sherwin-Williams and Carstab -- included in INFORM's study. In the early 1980s, MSD plant workers were becoming ill and, in a few cases, collapsing after exposure to organic vapors emanating from the wastewaters received at the sewage treatment plant. The vapors had the distinctly characteristic odor of trichlorobenzene, a hazardous substance used in large quantities at Sherwin-Williams for the manufacture of the synthetic sweetener, saccharin.

Upon investigation, MSD identified Sherwin-Williams as the source of the dangerous chemical. It issued a cease-and-desist order and subsequently altered the company's permit to limit the quantity of organic vapors that could be emitted from the company's wastewater discharges. Within several months, the operational changes and process improvements that reduced trichlorobenzene waste generation at the source, in combination with other waste management options, lowered discharges of trichlorobenzene to MSD by 25 to 30 percent. The specific nature of the changes were not reported to MSD nor would the company make them available to INFORM.

In another case, in response to restrictions imposed by the Union Sanitary District sewage treatment plant, Borden Chemicals took steps leading to significant waste reduction at its plant in Fremont, California. The District's stringent limit on phenol concentrations in Borden's wastewater was a burdensome requirement for Borden since phenol is a major raw material at the plant. Plant raw wastewaters contained phenol levels far in excess of the one to five milligrams per liter limit established by the sewage treatment plant.

Borden's initial response was to install a wastewater treatment system to lower the concentrations of phenol in its discharge to acceptable levels. However, problems with the treatment system, in concert with rising waste management costs and the stringent sewage treatment plant requirements, eventually led Borden to reduce phenol wastes at their sources through a variety of operational waste reduction practices.

Three other plants in this study adopted waste reduction practices in response to air emissions regulations controlling organic vapor emissions. USS Chemicals used a resin adsorber to reduce cumene emissions. Dow reduced vapor releases during material transfers and Merck introduced closed-loop process technology to prevent the loss of methylene chloride.

By increasing regulatory requirements for current waste generation and handling practices: Regulatory requirements for the disposal of solid hazardous wastes have been causing a rise in plant costs for disposing of them. If an industrial plant treats, stores (for more than 90 days) or disposes of hazardous solid wastes on-site, it must file as a Treatment, Storage or Disposal Facility (TSDF) under provisions of the Resource Conservation and Recovery Act (RCRA). The 1980 list of wastes regulated under RCRA contains over 300 individual chemicals and 80 combined wastestreams classified as hazardous (see Appendix A). The technical and administrative requirements for TSDFs are costly and complex. To avoid having to file as a TSDF, four INFORM study plants -- Borden Chemicals in California, Carstab Corporation and USS Chemicals in Ohio, and Exxon Chemicals in New Jersey -- reported changes in their management of hazardous wastes. Only the changes at the Borden plant, however, could be identified as involving waste reduction measures.

The Borden plant reported operational changes consisting of waste reduction practices that enabled the plant to close its evaporation pond. Although the pond had been built in 1978 before RCRA was fully implemented, its continued operation would have caused the Borden plant to be classified as a TSDF facility under the emerging requirements of the law. Borden's revised rinsing procedures helped eliminate the need for the evaporation pond by reducing the level of phenols in its wastewater.

To avoid TSDF classification, Exxon instituted wastewater treatment; and Carstab reported changes that allowed its plant to discontinue use of a treatment tank to separate toluene. Exxon's change was not waste reduction; insufficient details were given to evaluate whether Carstab's changes constituted waste reduction.

The effect of rising waste management costs resulting from increasing regulatory restrictions are also discussed below in the section on cost factors encouraging waste reduction.

<u>By increasing regulatory and public scrutiny</u>: In enforcing regulations, the active scrutiny of industry practices by government agencies can trigger waste reduction. Thus, CIBA-GEIGY instituted changes at its plant in Toms River, New Jersey, to reduce nitrobenzene wastes by an estimated 90 percent as a result of inspections and review by the U.S. EPA and New Jersey Department of Environmental Protection (NJDEP). The Toms River plant had frequently exceeded the limit NJDEP imposed of 76 pounds per day of nitrobenzene in its wastewaters. Continuing pressure from environmental agencies to eliminate permit violations was enough to induce the plant to adopt waste reduction for nitrobenzene.

In addition to focusing on nitrobenzene wastes, recent agency and public pressures have caused CIBA-GEIGY to re-evaluate its entire program of waste management:

--The U.S. EPA has been investigating groundwater quality at the Toms River plant site to see if the large quantities of hazardous wastes disposed of in the past have led to environmental problems that would require cleanup of the site.

--Current waste disposal practices in an on-site landfill have been scrutinized by the NJDEP to insure that they were in compliance with state regulations.

--Media coverage and public hearings throughout New Jersey focused tremendous public concern on CIBA-GEIGY's waste disposal practices.

--The environmental activist group, Greenpeace, selected the plant as the target for protest actions after learning that it was the only facility in New Jersey that discharged chemical wastes into the Atlantic Ocean.

--After the plant's permit to discharge its wastewaters into the Atlantic had expired and was subject to renewal, the plant was involved in public hearings and a NJDEP review of permit requirements.

In 1985, the U.S. EPA's investigation at this plant site found that, as a result of past and recent landfill practices, groundwater has become contaminated at the site. The NJDEP levied a $1.45 million fine against the plant -- the largest environmental fine ever imposed in New Jersey -- with CIBA-GEIGY to bear the eventual multi-million dollar costs of site clean-up. The plant's renewed wastewater discharge permit, along with other stipulations imposed by state officials, currently constitute the most stringent environmental controls ever required by the NJDEP. New Jersey's Division of Criminal Justice also began investigating whether to bring criminal charges against plant officials.

Under the impact of these events, CIBA-GEIGY's entire spectrum of waste generation, management, treatment and disposal at the plant has undergone re-evaluation. The combined regulatory, economic and public pressures confronting CIBA-GEIGY may well lead to additional waste reduction measures at the plant.

Three ways in which regulations indirectly inhibit hazardous waste reduction.

In the examples just described it is clear that regulations can promote waste reduction practices. Perhaps the more significant finding from this study, however, is that the regulatory structure permits alternative waste treatment and disposal methods that much more frequently have the effect of discouraging waste reduction.

This is possible because environmental laws regulate particular wastes in particular media rather than total wastes from a plant, regardless of types and where they are discharged. There are currently six hazardous air pollutants regulated under the federal Clean Air Act, 126 priority pollutants listed in the Clean Water Act and over 300 chemicals and 80 combined wastestreams defined as hazardous solid wastes under the Resource Conservation and Recovery Act. Thus, each law focuses on the different chemicals discharged to one medium and a hazardous waste regulated in one medium is not, therefore, necessarily regulated as hazardous if discharged to another.

As regulatory pressures have increased over the past years, the primary response among INFORM study plants has been to move wastes around, choosing the disposal media that are less regulated and, hence less expensive for the plants, rather than to search for ways to reduce these wastes at their source.

INFORM found that, by 1985, regulations had inhibited the search for waste reduction in at least three ways: a) by allowing access to no-cost or relatively low cost waste management alternatives; b) by continuing to provide regulatory exemptions; and c) by imposing regulatory fees that are too low to provide an incentive to reduce wastes.

<u>By allowing no-cost or low cost waste management alternatives</u>: The burning of hazardous wastes as fuel and the use of deep well injection for waste disposal were two legally permitted waste management options found to be cheaper and more attractive to INFORM's study plants than waste reduction. Both merit some detailed discussion.

The burning of hazardous wastes is one of the practices regulated by the federal Resource Conservation and Recovery Act (RCRA). Waste materials may be burned for either of two reasons. Some are burned solely to destroy them. These are subject to stringent regulatory requirements. Such wastes must be incinerated in highly effective burners that operate according to standards set by EPA to insure an extremely high degree of destruction. In addition to operating standards, hazardous waste incinerators are subject to record-keeping, monitoring, and financial security requirements established by federal and state authorities.

Other wastes are burned as fuels in order to recover usable energy. Wastes burned as fuels are exempt, as follows, from the requirements applied to other solid hazardous wastes:

--the waste generator need not notify federal or state agencies of the practice of burning wastes as fuels or of the quantity generated;

--the hazardous waste manifest system does not have to be used for waste shipments;

--administrative requirements such as facility closure plans, recordkeeping requirements, and financial safeguards governing hazardous waste incinerators do not apply;

--design, construction and operating standards for hazardous waste incinerators do not apply.

The fuel-use exemption offers an attractive option for managing hazardous wastes since it both saves money on fuel and the costs of waste disposal, and also avoids stringent and burdensome regulatory requirements. The exemption provides a compelling incentive for waste destruction rather than waste reduction.

The role that wastes burned as fuel can play in the economics of a plant's operations was seen at the USS Chemicals plant in Haverhill, Ohio. Tarry residue wastes from distillation towers in the plant's phenol manufacturing operations are generated at the rate of 175,000 gallons (about one million pounds) per month. These wastes have a heat value of 141,000 BTUs per gallon, about equal to conventional fuel oil, and have been used by the plant as supplemental boiler fuel since 1976.

With fuel oil prices at roughly a dollar per gallon, USS Chemicals saves about $175,000 per month on fuel costs by burning the phenolic wastes. Additional savings on a par with the fuel savings are realized when the plant avoids the costs of having to dispose of these materials at a commercial waste-handling facility.

The overall economic importance of avoiding these fuel costs and waste disposal costs is evident from an incident that threatened the continuation of this practice. The phenol tars were not considered by USS Chemicals to be legally classified as hazardous wastes due to the fuel-use exemption, and the practice of burning this material was not included in the plant's annual reports to the Ohio EPA on its hazardous waste generation.

When Ohio EPA inspectors became aware of the practice, the plant was ordered to discontinue burning the waste materials. USS Chemicals argued that, if it had to comply with the order, the plant would be forced to shut down its operations entirely, due to the absence of on-site capacity for storing the waste materials, and the lack of an alternative for disposing of them in an economically feasible manner. Ohio EPA eventually rescinded its order to stop burning the phenol tars pending further investigation of the situation.

It is by no means evident that USS Chemicals had a viable waste reduction option for the phenol tars. But it is clear that the economic advantages of producing wastes that are subsequently used as fuels eliminates the economic motivation for looking for waste reduction opportunities for this wastestream.

The practice of burning wastes as fuel does entail some costs due to the need to conduct additional monitoring and trial burns in order to insure thorough combustion of the waste material. For USS Chemicals, burning 175,000 gallons per month, the costs of monitoring are minor, relative to the overall savings due to the large volume of wastes. However, for Merck's plant in Rahway, New Jersey, the costs of burning 500 gallons per month of waste solvents as fuels became prohibitively expensive due to the increasingly stringent requirements the New Jersey Department of Environmental Protection imposed for continued and costly monitoring of the combustion products. Merck discontinued the practice in favor of off-site disposal even though the latter meant not only paying for disposal of the wastes, but having to assume the cost of buying additional fuel to replace the 500 gallons no longer being burned.

The Resource Conservation and Recovery Act's fuel-use exemption was deliberately designed to encourage the burning of hazardous wastes. It is therefore no wonder that, nationally, the practice of burning waste materials for fuel is believed to be widespread. A 1981 U.S. EPA report analyzed the impact of RCRA's emerging hazardous waste programs:

> "The Resource Conservation and Recovery Act of 1976 encourages the conservation of valuable resources, including energy, consonant with the protection of environmental quality. Recently promulgated hazardous waste regulations allow

the onsite use of wastes as fuels with few restrictions."[1]

It was therefore anticipated by U.S. EPA that a significant change in disposal practice would occur. Many wastes previously discarded would instead be burned in an attempt to qualify under the regulatory exemption. A further stimulus to incineration would be the escalating prices of the conventional fossil fuels that had been used almost exclusively for energy production. Any material that could supplement fossil fuels, at lower costs per BTU, could be seriously considered for use.

Deep well injection is another attractive waste management option. While the fuel use exemption is the only hazardous waste disposal option INFORM is aware of that can actually save a plant money, deep well injection is one option available that, although costing a plant money, is inexpensive enough to remain attractive.

From a regulatory point of view, deep well injection has been something of an environmental orphan. Such injection is widely practiced: 60 percent of all RCRA hazardous wastes are disposed of in deep wells and half this quantity is from the organic chemical industry (see Chapter 5). It is, however, not regulated under the major federal law for protection of clean water (the Clean Water Act). Nor is it regulated under the control of solid hazardous wastes (RCRA), although amendments to RCRA in 1984 and 1985 have placed underground injection increasingly under the control of this statute, and have led to detailed assessments by the U.S. EPA of the status of underground injection. These amendments, however, have not yet led to significant restraints on disposing of wastes in deep wells. Instead, deep wells come under the purview of the Safe Drinking Water Act of 1974, which establishes construction, operating, monitoring and closure standards for the wells, but places little prohibition on the types or quantities of waste materials that can be disposed of.

At USS Chemicals, more than 100,000 gallons per day of contaminated process wastewater are disposed of through deep well injection. The company's daily discharges contain 2,000 to 12,000 pounds of phenol along with other chemicals such as aniline and acetone which are used or produced at the plant. USS Chemicals considers the practice an attractive waste management option that is environmentally safe, easy to operate, and relatively low cost.

USS Chemicals would provide no detailed cost figures on this practice. However <u>Chemical Week</u>, a chemical industry magazine, reporting the extent to which deep well injection is practiced by industry, concluded that its use "is expected to increase because of deep well disposal's low cost relative to other waste disposal and treatment technologies."[2]

U.S. EPA officials in charge of underground injection have no cost data, but officials at Ohio EPA in charge of deep well injection estimated the typical cost of operating a deep well at $50,000 per year. Such wells are particularly attractive in Ohio because a large portion of the operating cost is electricity and Ohio utilities charge very low rates. If the Ohio EPA estimate is correct, then a deep well costs less than $150 per day to operate and may dispose of more than six tons of hazardous organic wastes daily -- at a cost to a plant of $25 per ton.

USS Chemicals officials have looked into two alternatives to deep well injection — recovering phenol from the wastewater prior to injection, and reducing phenol wastes at the source. They concluded that waste reduction, if possible, would be too costly and technologically difficult, and that recycling, although technologically straightforward, would be too costly to compete with continued injection.

<u>By continuing to provide regulatory exemptions</u>: The main thrust of Clean Air regulations has been to regulate volatile organic substances that play a part in the creation of smog and ozone pollution. This

125

focus on ozone pollution has resulted in exemptions from regulation for small volumes of chemicals or those not seen as contributing to the ozone problem.

The state of California has studied the impact of these air pollution regulations on the emissions of hazardous chemicals in the state. A 1982 report to the California Air Resources Board found that four storage tanks at the INFORM study plant owned by Dow in Pittsburg, California, were among the major industrial sources in the state of emissions of two potentially carcinogenic chemicals, carbon tetrachloride and perchloroethylene. The Dow tanks are regulated as sources of precursors to ozone, rather than as sources of carcinogens. They, therefore, are exempt from air emissions controls, as shown in Table 7-3.

<u>By imposing low regulatory fees</u>: Industrial plants pay fees to government agencies for pollution control permits and when discharging to sewage treatment plants. No INFORM study plant reported that they were high enough to induce them to search for ways to reduce wastes of the chemicals at source.

An unusual fee structure for air emissions in southern California provided a potential economic mechanism for encouraging waste reduction: the smaller the quantity of emissions coming from a plant, the smaller the fee. Only one of INFORM's study plants, Union Chemicals in La Mirada, California, was subject to these emission fees.

The South Coast Air Quality Management District (SCAQMD) requires plants within its jurisdiction to pay a fee for emissions of organic vapors in excess of 10 tons (20,000 pounds) per year. SCAQMD's fees are based on emission quantities as follows:

Table 7-3. Emissions Exempted from Controls at California Plants owned by Companies Represented in INFORM's Study

Plant and Source	Chemical	Maximum Emissions (lbs/yr)	Regulatory Status
Stauffer (San Pedro) 3 storage tanks	Ethylene Dichloride	23,724	Exempt since vapor pressure less than 1.5 psi
Allied (El Segundo) 1 storage tank	Carbon tetrachloride	5,500	Exempt since tank smaller than 39,630 gallons (150 cubic meters)
Dow* (Pittsburg) 4 storage tanks	Perchlorethylene and carbon tetrachloride	12,700	Exempt since tank smaller than 39,630 gallons and/or substances not classified as organic liquids
Du Pont (Antioch) 1 storage tank	Carbon tetrachloride	20,857	Exempt since substance not classified as organic liquid

* an INFORM study plant

Source: Science Applications Inc., "Inventory of Carcinogenic Substances Released into the Ambient Air of California: Phase II," prepared for the California Air Resources Board, Sacramento, CA, November 1982.

Chemicals	Fee per Ton of Emissions
Methane	$ 6.45
Methylene chloride, 1,1,1 trichloroethane, trifluoromethane and chlorinated-fluorinated hydrocarbons	$ 9.00
Other organic chemical vapors	$ 52.00

The "other organic chemical vapors" listed are the substances that can contribute to ozone formation and are subject to a higher fee as an effort to control ozone pollution. Chemicals such as methylene chloride and trichloroethane, which are regulated as toxic chemicals under clean water and hazardous solid waste laws, are given a lower priority in SCAQMD's fee system since these substances do not contribute to ozone formation.

The imposition of emission-based fees could provide a direct economic incentive for waste reduction. In Union Chemicals' case, however, the fee structure has no impact in encouraging the plant to reduce emissions below the current level because the plant pays no fees: Union's annual emissions fall below the 10 ton exemption limit.

None of INFORM's other study plants is subject to such a fee system. SCAQMD's counterpart in northern California, the Bay Area Air Quality Management District, does not use an emissions-based fee system, nor do the air agencies in New Jersey or Ohio. All three states have a permit fee system that charges companies for each permitted source of emissions and charges vary, depending on the size and type of source. But since these fees are related to the number of emission sources, they cannot be minimized by reducing emissions and therefore provide little, if any, economic incentive for waste reduction.

Fees imposed by sewage treatment plants are the other type of regulatory fees that most of the study plants pay. Seventeen of INFORM's 29 study plants dispose of their wastewaters -- often highly contaminated with industrial chemicals -- to publicly-owned sewage treatment plants, making this one of the most widely-used means of waste disposal among INFORM's study plants.

Sewage treatment plants not only impose fees on those plants discharging into the treatment facility, but also can impose restrictions on the types and quantities of materials discharged. Table 7-4 compares the discharge fees and limitations applied to the 17 INFORM study plants that discharge to local sewage treatment plants.

The only uniform practice among the sewage treatment plants is the manner in which the fees are structured: dischargers, with few exceptions, pay according to the amount of water discharged and the amount of solids and organic material present in the discharges. Otherwise the fees and controls vary enormously depending on the sewage treatment plant. Flow charges differ by a factor of 14, from $87 to $1,270 per million gallons. Other charges vary by a factor of 25 or more: from $12 to $396 per ton of organics, and from $42 to $1,136 for solids.

The combined impact of these variable charges produces a large difference in the costs of sewering wastes. A wastewater stream consisting of a million gallons of water, one ton of organics (measured as COD or BOD -- see Table 7-4), and one ton of suspended solids would cost $224 to discharge in Los Angeles County, and $1,673 to discharge to the Bayshore (NJ) Regional Sewerage Authority -- more than a seven-fold difference.

Restrictions on the types and quantities of materials that can be discharged were equally variable. Three

Table 7-4. Variable Fees and Restrictions at
Sewage Treatment Plants Servicing INFORM Study Plants

INFORM Study Plant	Sewage Treatment Plant	FEES				Effluent Limitations for Hazardous Organic Chemicals 3/
		Flow $/million gallons	Organics $/ton 1/		Solids $/ton 2/	
CALIFORNIA						
Borden	Union Sanitary	$ 660.00	$123.20 (COD)		$286.68	Phenols limited to 1-5 mg/l
Colloids 4/	West Contra Costa	$ 87.00	0		0	None because no federal guidelines established
Chevron, Stauffer	Richmond	$1002.00	$370.00(BOD) 5/		$300.00 5/	Phenols,chlorinated hydrocarbons,toluene,xylene, brominated organics,phenaldehyde,benzene--1 mg/l
Union	Los Angeles County	$ 126.00	$ 19.80 (COD)		$ 78.40	Max 102 mg/l total toxic organics
NEW JERSEY						
Atlantic,Fisher,Scher	Passaic	$ 175.00	$ 92.00 (BOD)		$ 42.00	Not reported to INFORM
IFF	Bayshore Regional	$ 141.00	$396.00 (BOD)		$1136.00	No benzene/toluene in flammable amounts
Max Marx	Joint Meeting	$ 98.14	$210.77 (BOD)		$ 85.83	None because no federal guidelines established
Merck	Linden Roselle	$ 475.00	$206.00 (BOD)		$ 326.00	Phenols limited to 10 mg/l
Merck	Rahway	$ 150.00	$150.00 (BOD)		$ 120.00	None because no federal guidelines established
Rhone-Poulenc	Middlesex 6/	$ 97.47	$ 11.58 (BOD)		$ 62.96	Not reported to INFORM
OHIO 7/						
American Cyanamid	Marietta	$800(avg)8/	$100.00(BOD)		$100.00	No toxic, explosive or hazardous substances
Carstab, Sherwin-Williams	Cincinnati	$800(avg)8/	9/		9/	No toxic substances
Frank Enterprises	Columbus	$1269.88	$200.00 (BOD)		$200.00	None because no federal guidelines established
Perstorp Polyols	Toledo	$1223.10	$140 (BOD) or $60.00(COD) 10/		$60.00 10/	Phenols limited to 30 mg/l

1/ Organic discharges are indicated by biological oxygen demand (BOD) or chemical oxygen demand (COD), commonly used indices of organic chemical contamination.
2/ Solids discharges are indicated by conventional testing for suspended solids (SS), a commonly used index of solids content.
3/ Limitations are for hazardous organic chemicals only. Limits on metals, other organics, temperature or pH are not listed.
4/ Colloids is classified as a "Commercial Discharger" and the fees shown are for this classification. Facilities classified as "Industrial Discharger" pay according to a different fee schedule.
5/ Fees are charged only if concentration of the contaminant exceeds a minimum limit established by the treatment plant.
6/ Middlesex uses a complex system of debt service surcharges to levy additional fees. These can more than double the actual cost of discharges to this treatment plant.
7/ The Ohio treatment plants base their flow charges on amount of water used rather than amount of wastewater discharged.
8/ Actual flow fee rates decrease with increasing quantities of flow.
9/ Cincinnati uses a complex system of debits and credits based on concentration of COD and solids to assess charges.
10/ Toledo has a "High Strength Discharge Program" which assesses the Perstorp plant fees as shown.
Source: All information based on a 1985 telephone survey of the sewage treatment plants.

sewage treatment plants impose limits on phenol discharge concentration from INFORM's study plants: Stauffer and Chevron are limited to one mg/l phenol; Borden to one to five mg/l; and Perstorp to 30 mg/l. Other sewage treatment plants impose no particular restrictions on hazardous chemical discharges while some have identified specific contaminants that are prohibited or severely restricted. At Borden Chemical and Sherwin-Williams, sewage treatment plant restrictions were instrumental in spurring waste reduction, as described above.

The wide variation in fees paid to sewage treatment plants were not reported as an influence at any of INFORM's study plants. Thus, while the sewage treatment plant limitations were cited as a spur to waste reduction at two plants, not even the highest fees acted to stimulate waste reduction at any of INFORM's study plants.

Regulatory oversight is highly sporadic and inconsistent.

As indicated earlier in this chapter, regulations can be an effective incentive to waste reduction. However, the overall experience of INFORM's study plants shows that the influence of regulations is highly sporadic and inconsistent.

Actual implementation of regulatory programs varies greatly. At Sherwin-Williams, an individual hazardous chemical, trichlorobenzene (TCB), was singled out for intensive regulatory scrutiny. Until chemical emissions caused the collapse of sewage treatment plant personnel and until the sewage treatment plant could identify Sherwin-Williams as the source of the TCB discharges, there was no regulatory pressure to eliminate the company's TCB discharges. Once pressure was applied, it took Sherwin-Williams just a matter of months to solve the problem. Although Sherwin-Williams handles more than 15 other hazardous chemicals, no similar pressure by regulatory agencies has been applied to reduce the wastes stemming from their use at the plant.

Borden also responded to stringent limits, in its case to restrictions on phenol discharges, with waste reduction measures that led to a dramatic reduction of 93 percent in the amount of resin wastes. Perstorp's plant in Toledo, Ohio, is also subject to phenol limits but, since the plant does not use phenol in its operations, the restrictions have no impact on their waste management. The hazardous chemicals that Perstorp does handle -- formaldehyde and acetaldehyde -- are not restricted at all by the receiving sewage treatment plant and, as a result, there is no regulatory pressure to pursue waste reduction.

During the course of INFORM's research, regulatory attention at times seemed almost arbitrary, focusing on individual chemicals, certain processes, or selected plants while virtually ignoring others. Sometimes individual hazardous chemicals were singled out for attention while others were ignored, as was the case at Sherwin-Williams and Borden. The South Coast Air Quality Management District charges $52.00 per ton for emissions of hexane, a non-hazardous chemical, but only $9.00 per ton for methylene chloride, a hazardous chemical according to the criteria used in this study. Federal and state environmental agencies placed enormous pressure on CIBA-GEIGY to reduce nitrobenzene discharges from its Toms River plant in New Jersey -- a pressure that led to waste reduction practices for this chemical -- but gave little detailed attention to the hundreds of other chemicals handled at the plant.

Entire facilites are subject to an equally variable degree of regulatory pressures. The enormous variation in the costs of discharging wastes to sewage treatment plants and the degree of restrictions on the discharges has already been documented. Sewage treatment plant authorities in California regulated discharges from Union Chemical's plant based on a knowledge of its latex manufacturing operations, but were not even aware that the plant had solvent-blending operations as well. Environmental officials in Ohio, after searching for information on the Smith & Wesson Chemical plant in

Rock Creek, came up empty-handed and notified INFORM that the plant was probably no longer in business. After much prodding, local air officials did eventually locate some active air permit records for this facility. Even though Du Pont's Chambers Works plant in New Jersey did not cooperate in this study, INFORM was able to compile a 31-page profile on the plant based on information from records kept by state and federal environmental agencies. For another New Jersey facility, J.E. Halma, the profile is only two paragraphs, reflecting an almost complete absence of information on this company in public files.

INFORM found that where regulatory oversight was given to a facility, or to a particular process at a facility, or to a single chemical, this attention was critical in prompting waste reduction practices. Where regulatory pressures were seen to be minimal, potential waste reduction opportunities were not being taken advantage of.

The Impact of Cost Factors on Waste Reduction Practices

Every one of the 13 plants that INFORM interviewed indicated that its level of hazardous waste reduction was influenced by cost factors. Materials cost savings and reduced waste management costs, many of course related to meeting regulatory requirements, stimulated hazardous waste reduction, whereas implementation costs tended to inhibit source reduction. Cost factors are discussed in this section along with specific examples at INFORM study plants of how the various costs affected waste reduction practices.

<u>Potential cost savings in two areas encouraged waste reduction.</u>

Both the cost of lost materials and the cost of waste management can be lowered by waste reduction practices. Any practice that prevents material from becoming a waste reduces the amount (and dollar value) of lost materials and reduces the overall quantity (and potential cost) of wastes requiring management. If the reduced costs outweigh the costs of implementing waste reduction, then the plant comes out ahead from a strictly economic point of view. The savings from waste reduction practices that could be identified at INFORM's study plants are shown in Table 7-5.

Savings of material costs: INFORM study plants reported that waste reduction changes were implemented to increase overall efficiency and reduce the costs of material losses at the plant but, at the same time, these changes reduced wastes.

Perstorp, for example, reported that costs of material losses were one of the key reasons why it undertook modifications at its Toledo plant. These led to a 50-70 percent reduction in the amount of chemicals lost in the plant's wastewater and other reductions in air emissions. Perstorp reported that the economic benefits resulting from waste reduction efforts, along with other improvements made in energy efficiency and product quality, have made it possible for the plant to operate profitably, which was not the case when Pan American Chemicals, the previous owner, operated the plant.

Atlantic Industries also reported that economic and competitive pressures in the industry were prompting the plant to increase its efforts at waste reduction through process refinements. For Atlantic's major dye products, a primary objective of process development research is to reduce the amount of excess reactants in use and increase product recovery as a means of improving the overall economics of the process while at the same time reducing wastes.

As an example, Atlantic cited improvements in a diazo dye manufacturing process that resulted in an eight percent increase in yield and a consequent annual reduction of 55,000 pounds of organic chemical wastes and 250,000 pounds of inorganics along with a 500,000 gallon overall reduction in water use. These waste reduction measures were primarily motivated by the

Table 7-5. Material and Cost Savings of Waste Reduction Practices

INFORM Study Plant	Waste Reduction Practice (Cost)	Raw Material Savings	Product Savings	Waste Reduction	Disposal Cost Savings
CALIFORNIA					
Borden	Recycle resin manufacturing rinsewater and other improvements	--	--	93% organic materials in wastewater; 325 cubic yards/yr of sludge	Sludge: $48,750 (based on $150/cu yd)
Stauffer	DEVRINOL process change	--	$200,000/yr	--	--
	Kerosene reuse	800 gallons and $1,000/yr	--	--	--
	Replace packed seals with mechanical seals	2,600 gallons and $37,000/yr	--	--	--
	VAPAM process change	--	--	535 drums/yr	$28,085/yr
NEW JERSEY					
Atlantic	Process change in diazo dye manufacturing	--	8% improved product yield	55,000 lbs/yr organics; 250,000 lbs/yr inorganics; 500,000 gals/yr water use	--
Exxon	16 floating roofs at capital cost of $5,000 to $13,000 each; annual operating costs of $500 to $1000 each	681,810 lbs/yr (5 million lbs since 1974) $205,305/yr	--	90%	--
Merck	Internal recovery of isoamyl alcohol and methylene chloride in new Primaxin process	2,610,000 lbs/yr methylene chloride and 423,000 lbs/yr isoamyl alcohol	--	--	--
	Internal recovery of acetone in new Primaxin process	229,600 lbs/yr of acetone	--	--	$47,750/yr saved in sewer charges
OHIO					
Monsanto	New polystyrene process	--	5% improved product yield	99% or 59,800 lbs/day of process chemicals	--
USS Chemicals	Cumene adsorption system	715,000 lbs/yr $178,000/yr	--	80%	--
	Condenser for cumene at cost of $5,000	400,000 lbs and $100,000/yr	--	--	--

-- means data not reported to INFORM

desire to decrease material losses by improving the overall yield of the process. But Atlantic also realized savings in waste management costs since fewer contaminants in the wastewater stream meant smaller fees payable to the local sewage treatment plant.

USS Chemicals also reported that dramatic changes in the economics of phenol production were a major factor in motivating the company to look more carefully at the costs of material losses. From 1969, when the first phenol production unit was built, to 1979 when the Phenol II unit went into operation, the cost of cumene, the major raw material used in phenol manufacture, rose from five cents to 30 cents a pound. The rise prompted the plant to install resin adsorber systems that prevented the loss of 715,000 pounds per year -- almost $200,000 worth -- of cumene air emissions.

The economic incentives for decreasing material losses through waste reduction are related not only to the amount and value of the materials lost, but to the amounts lost relative to the scale of operations at a facility. This is particularly evident at USS Chemicals where the large volumes of chemicals produced at the plant can easily mask material losses of hundreds of thousands of pounds.

Even at the time when USS Chemicals was focusing attention on the economic and technical feasibility of using resin adsorbers, a valve in its original phenol manufacturing unit was leaking cumene into the atmosphere at a rate of 400,000 pounds per year (a material loss to the plant of about $100,000). However, for the enormous scale of operations at USS Chemicals, this loss represented only 0.06 percent of the 700 million pounds of cumene used annually at the plant.

The USS Chemicals loss of almost half a million pounds of raw material went undetected in both the material and economic record-keeping tallies at the plant -- it was just not a large enough quantity to be noticed. The plant can keep track of its materials to within plus or minus one percent, a level of accuracy that it considers state-of-the-art for the chemical industry. Nevertheless, this still leaves the possibility of millions of pounds of materials going unaccounted for. Not until an employee noticed the odor of cumene, identified the source of the leak and suggested use of a surplus condenser as a means of reducing this leak was the loss prevented.

Exxon also handles similarly large quantities of chemicals -- more than a billion pounds of raw materials per year -- and the plant's waste reduction savings, although substantial, represent only a tiny fraction of the overall quantity of materials in use. The plant has installed 16 floating roofs to minimize loss of organic chemical vapors from storage tanks. Each tank, equipped with a floating roof, typically handles from 50 million to 150 million pounds of chemicals per year; each floating roof saves less than 1/10 of one percent of this annual amount.

These small percentage savings, however, add up to large material quantities. The 16 floating roofs, since their installation in 1974, have prevented the loss of five million pounds of chemical vapors representing a current savings of $200,000 per year. In comparison, capital costs for all the roofs were less than $200,000 and annual operating and maintenance costs are less than $16,000. Despite the savings achieved by using the floating roofs, Exxon's initial adoption of this waste reduction practice was prompted by regulatory requirements of the New Jersey Department of Environmental Protection rather than by economic considerations.

Savings in waste management costs: Proper waste management typically involves equipment, materials, staff, transportation, energy, administration, and the use of commercial waste handling services, all of which add to the overall costs of a plant's operations. Waste management costs were cited as an important factor influencing the adoption of waste reduction practices by six INFORM study plants.

Disposal cost savings were in some cases cited as the primary motivation for initiating a waste reduction practice; in other cases they represented the fringe benefits of a practice initiated for some other reason, such as regulatory pressures.

One example is the Borden Chemical plant in Richmond, California, which until 1981 was generating an estimated 350 cubic yards (a cubic yard is roughly equal to one ton) of phenolic resin sludge, chiefly from its on-site treatment of wastewaters in an evaporation pond. The pond, built in 1978, was seen by Borden as the simplest and most cost-effective means of handling its organic wastes. But potentially costly problems began surfacing by 1980. Borden's higher than expected rate of sludge generation and unusually heavy rainfalls were causing the pond to fill to capacity sooner than anticipated. The costs of enlarging the pond were considered prohibitive. Additionally, emerging regulations under the federal Resource Conservation and Recovery Act would require Borden to file as a hazardous waste "Treatment, Storage or Disposal Facility" if the company continued to operate the pond. This raised the spectre of administratively complex and expensive operational, monitoring and record-keeping requirements. As an added economic blow, the costs of disposing of the sludges accumulating in the pond were rising rapidly. Based on 1982 prices, Borden would have to spend over $100,000 to haul the sludges in the pond to a commercial waste handling facility for disposal.

Faced with these rising waste management and disposal costs, Borden's plant manager (in the extra time he had available due to a slump in business) began looking at ways to reduce the generation of phenol wastes at source, to see if it was possible to eliminate use of the troublesome evaporation pond altogether. Through a combination of operational changes involving new procedures for rinsing filters, rinsing reactor vessels, and increasing employee awareness of how to prevent small, but significant losses of materials, Borden achieved its goals. Phenolic wastes were reduced by 93 percent and use of the evaporation pond was discontinued. Total sludge generation from resin manufacturing was reduced from 350 to 25 cubic yards per year which, at disposal costs of $150/cubic yard, meant annual savings in waste disposal costs of $48,750.

At Monsanto's plant, the hazardous solid wastes, ranging from 400,000 to more than a million pounds annually, are sent to off-site commercial disposal facilities in Ohio and Alabama. Monsanto's plant manager reported that in order to reduce disposal costs, they trained trash specialists who now segregate the plant's hazardous solid wastes from other trash, such as computer paper or cardboard boxes. This sorting ensures that non-hazardous trash is not contaminated with hazardous materials, even in small amounts. Under federal regulations such contamination would require that the entire trash load be handled as hazardous waste.

Two Cost Factors Inhibited Waste Reduction.

INFORM found that waste reduction practices were not actively pursued for some types of wastes because there were either no forseeable cost advantages, or possible cost savings were small and uncertain enough to discourage taking the initiative. The factors that were seen to partially or completely offset potential cost savings of waste reduction are discussed below.

Fixed costs an obstacle: The management at Atlantic Industries reported that the fixed costs of operating blowers that are part of the plant's air pollution control equipment could not be substantially reduced through waste reduction practices. The blowers would still have to operate at the same capacity in order to trap even reduced amounts of dust, according to Atlantic, and energy costs would not be affected. Atlantic reported to the U.S. Department of Commerce (in the government's 1984 Survey of Pollution Abatement Costs and Expenditures) approximately $321,000 in costs for environmental management, the largest proportion of this amount going to control air emissions. Similarly, Atlantic told INFORM that a large portion of its costs for wastewater management was the $25,000 spent to monitor its wastewater discharges. Again, this fixed cost would not be affected by waste reduction since monitoring would be required even at lower levels of waste generation.

Cost disparity between chemicals: Merck reported that an economic obstacle to focusing company attention on reducing solvent wastes was the large disparity between the sale price of its products and the costs of the solvents used in its manufacturing process. Solvents typically cost Merck 20 to 25 cents per pound while the pharmaceutical chemicals manufactured by Merck sell for tens or even hundreds of dollars per pound (the company did not provide prices for specific chemicals). Merck reported that most of its efforts at improving plant efficiency were directed at increased recovery of the much more valuable product chemicals rather than the cheaper solvents in its wastes.

In addition, Merck cited the need to maintain the highest degree of quality control of their products as an explanation for their reluctance to tamper with process conditions in the interest of reducing solvent wastes. Eventually, the combined impact of rising solvent prices, increasing costs of waste management, limited disposal options, and more stringent regulations of solvent air emissions prompted Merck to find ways of reducing solvent wastes in newly-developed processes.

The Impact of Liability on Hazardous Waste Reduction

Liability -- the legal, financial and possibly criminal responsibility for problems caused by hazardous wastes as well as their far broader potential public health impacts -- has become a major consideration for companies handling hazardous chemicals and disposing of hazardous wastes. Under increasingly stringent legal definitions of liability, some companies have found themselves held accountable for large-scale environmental problems. Further, individual corporate officers have, in many cases, become liable for criminal prosecution. The costs of cleanup, compensation for damages, and litigation have the potential of putting firms out of business. Corporate concerns about liability were seen in INFORM's study plants to both encourage and discourage waste reduction efforts.

Liability Risks that Encourage Waste Reduction.

Liability risks were a factor leading to waste reduction at Borden Chemical. The plant's reliance on an evaporation pond to treat and store hazardous chemical wastewaters and sludges carried with it the possibility of a leak in the pond for which Borden would be responsible. The specter of the potentially enormous costs of remedial action, along with possible government and citizen lawsuits, was a strong inducement to search for an alternative means of waste management. Borden launched a search for waste reduction practices that finally enabled the plant to eliminate the use of the pond.

CIBA-GEIGY reported that its primary motivation for developing a mercury-free dye manufacturing process was the possible "profound effect" on industry of future serious environmental contamination and poisoning problems involving mercury. As a response to the industry-wide concerns about mercury, CIBA-GEIGY spent over 10 years developing a mercury-free dye manufacturing process which is now in use at its Toms River, New Jersey, plant. The new process is more costly than the original one that used mercury, but the company considers the added certainty that the new process will not lead to future problems to be well worth the extra cost.

The widespread mercury contamination that occurred in the 1960s and 1970s in Sweden, Japan and Iraq caused thousands of illnesses and hundreds of deaths. In some cases, industrial discharges of mercury wastes that were thought to present no dangers to health had inadvertently been introduced into the food chain through chemical transformations in the environment and had poisoned people eating mercury-contaminated foods.

Liability Risks That Discourage Waste Reduction.

Liability risks can also steer waste management decisions toward well-known and traditionally-used technologies regarded as environmentally secure that avoid the need to pursue alternatives that may be more costly and complex -- including waste reduction.

The Dow Chemical plant in Pittsburg, California, is moving increasingly away from landfilling its hazardous wastes toward incineration as a means of destroying them. Dow officials cite a concern over the long-term liability of wastes disposed of on land, along with a commitment to environmental stewardship, as the reasons for preferring incineration to waste disposal methods such as landfilling. Dow reported that incineration is significantly more expensive than landfilling its wastes and that the company has had to add an internal company surcharge of $215 per drum (approximately $500 per ton) on hazardous wastes going to landfills to equalize the cost of the two disposal practices.

Different companies come to different conclusions about which waste management practices to select after considering liability and more general environmental concerns. Dow's corporate environmental policy explicitly rejects the use of deep wells as a means of disposing of hazardous wastes -- Dow abandoned the practice more than 10 years ago. USS Chemicals, in contrast, has chosen deep well injection because the company believes it is an environmentally safe practice. Although USS Chemicals' Haverhill plant was the only plant in INFORM's study to rely on deep well injection, many of the other companies included in this study make active use of deep well injection at other (non-study sample) plants. One of these companies, Du Pont, has 34 active wells nationwide. These wells contain 13 percent of the total volume of wastes disposed of by deep well injection in the U.S.

Conclusion: Waste Reduction is Usually the Last Choice

An inevitable conclusion to be drawn from the sum of experiences and choices seen at INFORM's 29 study plants is that waste reduction alternatives were seldom considered until circumstances virtually forced plants to review their waste management practices. Recycling, treatment and disposal options prevailed.

One reason for this is that some of the options are costless or very inexpensive. Regulations were a major factor found in this study to influence these costs and, thus, a plant's decision to reduce wastes. Table 7-6 explains how regulations help determine the relative costs of the most frequently used waste treatment and disposal options.

The range of costs associated with the practices is striking. Sending a ton of organic wastes to a commercial hazardous waste incinerator can cost as much as $1,000 for state-of-the art thermal destruction (the type that would be used, say, to destroy chlorinated organic chemical wastes). But burning waste material as a supplemental fuel virtually eliminates disposal expenses and can save a plant about $250 per ton in lowered fuel bills. Other waste management options for the same one ton of organic wastes can range from almost zero for air emissions to low costs for disposal if the wastes are simply discharged in deep wells or landfills, to fees varying from less than $20 to almost $400 (plus additional charges for wastewater flow) for discharge to a sewage treatment plant.

Burning wastes in boilers clearly is advantageous to many plants because fuel costs are reduced, as are the costs of waste disposal. But, as was the case at Merck, where burnable waste volumes are small and monitoring requirements are stringent, the economic advantages can be lost. Air emissions have little

Table 7-6. Treatment and Disposal of Hazardous Wastes: Regulations and Costs

Regulatory Status	Nature of Costs	Net Costs to Plant
BURNED IN ON-SITE BOILERS		
Exempted from most federal and state hazardous waste regulations if burned as a supplemental fuel. Monitoring and testing requirements are not uniform, but are decided on a case-by-case basis.	Costs of monitoring and testing depend on the requirements imposed. These costs tend to be relatively fixed, that is, they do not increase as the volume of wastes burned increases.	NET SAVINGS to the plant provided that large enough volumes are burned so that reduced fuel bills offset the costs of monitoring and testing requirements. Fuel savings can amount to $250 per ton of waste burned (based on replacement cost of fuel oil priced at $1 per gallon).
EVAPORATED TO THE ATMOSPHERE		
Regulations place an upper limit on the amount that can be emitted but can allow hundreds of tons to escape from a single plant. This limit is not a fixed amount for a given source or for specific chemicals, but can vary widely depending on the substance, the nature of the source, the industrial category, and local air quality conditions.	—Emissions-based fees range from $6.45 to $52 per ton (Southern California only). —Modest cost for air pollution control permits ranging from $50 to $1000 per source at INFORM's study plants. These permit fees are fixed since they must be paid irrespective of the quantity of emissions to the atmosphere.	Close to ZERO COST for waste disposal. Relatively small permit fees.
UNDERGROUND INJECTION WELLS		
Currently regulated under the Safe Drinking Water Act with virtually no restrictions on the types or quantities of materials that can be discharged. Restrictions are likely to increase due to 1984 amendments to the Resource Conservation and Recovery Act (RCRA).	—Currently small permit fees and monitoring costs. —Operating costs are generally considered modest. —These costs may increase under the 1984 RCRA amendments.	SMALL to MODEST disposal costs. Ohio EPA estimates $50,000 per year per well.

Table 7-6 (cont.). Treatment and Disposal of Hazardous Wastes: Regulations and Costs

Regulatory Status	Nature of Costs	Net Costs to Plant
DISPOSAL IN LANDFILLS		
Many landfill facilities can currently accept a wide variety of hazardous waste even in the absence of final approval as a hazardous waste facility because they have interim status. The 1984 RCRA amendments restrict the types of materials that can be landfilled and are expected to reduce the numbers of landfills in active use.	Costs stem from permit fees, monitoring and testing, transportation and actual operating costs. These have all been small in the past but are changing as the regulatory framework falls into place. Detailed cost figures are not available.	SMALL to MODEST costs at present. The Stauffer study plant pays roughly $250 per ton for landfilling, but costs of $75* per ton are considered typical. All landfilling costs are expected to rise in the near future.
DISCHARGED TO SEWAGE TREATMENT PLANTS		
Federal pretreatment guidelines for organic chemical plant discharges to sewage treatment plants (STPs) have not yet been issued. Each STP establishes its own guidelines, and restrictions range from prohibition to virtually no limitations on the types or quantities of chemicals that can be discharged.	Fees charged by each STP based on its annual operating costs are usually scaled to the quantity of water, solids, and organic material discharged by an industrial plant.	Costs are HIGHLY VARIABLE. Fees for INFORM's study plants range from $11.58 to $396 per ton of organic contaminant plus additional charges for the net flow of wastewater from the plant.
BURNED IN A HAZARDOUS WASTE INCINERATOR		
Burning hazardous wastes (except as a supplemental fuel—see above) is strictly regulated in terms of operating conditions as well as monitoring, permit and reporting requirements.	--Fees charged by the incinerator operator. --Transportation costs.	Considered one of the HIGHEST cost disposal options, although very limited data are available. Costs of $1000* per ton are typical for difficult-to-destroy wastes such as chlorinated organics. The Dow study plant adds a surcharge of $215 per drum (about $1000 per ton) on wastes from its own plants in order to equalize the costs of landfilling and incineration.

* See, for example, "Managing Hazardous Wastes: 1985," Supplement to CHEMICAL WEEK magazine, August 21, 1985. Other figures are from discussions in the text of this chapter and from the profiles in Part II of this study.

cost associated with them (other than the loss of material itself) and reducing them may not be considered worth the effort in the absence of a regulatory requirement to do so.

The low costs of landfilling wastes may make this seem an attractive option for some plants, although more long-term concerns about liability may give other plants pause.

USS Chemicals, in reviewing alternatives for its deep well disposal operations, found reduction at the source to be technologically difficult and the recycling of phenol wastes, although possible, to be more costly than continuing to inject the wastes into the ground. Disposal to the environment was the option of choice for USS Chemicals. The other, ostensibly more desirable procedures were not chosen because of the expense of destroying or recycling these wastes, and the technological obstacles to waste reduction at the source.

It was also observed that four plants missed the savings presented by waste reduction until they were virtually forced by government regulations or operating problems to review their waste management practices. Borden, needing to meet a stringent phenol limit in its wastewater discharges, turned first to treatment methods (an evaporation pond). Only after treatment failed to solve the problem did it give serious attention to reducing wastes at the source, finding relatively straightforward options for eliminating most of its phenolic resin wastes.

Exxon did not install floating roofs until the state regulations were being drafted requiring the roofs to be used; only after the regulations were in place did the company find the roofs to be cost-effective. Operational problems at Exxon's treatment plant spurred it to reduce organic chemicals in its wastewater.

Sherwin-Williams implemented waste reduction of trichlorobenzene discharges only after their route of discharge was closed off by authorities at the sewage treatment plant receiving the company's wastewaters. Exxon and Borden (and perhaps Sherwin-Williams as well) implemented waste reduction measures that saved rather than cost the plant money.

There were no technological obstacles to having implemented waste reduction at these plants years earlier. Yet, it took changes (and relatively dramatic changes) in the regulatory and economic climates of these plants before they recognized the waste reduction steps that could save money. Other company plants' operations have not yet felt the same urgency to adopt waste reduction practices, but it is more than likely that similar regulatory and economic developments will, sooner or later, persuade them to do so.

References

1. Industrial Environmental Research Laboratory, <u>Assessment of Hazard Potential from Combustion of Waste in Industrial Boilers</u>, EPA 600/S7-81-108, U.S. Environmental Protection Agency, Research Triangle Park, NC, August 1981.

2. "How safe is deep-well disposal of waste?" <u>Chemical Week</u>, November 21, 1984, p. 34.

Part II

THE PLANT PROFILES

Introduction

The following profiles of the 29 organic chemical plants that INFORM has studied represent all that INFORM was able to learn about plant operations, waste management, waste reduction initiatives and their results. Each of the 29 companies was sent a copy of its profile for review and was asked to supply comments, additions or corrections.

The information in the profiles is largely self-explanatory. There are ambiguities and outright gaps, however, in much of the existing data on waste reduction. Because of these, the clarifications and caveats below provide a necessary background to the data presented in the profiles.

Organization of the Profile Texts

The profile texts follow a uniform format consisting of: a brief SUMMARY; a review of PLANT OPERATIONS that includes a description of production as well as the plant's relationship to its parent company; and an accounting of WASTE GENERATION AND WASTE REDUCTION that is subdivided into sections on air emissions, wastewater discharges and solid wastes. The OBSERVATIONS section discusses broad-based factors that exercise a significant influence on plant operations, particularly on waste generation and waste reduction. For the six plants where information is scarce, the format has been condensed (see, for example, J.E. Halma Co.).

Boldface type is used to identify hazardous chemicals, that is, chemicals included in the lists of Appendix A. For example, **phenol** is a hazardous chemical, acetic acid is not. A waste containing a chemical in boldface print does not necessarily present a toxic hazard, since factors such as chemical concentration and the manner of disposal must be considered as well.

The SOURCES OF INFORMATION for the profiles are listed following the text. It was not uncommon for the records of one environmental agency to appear in the files of another, so that, although the source of information may be a memo from a state water quality agency, the location of that memo may well have been a federal solid waste file. Both the source and the location are noted.

Profile Tables

Two types of tables follow the source listing -- one on waste reduction practices, the other on waste generation and disposal. The two share several types of information-reporting categories:

The "Type of Waste" column identifies the air emissions, wastewater discharges or solid waste materials reported to be generated at the plant. Pure chemical wastes are referred to by their chemical names only, while mixed streams are referred to as "wastes." A listing of "10,000 pounds of benzene" means a company is discarding that quantity of the chemical. On the other hand, "10,000 pounds of benzene waste" means that quantity of material was discarded but the actual quantity of benzene is not specified.

Because of the enormous variations in how wastes are reported in environmental files, INFORM included, wherever possible, the actual chemical identities of waste materials. In many cases, however, there was no specific information other than the vague and incomplete descriptions contained in environmental files.

Where possible, wastes are listed under the appropriate sub-heading of either "Hazardous" or "Non-Hazardous" according to INFORM's criteria -- that is, according to whether or not the chemical is included in the listings in Appendix A. Where official descriptions were vague, wastes are listed as either "Likely to Contain Hazardous Chemicals" or "May Contain Hazardous Chemicals," based on INFORM's evaluation of the types of materials handled at the plant and the ways in which wastes are generated. A listing that cannot definitely be classified as hazardous or non-hazardous, for example waste solvents, would appear under the "May Contain..." heading. If a wide variety of hazardous solvents are known to be used at a plant, however, the listing waste solvents would appear under the "Likely to Contain..." heading.

Organic chemicals are listed in alphabetical order, followed by inorganics. Exceptions are made in order to keep broad categories of wastes with a common origin or fate together on the table.

A single chemical can have more than one name. Tetrachloroethylene is also widely known as perchloroethylene, but may also appear as tetrachloroethene, ethylene tetrachloride, or simply perc. It was not possible for INFORM to provide synonyms for the various chemicals, so all chemicals are listed as companies reported them. The absence of formal numerical prefixes can also create confusion: 1,2 dichlorobenzene and 1,4 dichlorobenzene are different chemicals, yet may appear in reports without their identifying numbers. INFORM has not undertaken the massive and probably impossible task of resolving all potential ambiguities caused by variations in nomenclature.

The "Waste Generation or Disposal" column summarizes the source of the waste material and its ultimate disposition as air emissions, wastewater discharges, burial in a landfill, and so on. Indeterminate descriptions such as "Sent off-site for disposal" or "Stored on-site" reflect the lack of detailed information provided in government records or by the companies on the fate of the listed waste.

Because the same amount of waste may be reported more than once under another or the same "Type of Waste" heading, figures in the "Quantity of Wastes" column cannot necessarily be totalled. For instance, Du Pont reported generating 5,388,766 pounds of "non-halogenated solvent wastes," a figure which may or may not include individual solvent wastes also reported by Du Pont, such as toluene (85,820 pounds) or methyl alcohol (173,837 pounds). In addition, the figures on quantities of wastes do not always represent actual quantities generated. In some cases, they may be estimates by either the company or the government. In others, they may represent the maximum amount of waste that is allowed to be generated as stipulated by permits. Where the figures are known to represent a maximum, this is indicated in the tables.

The units of waste generation reported as pounds per year may not represent an average annual amount of waste generated, due to the year-to-year variations in production at many plants, as well as sporadic incidents such as spills. The quantities of wastes are reported, for the most part, in the units found in the files; only figures of weight (tons, kilograms, etc.) were routinely converted to pounds. On several occasions, when information was deemed complete enough, computations of waste discharges were made by INFORM; these are clearly identified in the profiles by footnotes.

The dates of waste generation reported in the tables are, where possible, the actual dates of generation. However, this is not consistently so. Sometimes, only the information source itself had a date, but did not indicate the actual date of waste generation (and, some data referred to "typical" waste generation without specifying a particular time frame). At other times, even a date for an information source is hard to pin down. Air permit files, for instance, consist of hundreds of individual paper sheets, each submitted at different dates. Because these files are intended to be up-to-date, INFORM assigned the year in which the files were received as the appropriate date to use in citing them in the tables. It is possible though, that the actual information listed was submitted years earlier.

Lastly, individual waste generation figures represent many different scales of operation and different sources of waste at any given plant. For instance, Exxon's profile table lists three emissions figures for phenol:

```
Emissions from 5 process sources     3.6 lbs/yr
Emissions from 3 storage tanks   700-800 lbs/yr
Plant-wide emissions               2,230 lbs/yr
```

The emissions from eight discrete sources (five process sources, three tanks) do not add up to Exxon's plant-wide emissions figures. Hence, either (1) the list of discrete emissions is not complete or (2) the emissions have changed in the time between reporting periods; or (3) the information in one or more of the data sources is inaccurate. The figures are simply presented as reported.

The "Sources of Information" column gives the number of the information sources listed at the end of the text of each profile.

Wastestreams that have been affected by waste reduction practices are identified on the "Waste Generation and Disposal" table with a solid dot in the left-hand margin, and reported in more detail on a separate Waste Reduction table.

The "Waste Reduction" table contains two additional columns. The "Waste Reduction Practice" column identifies the actual change made to implement source reduction, the category or categories to which it belongs (product, process, equipment, chemical substitution, or operational) and the date the change was implemented. The "Results" column summarizes what could be learned about the impact of the practice in terms of material savings and economic costs and benefits.

The Waste Reduction table summarizes as clearly as possible the before-and-after picture of waste generation. Where, however, information on waste reduction came from different sources than those on waste generation and the two were not clearly cross-referenced, the relationship between waste generation data and waste reduction practices was not always apparent. The text of the profiles contains more detail and discussion.

American Cyanamid Company

P.O. Box 388
1405 Greene Street
Marietta, Ohio 45750
(614) 374-7171

SUMMARY

American Cyanamid's plant (Marietta, Ohio) manufactures water-soluble organic dyes, ultraviolet absorbers, explosives and organic intermediates. Air emissions from the plant include 36,540 pounds per year of volatile organic chemicals. Process and other wastewaters containing **chlorobenzene** are treated and discharged to the City of Marietta Wastewater Treatment Plant. Non-contact cooling water is recycled and excess is discharged from a holding pond into Duck Creek. Solid waste streams are disposed of through a variety of methods, including landfilling, recycling and incineration, and are also stored on-site in surface impoundments. Hazardous components of these solid wastes include **coumarin, chlorobenzene** and **zinc.** There is no indication from available information that Cyanamid has adopted any waste reduction techniques.

PLANT OPERATIONS

American Cyanamid Company operates a 128-employee organic dye manufacturing plant in Marietta, a city on the Ohio River in the southeastern corner of Ohio. This is one of the seven major U.S. manufacturing plants of American Cyanamid's Chemicals Group, which operates a total of 36 facilities.

NOTE: Chemicals in **boldface** type are hazardous, according to INFORM's criteria (see Appendix A).

Cyanamid's main business areas are medical, agricultural, chemical, consumer and Formica brand products. The Chemicals Group consists of three divisions: Chemical Products, Industrial Products and Polymer Products. The Chemical Products Division, to which the Marietta plant belongs, manufactures catalysts, dyes, rubber chemicals, sulfuric acid and organic chemicals for polymer manufacture. The Chemicals Group's worldwide sales were $973 million in 1983; sales for the entire corporation were over $3.5 billion in 1983.

The Marietta plant, established in 1915, manufactures water-soluble red, yellow, purple and green organic dyes, ultraviolet absorbers (used as additives by plastics and paint manufacturers to prevent decomposition from ultraviolet light) and explosives. Its 75 different chemical processes, carried out in batch operations, annually yield 5.5 million pounds of organic chemical products.

WASTE GENERATION AND WASTE REDUCTION

AIR EMISSIONS

Generation

The Marietta plant discharges 36,540 pounds per year of volatile organic chemicals from three air emissions sources, according to Ohio EPA air permit files. Little information was available in these files as to the type and quantity of wastes being emitted from most of the plant's 23 air emissions sources. Pollutants from other process air emissions sources include dye dust and trinitrotoluene dust, both of which are described as being discharged at "negligible" rates.

The plant's 1983 City of Marietta Wastewater Treatment Plant Industrial Waste Questionnaire cited two air pollution control methods used at Cyanamid: particulates are recovered in dust collectors, and acid gases are scrubbed and then discharged to the plant's neutralization pond for treatment prior to discharge to the City of Marietta Wastewater Treatment Plant.

WASTEWATER DISCHARGES

Generation

According to the plant's 1983 Industrial Waste Questionnaire, process and other wastewaters from the plant are discharged at the rate of 353,000 gallons per day to the City of Marietta Wastewater Treatment Plant; Cyanamid discharges a total of 121 million gallons per year of wastewater to the treatment plant. These wastes are treated in two on-site surface impoundments prior to discharge. Individual batch process wastestreams are sent to the first of these two large basins for neutralization. Wastewater is then sent to the second basin for settling, equalization and further neutralization. One-quarter to one-third of this annual effluent to the sewer is rainwater which washes into the ponds.

Cyanamid reported that its wastewater discharge to the treatment plant contains the following contaminants: dyes, coloring agents and related intermediates; organic solvents and thinners; alcohols; aldehydes and ketones; soaps, surfactants and detergents; **benzene** derivatives; and chlorinated organic compounds. Specific contaminants include **chlorobenzene,** 900 pounds per year, out of a total 88,600 pounds annually used on-site and a variety of heavy metals, including **chromium, copper, lead, nickel** and **zinc.**

According to Cyanamid's 1981 wastewater discharge permit renewal application, non-contact cooling water is discharged to an on-site holding pond after initial use and then 1.3 million gallons of this water is recycled back to the plant processes each day for reuse (as process or cooling water). Excess cooling water (a total of 266,000 gallons per day) is discharged to Duck Creek as the holding pond begins to fill up. A November 1975 letter from Cyanamid to the Ohio Environmental Protection Agency explains that this discharge contains **copper**, because copper sulfate is added to the pond to prevent algae growth.

SOLID WASTES

Generation

Cyanamid handles and disposes of a wide variety of organic and inorganic solid waste streams through the following techniques: on-site storage in surface impoundments; off-site landfilling in Ohio, South Carolina and Alabama; on-site and off-site recycling; and incineration at American Cyanamid's West Virginia plant. The February 1981 Industrial Waste Survey indicates that some of Cyanamid's wastestreams are temporarily being incinerated until it selects permanent disposal methods and sites for these wastes.

According to the plant's 1981 Industrial Waste Survey, the largest of Cyanamid's hazardous solid waste streams include 36,000 pounds per year of **coumarin** clarification cake containing **zinc,** and 20,000 pounds per year of **chlorobenzene** wastes. The single largest solid waste stream is 100,000 pounds per year of **auramine** tar cake.

SOURCES OF INFORMATION

SOURCE	LOCATION
1. Air Permit files, 1984	Ohio Environmental Protection Agency (OEPA) Southeast District Office (SEDO) Division of Air Pollution Control 2195 Front Street Logan, OH 43138 (614) 385-8501
2. Correspondence from American Cyanamid to Ohio Environmental Protection Agency, Southeast District Office, January 1978	OEPA, SEDO Division of Air Pollution (see #1)
3. Wastewater Discharge Permit Renewal Application, March 1981	U.S. Environmental Protection Agency (U.S. EPA), Region V Water Division 230 South Dearborn Chicago, IL 60604 (312) 886-6112
4. City of Marietta Wastewater Treatment Plant Industrial Waste Questionnaire, April 1983	OEPA Pretreatment Division 361 East Broad Street Columbus, OH 43215 (614) 462-6787
5. OEPA memo from SEDO to the central office, January 1983	OEPA Division of Hazardous Materials Management Permits and Manifest Records Section 361 East Broad Street Columbus, OH 43215 (614) 466-1586

6. Industrial Waste
   Survey, February 1981

   OEPA, SEDO
   Division of Hazardous
     Materials Management
   2195 Front Street
   Logan, OH 43138
   (614) 385-8501

7. Revised Industrial
   Waste Survey,
   October 1981

   OEPA, SEDO
   Division of Hazardous
     Materials Management
   (see #6)

8. RCRA Part A
   Application, 1980

   U.S. EPA, Region V
   Waste Management Branch
   230 South Dearborn
   Chicago, IL 60604
   (312) 886-6134

9. Off-site manifest
   approvals, April 1980

   OEPA
   (see #5)

10. Treatment, Storage
    and Disposal
    Hazardous Waste
    Annual Report, 1981

    OEPA
    (see #5)

11. Treatment, Storage
    and Disposal
    Hazardous Waste
    Annual Report, 1982

    OEPA
    (see #5)

12. Generator Annual
    Hazardous Waste Report,
    1981

    OEPA
    (see #5)

13. Correspondence from
    American Cyanamid to
    the Ohio Environmental
    Protection Agency,
    November 1975

    OEPA, SEDO
    Office of Water Pol-
      lution Control
    2195 Front Street
    Logan, OH 43138
    (614) 385-8501

AMERICAN CYANAMID

WASTE GENERATION & DISPOSAL

Type of Waste ● Indicates Wastes Affected by Reduction	Generation or Disposal	Quantity of Waste	Sources of Information
Air Emissions			
MAY CONTAIN HAZARDOUS CHEMICALS Dye dust	Emissions from 5 sources	?* (1984)	(1)
Volatile organic chemicals from dye and ultraviolet absorber manufacture	Emissions from 3 sources	36,540 lbs/yr (1978)	(1,2)
NON-HAZARDOUS Trinitrotoluene dust	Emissions from 1 source	?* (1984)	(1)
Wastewater Discharges			
Non-contact cooling water, including: HAZARDOUS	Discharged from a holding pond into Duck Creek	266,000 gals/day (1981)	(3)
Copper	"	.74 lbs/day (1981)	(3)
Process and other wastewaters, including:	Treated and discharged to the City of Marietta Wastewater Treatment Plant	353,000 gals/day 121 million gals/yr (1983)	(4)
		80,000,000 gals/yr (1981)	(7)
HAZARDOUS Chlorobenzene	"	900 lbs/yr (1983)	(4)
Chromium	"	.013 ppm (1983)	(4)
Copper sulfate	"	110 lbs/yr (1983)	(4)
Lead	"	.175 ppm (1983)	(4)
Nickel	"	.158 ppm (1983)	(4)

* Emissions listed as "negligible" in Ohio Environmental Protection Agency air permit files.

AMERICAN CYANAMID

WASTE GENERATION & DISPOSAL

Type of Waste ● Indicates Wastes Affected by Reduction	Generation or Disposal	Quantity of Waste	Sources of Information
Wastewater Discharges			
Process...other wastewaters (cont.)	Treated and discharged to the City of Marietta Wastewater Treatment Plant		
HAZARDOUS (cont.)			
Zinc	"	4,740 lbs/yr (1983)	(4)
Zinc chloride	"	41,530 lbs/yr (1983)	(4)
NON-HAZARDOUS			
Aluminum	"	1.25 ppm (1983)	(4)
Ammonia	"	3,000 lbs/yr (1983)	(4)
Iron	"	1.56 ppm (1983)	(4)
Solid Wastes			
HAZARDOUS			
Aniline-containing quinaldine tar	Disposed of in an off-site landfill	5,000 lbs/yr (1981)	(6,7)
Asbestos insulation	Disposed of in an off-site landfill	2,000 lbs/yr (1981)	(6,7)
Auramine tar cake	Disposed of in an off-site landfill	100,000 lbs/yr (1981)	(6,7)
Auramine waste	Recycled at Chemical Waste Management Inc., Emelle, AL	600 lbs/yr (1981)	(6,7)
Chloro anthraquinone waste containing chlorobenzene and methanol	Disposed of in an off-site landfill	1,200 lbs/yr (1981)	(6,7)
Chloro anthraquinone off-grade product	"	? (1981)	(7)
Chlorobenzene waste from dye manufacture	Disposed of in an off-site landfill at SCA Chemical Services Co., Pinewood, SC	4,000 lbs/yr (1980, 1981)	(7,9)
Chlorobenzene wastes	"	20,000 lbs/yr (1981)	(6,7)
Coumarin waste containing zinc	Disposed of in an off-site landfill	36,000 lbs/yr (1981)	(6,7)
Crotonaldehyde	Stored on-site	500 lbs/yr (1980)	(8)

AMERICAN CYANAMID

WASTE GENERATION & DISPOSAL

Type of Waste ● Indicates Wastes Affected by Reduction	Generation or Disposal	Quantity of Waste	Sources of Information
Solid Wastes			
HAZARDOUS (cont.)			
Methyl anthraquinone waste containing chlorobenzene	Disposed of in an off-site landfill	600 lbs/yr (1981)	(6,7)
Methyl ethyl ketone	Stored on-site	50 lbs/yr (1980)	(8)
Mononitrobenzene wastes:			
From quinaldine	Recycled	? (1981)	(7)
Residue	Disposed of an in off-site landfill at Chemical Waste Management, Inc., Emelle, AL	1,000 lbs/yr (1981)	(6,7)
Still bottoms	Disposed of in on-site neutralization pond	9,000 lbs/yr (1981)	(6,7)
Naphthalene	Stored on-site	200 lbs/yr (1980)	(8)
Naphthalene waste	Recycled at Chemical Waste Management, Inc., Emelle, AL	10,000 lbs/yr (1981)	(6,7)
Phthalic anhydride	Stored on-site	500 lbs/yr (1980)	(8)
Polychlorinated biphenyl wastes	Sent off-site for disposal	0-5,000 lbs/yr (1981)	(6,7)
Resorcinol	Stored on-site	500 lbs/yr (1980)	(8)
Resorcinol waste	Disposed of in an off-site landfill	0-3,000 lbs/yr (1981)	(6,7)
Solvents, halogenated	Stored in an on-site impoundment	15,000 lbs/yr (1981)	(10)
Wastes from solvent recovery	"	11,000 lbs/yr (1982)	(11)
Vanadic acid, ammonium salt	Stored on-site	100 lbs/yr (1980)	(8)

165

AMERICAN CYANAMID

WASTE GENERATION & DISPOSAL

Type of Waste • Indicates Wastes Affected by Reduction	Generation or Disposal	Quantity of Waste	Sources of Information
Solid Wastes			
HAZARDOUS (cont.)			
Vanadium pentoxide	Stored on-site	100 lbs/yr (1980)	(8)
LIKELY TO CONTAIN HAZARDOUS CHEMICALS			
Paint wastes and solvent	Incinerated at American Cyanamid, Willow Island, WV	55 gals/yr (1981)	(6,7)
Trash containing hazardous materials	Disposed of in an off-site landfill	5,000 lbs/yr (1981)	(6,7)
MAY CONTAIN HAZARDOUS CHEMICALS			
Lagoon sludge	?	? (1981)	(7)
Rhodamine effluent sludge	Disposed of in an off-site landfill	15,000 lbs/yr (1981)	(6,7)
Solid wastes, miscellaneous	Disposed of an in off-site landfill, recycled, incinerated or recovered	At least 282,000 lbs/yr (1981)	(6,7,9)
Waste solvent, used for maintenance	Incinerated at American Cyanamid, Willow Island, WV	110 gals/yr (1981)	(6,7)
NON-HAZARDOUS			
Chinoline waste	Disposed of in an off-site landfill	60,000 lbs/yr (1981)	(6,7)
Corrosive wastes, unspecified	Stored on-site	84,500 lbs/yr (1980)	(8)
		59,135 lbs/yr (1981)	(10)
		10,135 lbs/yr (1982)	(11)
	Sent off-site for disposal at CECOS/CER Co., Williamsburg, OH	68,433 lbs/yr (1981)	(12)
	Sent off-site for disposal at Chemical Waste Management, Inc., Emelle, AL	79,659 lbs/yr (1981)	(12)

AMERICAN CYANAMID

WASTE GENERATION & DISPOSAL

●	Type of Waste Indicates Wastes Affected by Reduction	Generation or Disposal	Quantity of Waste	Sources of Information
	\multicolumn{3}{c	}{**Solid Wastes**}	

	Type of Waste	Generation or Disposal	Quantity of Waste	Sources of Information
	NON-HAZARDOUS (cont.)			
	Dimethoxy benzene tar	Disposed of in an off-site landfill	4,000 lbs/yr (1981)	(6,7)
	Ignitable wastes, unspecified	Stored on-site	21,000 lbs/yr (1980)	(8)
			10,100 lbs/yr (1981)	(10)
			722 lbs/yr (1982)	(11)
		Sent off-site for disposal at CECOS/CER Co., Williamsburg, OH	992 lbs/yr (1981)	(12)
		Sent off-site for disposal at Chemical Waste Management, Inc., Emelle, AL	2,420 lbs/yr (1981)	(12)
	Methyl isobutyl carbinol from brighteners	Recycled at Chemical Waste Management, Inc., Emelle, AL	10,000 lbs/yr (1981)	(6,7)
	B-Napthyl quinaldine	Disposed of in an off-site landfill	30,000 lbs/yr (1981)	(6,7)
	Nitro diphenyl amine waste	"	4,000 lbs/yr (1981)	(6,7)
	Red BG distillate containing ethyl acetate	Incinerated at American Cyanamid, Willow Island, WV	At least 40,000 lbs/yr (1981)	(6,7)
	Rhodamine wastes, miscellaneous	Mostly recycled at Chemical Waste Management, Inc., Emelle, AL; rest is disposed of in an off-site landfill	At least 6,000 lbs/yr (1981)	(6,7)
	Sludge containing iron oxides	Disposed of in an off-site landfill	? (1981)	(7)
	Waste oils	Incinerated at American Cyanamid, Willow Island, WV	550 gals/yr (1981)	(6,7)

Atlantic Industries

10 Kingsland Road
Nutley, New Jersey 07110
(201) 235-1800

SUMMARY

Atlantic Industries' plant (Nutley, New Jersey) manufactures hundreds of dyes from raw materials that include **aniline, phenol,** and **formaldehyde.** Relatively small amounts of raw materials and finished dye products are lost as air emissions, although the plant produces sizable wastewater and solid wastestreams. Air emissions of product dust have been reduced by unspecified changes. Through process refinements, Atlantic reduced organic and inorganic contaminants in the wastewater from a diazo dye manufacturing process, improving product yield by eight percent. The adoption of other waste reduction practices at the plant was inhibited by several factors cited by Atlantic: waste generation is already quite minimal; the highly variable nature of batch process dye manufacture is not conducive to reducing waste at the source; and there is little economic incentive to pursue waste reduction initiatives due to the fixed costs of waste management.

PLANT OPERATIONS

Atlantic Industries' dye manufacturing plant, located in a residential area of Nutley, New Jersey, was built in 1939 and currently employs over 200 people. Atlantic Industries markets a wide variety of organic dyes which are used primarily to dye textiles, paper and leather.

NOTE: Chemicals in **boldface** type are hazardous, according to INFORM's criteria (see Appendix A).

Total production at the Nutley plant is 5 to 10 million pounds of dyes per year. The mix of products is enormously variable, depending on customer demand; 300 to 400 dyes are manufactured each year out of a total of more than 700 that the company can produce, and are marketed as components of over 2,500 products. Hundreds of chemicals are used as raw materials depending on the types of dyes in production.

Production at the plant is 100 percent batch (as is typical in the dye industry) and is carried out in about 60 batch-reaction vessels which are in almost constant use. Plant management reported to INFORM that "it is rare to find an empty tank" at the Nutley plant.

WASTE GENERATION AND WASTE REDUCTION

AIR EMISSIONS

Generation

The plant emits relatively small amounts of raw materials and dyes as air emissions. Emissions of **aniline,** for instance, were 110 pounds in 1978, or less than 0.05 percent of the 224,000 pounds of **aniline** used at the plant. Other emissions include 75 pounds per year of **phenol** and 35 pounds of **chromium,** according to information from the New Jersey Department of Environmental Protection (NJDEP) for 1978, the most recent year emissions figures are available.

Reduction

INFORM has learned of one waste reduction practice affecting air emissions at Atlantic. The plant reported that it has been able to reduce the amount of product dust generated during drying, grinding and blending operations. However, further analysis of the impact of this waste reduction practice is not possible, since Atlantic did not provide details about the nature of this modification.

Plant management also reported to INFORM that the enormous variety of materials used and manufactured at the plant makes it infeasible to reuse the particulate materials trapped by air pollution control equipment (for example, filters) because of the many types of chemicals that are inevitably mixed together.

WASTEWATER DISCHARGES

Generation

The plant generates 300,000 to 400,000 gallons per day of wastewater, chiefly from process water and rinsewaters of reaction vessels and related equipment. The water is discharged to an on-site neutralization and settling tank where some of the solid materials settle out (see Solid Wastes section) and are then discharged to a sewage treatment plant operated by the Passaic Valley Sewerage Commission. According to NJDEP file information, these annual discharges contain 5,000 pounds of **acetonitrile;** 4,700 pounds of **aniline;** 5,000 pounds of **formic acid;** 6,000 pounds of **phenol;** 1,000 pounds of **copper;** 700 pounds of **chromium**; and numerous other organic and inorganic contaminants in smaller quantities.

Reduction

Atlantic told INFORM that a primary objective of all process research is waste reduction, through such methods as reducing excess reactants, changing process chemical concentrations or changing the techniques used to separate products from other reactants. It explained that "constant process maintenance and process improvement programs," since 1981, have reduced plant water use from 750,000 to 300,00-400,000 gallons per day, and have also reduced the level of organic and other contaminants discharged in the plant's wastewater.

In order to illustrate these reduction efforts, Atlantic provided one example of a waste reduction practice affecting its wastewater discharge. The plant implemented changes in a diazo dye manufacturing process which have improved product yield by eight percent, reduced organics in wastewater by 55,000 pounds per year, reduced inorganics in wastewater by 250,000 pounds per year and reduced water use by 500,000 gallons per year.

These process changes included increased process chemical concentrations, a lower reaction temperature and a new method for combining dye components, all of which contributed to improving yield and lowering waste generation. Atlantic told INFORM that these changes, implemented over the course of a year, were motivated by three needs recognized by both plant and Research and Development management: to reduce water use, to reduce the plant's waste load and to improve product yield. These goals were necessary in order to lower costs and remain competitive.

SOLID WASTES

Generation

Sludges accumulate in the settling tank as solid wastes and are shipped off-site to a commercial hazardous waste disposal facility in South Carolina. About 40,000 pounds of sludge are generated annually, according to NJDEP records, which also indicate that the sludges contain relatively small quantities of hazardous contaminants, including 300 pounds each of **aniline, p-nitroaniline,** and **chromium.**

Atlantic told INFORM that it is replacing all drums it uses for dye handling at the plant with reusable plastic drums. However, an absence of details precludes any evaluation of the nature or impact of this procedure.

OBSERVATIONS

Plant management at Atlantic reported to INFORM that several factors relating to the technology and economics of dye manufacture act as disincentives for the company to reduce their waste generation below the current levels that they already consider to be quite minimal.

Diversity of Operations: The enormous variability in the numbers and types of products manufactured at Atlantic was seen by the plant management as limiting waste reduction opportunities. Thorough and complete rinsing of equipment when switching production from one product to another is absolutely essential in the dye business because even very minor amounts of contamination can discolor a dye, making it unsuitable for commercial use. There is no opportunity to minimize the amount of material entering the wastewater stream during these washing operations. Furthermore, the quantity of materials involved in any one rinse is too small to warrant an effort to recover materials from the wastewater.

Because the types of raw materials and final products in use are constantly changing at the plant, and very few of these materials are handled in high volumes, the plant can only operate efficiently if each piece of equipment is capable of handling a variety of materials. When material handling varies as frequently as it does at Atlantic, there is no opportunity to rely on dedicated equipment -- equipment such as tanks, pumps or piping that is used to handle only a single chemical -- and, hence, no opportunity to eliminate the constant rinsing and waste generation associated with the use of non-dedicated equipment.

Fixed waste management costs: Atlantic spent $321,000 for pollution abatement, it reported to the Census Bureau in the 1984 Survey of Pollution Abatement Costs and Expenditures. The following table, based on data that Atlantic gave to INFORM, compares abatement costs with the amount of waste abated.

Type of Waste	Cost of Abatement ($/yr)	Amount of Waste Abated (lbs/yr)
Air	$159,000	32,000
Water	76,000*	56,000
Solid	86,000	3,252,000
TOTAL	$321,000	3,340,000

* Includes sewage treatment plant fees.

Plant management reported to INFORM that there was little opportunity for reducing waste management costs through waste reduction. The control of air emissions, as indicated above, is the most costly aspect of pollution control at Atlantic, due to the high energy costs of operating equipment such as blowers. Although less than one percent of the abated wastes are air pollutants, air pollution control accounts for almost half of all the plant pollution abatement costs. These costs are relatively fixed since energy use would not be affected by waste reduction; the air pollution control equipment would still have to be operated even if emissions could be reduced at the source.

Similarly, the costs of wastewater management are largely dictated by the expense of regularly monitoring water quality at a cost of $25,000 per year -- an expense that would not be affected by waste reduction practices.

Ninety seven percent of the abated wastes are solid wastes, yet they only account for 27 percent of total plant pollution abatement costs. The costs of commercially disposing of hazardous waste sludges could be minimized by waste reduction practices. However, Atlantic reported that the potential savings are not compelling enough for Atlantic to explore means of reducing the volume of sludge generation.

Small cost of material losses: For many raw materials in use at the plant, Atlantic reported to INFORM that there is little potential economic advantage to minimizing wastestreams that the company already considers to be quite small. Processes involving **aniline**, a major raw material at Atlantic (and in the dye industry in general), are already highly efficient at converting this material to a final dye product and the amount lost as waste is so small that there is no significant economic pressure to further reduce the quantity of **aniline** wastes.

Not all raw materials can be processed so efficiently, however. Some chemicals must be present in excess amounts in order for the manufacturing process to operate effectively; a large portion of these chemicals ends up in the wastestream. For instance, almost 10 percent of the **phenol** and **copper** used at Atlantic are disposed of in the plant's wastewater. Yet, the costs of these lost materials are small enough that there is, again, little economic incentive to pursue waste reduction at the source. For instance, 6,000 pounds of **phenol** discharged as waste by Atlantic represents a loss to the plant of about $2,000 per year. Atlantic explained that reducing wastes from some of these processes would be detrimental to product quality and process yield.

Atlantic told INFORM that it has stopped using a number of chemicals based on concerns for the health and safety of its workers and neighbors. For example, within 24 hours of learning from an unpublished report in 1969 that **benzidine** was a suspected cause of cancer in dye-industry workers, management decided to stop using this chemical. At that time, benzidine dyes accounted for 30 to 40 percent of Atlantic's production. Similarly, the plant stopped using **alpha naphthylamine** based upon the Occupational Safety and Health Administration's reports of its potential carcinogenicity. The use of **phosgene,** which was being used in 15 percent of Atlantic's production processes, was discontinued in 1975 because of health and safety concerns. In particular, Atlantic cited a fatal accident involving release of **phosgene** at the Toms River Chemical Co. (now the CIBA-GEIGY plant in Toms

River, New Jersey) as a motivation for discontinuing use of this chemical.

SOURCES OF INFORMATION

SOURCE	LOCATION
1. New Jersey Industrial Survey, 1978 data	New Jersey Department of Environmental Protection (NJDEP) Office of Science and Research CN402 Trenton, NJ 08625 (609) 292-6714
2. Passaic Valley Sewerage Commission user survey, July 1981	NJDEP Division of Water Resources CN209 Trenton, NJ 08625 (609) 292-5602
3. RCRA Part A Application, November 1980	U.S. Environmental Protection Agency, Region II Permits Administration Branch 26 Federal Plaza New York, NY 10278 (212) 264-5602
4. Hazardous Waste Manifest Records, 1981	NJDEP Division of Waste Management CN027 Trenton, NJ 08625 (609) 292-9879
5. Information supplied by Atlantic to INFORM	

WASTE REDUCTION

ATLANTIC

Type of Waste	G-Generation D-Disposal Q-Quantity (year)	Waste Reduction Practice	Results	Sources of Information
MAY CONTAIN HAZARDOUS CHEMICALS Diazo dye manufacturing wastes	G - ? D - ? Q - ? (1985)	Changes in diazo dye manufacturing process PS (year unknown)	Product yield improved by 8%; organics in wastewater reduced by 55,000 pounds per year; inorganics in wastewater reduced by 250,000 pounds per year; and water use reduced by 500,000 gallons per year	(5)
NON-HAZARDOUS Product dust	Dust vented to scrubber Q - ? (1985)	Type of practice not reported ? (year unknown)	Less dust is generated during drying, grinding and blending operations	(5)

Key to Waste Reduction Changes
  PR – Product
  PS – Process
  EQ – Equipment
  CH – Chemical Substitution
  OP – Operational

ATLANTIC

WASTE GENERATION & DISPOSAL

● Type of Waste Indicates Wastes Affected by Reduction	Generation or Disposal	Quantity of Waste	Sources of Information
Air Emissions			
HAZARDOUS			
Aniline	Plant-wide emissions	110 lbs/yr (1978)	(1)
Formaldehyde	"	1 lb/yr (1978)	(1)
Phenol	"	75 lbs/yr (1978)	(1)
Chromium	"	35 lbs/yr (1978)	(1)
Zinc	"	1.5 lbs/yr (1978)	(1)
NON-HAZARDOUS			
p-Amino azobenzene	"	30 lbs/yr (1978)	(1)
Assorted dyes*	"	29 lbs/yr (1978)	(1)
● Product dust	Dust vented to scrubber	? (1985)	(5)

* Represents emissions of nine dyes from the New Jersey Department of Environmental Protection's Industrial Survey because of their potentially hazardous nature.

178

ATLANTIC

WASTE GENERATION & DISPOSAL

Type of Waste ● Indicates Wastes Affected by Reduction	Generation or Disposal	Quantity of Waste	Sources of Information
Wastewater Discharges			
Process, non-contact cooling and other wastewaters, including: HAZARDOUS	Discharged to the Passaic Valley Sewerage Commission	300,000–400,000 gals/day (1985)	(1,5)
Acetonitrile	"	5,000 lbs/yr (1980)	(3)
Aniline	"	4,700 lbs/yr (1980)	(3)
		240 lbs/yr (1978)	(1)
Cresols	"	200 lbs/yr (1980)	(3)
3,3-Dimethoxybenzidine	"	50 lbs/yr (1980)	(3)
3,3-Dimethylbenzidine	"	50 lbs/yr (1980)	(3)
Formaldehyde	"	200 lbs/yr (1980)	(3)
		10 lbs/yr (1978)	(1)
Formic acid	"	5,000 lbs/yr (1980)	(3)
p-Nitroaniline	"	300 lbs/yr (1980)	(3)
Phenol	"	6,000 lbs/yr (1980)	(3)
Resorcinol	"	300 lbs/yr (1980)	(3)
Toluene diamine	"	80 lbs/yr (1980)	(3)
o-Toluidine hydrochloride	"	90 lbs/yr (1980)	(3)
Antimony	"	.20 mg/l (1981)	(2)
Arsenic	"	.025 mg/l (1981)	(2)

ATLANTIC

WASTE GENERATION & DISPOSAL

●	Type of Waste Indicates Wastes Affected by Reduction	Generation or Disposal	Quantity of Waste	Sources of Information
	\multicolumn{3}{c}{**Wastewater Discharges**}	
	Process...wastewaters (cont.) HAZARDOUS (cont.)	Discharged to the Passaic Valley Sewerage Commission		
	Cadmium	"	.056 mg/l (1981)	(2)
	Chromium	"	.182 mg/l (1981)	(2)
			700 lbs/yr (1980)	(3)
			50 lbs/yr (1978)	(1)
	Copper	"	2.3 mg/l (1981)	(2)
			1,000 lbs/yr (1978)	(1)
	Lead	"	.30 mg/l (1981)	(2)
	Mercury	"	.0007 mg/l (1981)	(2)
	Nickel	"	.48 mg/l (1981)	(2)
	Selenium	"	.04 mg/l (1981)	(2)
	Silver	"	.07 mg/l (1981)	(2)
	Zinc	"	.51 mg/l (1981)	(2)

ATLANTIC

WASTE GENERATION & DISPOSAL

Type of Waste ● Indicates Wastes Affected by Reduction	Generation or Disposal	Quantity of Waste	Sources of Information
Wastewater Discharges			
Process...wastewaters (cont.) NON-HAZARDOUS	Discharged to the Passaic Valley Sewerage Commission		
p-Amino azobenzene	"	200 lbs/yrs (1978)	(1)
Assorted dyes*	"	204 lbs/yr (1978)	(1)
MAY CONTAIN HAZARDOUS CHEMICALS			
● Diazo dye manufacturing wastes	?	? (1985)	(5)
Solid Wastes			
Sludge from on-site treatment tank, including: HAZARDOUS	Sent off-site for disposal at SCA Chemical Services Co., Pinewood, SC	40,000 lbs/yr (1981)	(4)
Aniline	"	300 lbs/yr (1980)	(3)
3,3-Dimethoxybenzidine	"	50 lbs/yr (1980)	(3)
3,3-Dimethylbenzidine	"	50 lbs/yr (1980)	(3)
p-Nitroaniline	"	300 lbs/yr (1980)	(3)
Toluene diamine	"	20 lbs/yr (1980)	(3)
o-Toluidine diamine	"	10 lbs/yr (1980)	(3)
Chromium	"	300 lbs/yr (1980)	(3)

* Represents discharges of nine dyes from the New Jersey Department of Environmental Protection's Industrial Survey because of their potentially hazardous nature.

Bonneau Dye Corporation

35520 Schneider Court
Avon, Ohio 44011
(216) 934-4800

Bonneau Dye Corporation is located in Avon, Ohio, about fifteen miles west of Cleveland. This two-employee plant manufactures custom blend dye formulations and specialty chemicals for candle-making and other crafts. A total absence of data in public records precluded analysis of Bonneau's waste generation and handling methods. There is no indication from available information that Bonneau has adopted any waste reduction techniques.

Borden Chemical Company

41100 Boyce Road
Fremont, California 94538
(415) 657-4500

SUMMARY

The Borden Chemical Company's plant (Fremont, California) manufactures resins and adhesives for industrial use. **Phenol, formaldehyde** and urea are the major raw materials for resin manufacture; **methylene chloride, toluene** and other organic solvents are used in the adhesives operations. Since 1981, operational changes in Borden's resin operations -- particularly revised equipment-rinsing procedures -- have reduced the amount of organic materials discharged in its wastewater by 93 percent. This result, in turn, has eliminated most of the hazardous solid wastes generated from treating resin wastewaters. In contrast, the plant's adhesives operations, which are far more complex, have achieved only a limited amount of hazardous waste reduction. Two key factors that prompted Borden's efforts to reduce its generation of hazardous wastes were (1) strict regulatory limits on **phenol** discharges in its wastewater and (2) management's desire to eliminate the plant's reliance on an on-site evaporation pond, which was designed to treat the plant's wastewater but which posed regulatory, logistical, and financial problems.

PLANT OPERATIONS

Borden Chemical Company's 50-employee Fremont, California plant, located in an industrialized area

NOTE: Chemicals in **boldface** type are hazardous according to INFORM's criteria (see Appendix A).

southeast of San Francisco, has been in operation since 1959. Borden Chemical is a division of Borden, Inc., a large international corporation best known for dairy products and such adhesives as Elmer's Glue. The Chemical Division's 1982 sales were $680 million, representing 16 percent of Borden, Inc.'s total revenues.

The Fremont plant produces four major product lines through batch manufacturing processes:

-- UREA-FORMALDEHYDE AND PHENOL-FORMALDEHYDE RESINS, used primarily in the manufacture of plywood and particle board for the housing industry. Major raw materials used for resin production are **formaldehyde** (225 million pounds per year produced on-site from **methanol), phenol** and urea.

-- WATER-BASED ADHESIVES, used primarily by local food and wine industries to secure labels to cans and bottles and for packaging. For example, Gallo Wineries, a major client, purchases 30,000 pounds of adhesives a month. This product line includes 60 formulations based either on polyvinyl acetate or starch. **Toluene, 1,1,1-trichloroethane** and other organic solvents are used in some formulations.

-- SOLVENT-BASED ADHESIVES, used for industrial applications, contain a rubber base and various solvents, including **toluene,** acetone, **methyl ethyl ketone,** hexane, **methylene chloride** and **trichloroethane.**

-- HOT-MELT ADHESIVES, used for high-speed packaging and labeling applications. Major raw materials used to formulate these adhesives include mostly wax and wood rosin derived substances.

In 1982, for economic reasons, the plant permanently discontinued its printing ink manufacturing operations.

WASTE GENERATION AND WASTE REDUCTION

AIR EMISSIONS

Generation

The Fremont plant has 96 sources of air emissions, chiefly from 73 storage and handling tanks, according to air permit files of the Bay Area Air Quality Management District (BAAQMD). Annual emissions for specific chemicals are **methanol,** 18,250 pounds; **trichloroethane,** 11,600 pounds; and **formaldehyde,** 800 pounds. A 1983 U.S. EPA emissions survey reported that Borden's **formaldehyde** emissions are 170,400 pounds a year.*

According to BAAQMD files, Borden's annual emissions for general categories of organic vapors (which may include the above-listed hazardous chemicals) are as follows: total organic emissions, 116,580 pounds; fugitive organic emissions (caused by leaking pipes, valves, pumps, and other plant equipment), 22,300 pounds; and unspecified adhesives emissions, 53,000 pounds. Borden considers the BAAQMD estimates to be inaccurate, but did not supply INFORM with more reliable emissions data.

---

* The U.S. EPA's **formaldehyde** emissions figures are 200 times larger than those reported by the BAAQMD -- a discrepancy that could be partially explained by the fact that (1) in the EPA's opinion, the estimation techniques used are unreliable and/or (2) a separate set of BAAQMD emissions data referring to unspecified organic chemicals could include large amounts of **formaldehyde.** Borden strongly rejected EPA's estimate, but did not provide INFORM with a more accurate figure.

WASTEWATER DISCHARGES

Generation

Borden generates an average of 36,385 gallons per day of wastewater, originating from frequent rinsing of reaction vessels, filters, and other equipment, along with cooling water. The wastewater is discharged to the sewage treatment plant operated by the Union Sanitary District (USD), which restricts the levels of two chemicals in the wastewater -- **zinc** concentrations are limited to three milligrams per liter (mg/l) and **phenol** concentrations cannot exceed a limit of one to five mg/l, depending on USD's capacity to handle its total load of **phenol** from all dischargers.

Reduction

Borden told INFORM that several waste reduction practices adopted at the plant have lowered the amount of organic contaminants in its wastewater and that these changes ultimately reduced the generation of solid waste. Descriptions of these practices and their net results appear in the following section.

SOLID WASTES

Generation

Until the introduction of waste reduction changes, the Fremont plant annually disposed of 500,000 pounds of **resin wastes** and 200,000 pounds of **adhesive wastes,** according to information given by Borden to the U.S. EPA in 1980. These wastes, which are transported to a commercial hazardous waste disposal facility, result chiefly from sludges generated during on-site treatment of the plant's wastewater in an on-site evaporation pond. Borden reported to INFORM that in addition to off-site disposal, other solid wastes were generated and stored on-site in the plant's evaporation pond, but did not specify the quantity of wastes in storage.

Reduction

Borden's plant manager reported that since 1981, major changes in waste generation and waste management practices have reduced the quantity of organic materials discharged in the plant's resin manufacturing wastewater by 93 percent. This reduction enabled Borden to stop using an on-site treatment facility known as an evaporation pond which had generated resin sludges containing **phenol** and other process chemicals, and has eliminated most of the plant's **phenolic** resin sludges. Whereas when the pond was operating, generation of resin sludge totalled 350 cubic yards a year, current sludge generation is only 25 cubic yards a year. Based on the current cost for off-site disposal of these sludges of $150 a cubic yard, reduced sludge generation has led to $48,750 per year in disposal cost savings.

Before 1978, Borden sent its **phenolic** wastewater off-site in order to comply with the USD's strict limits on **phenol** discharges to its sewage treatment plant.*
In 1978 Borden installed a large-capacity (1,250,000 gallon) on-site evaporation pond that concentrated the wastewater into solid sludges for temporary storage and eventual transport to a commercial waste-handling facility. Borden viewed the pond as a "passive means of waste disposal which requires no support facilities, materials or personnel,"** -- unlike smaller-scale treatment systems, which are labor-intensive and require chemical additives and energy. In Borden's opinion its pond could provide the necessary level of wastewater treatment at the lowest cost to the plant.

---

* According to the plant manager, similar plants owned by Borden Chemical in other states have **phenol** limits as high as 35 mg/l, in contrast to Fremont plant's limit of one to five mg/l.

** Letter from Borden's Production Superintendent to the California Regional Water Quality Control Board, October 18, 1977.

However, the evaporation pond produced problems that spurred Borden to begin exploring waste reduction possibilities:

-- The pond filled up faster than Borden expected as a result of high rates of sludge generation and unusually heavy rainfall.

-- Borden found that the cost of enlarging the pond, in order to accommodate large amounts of sludge and rainwater, was prohibitive.

-- Continued operation of the pond would cause the plant to be classified as a hazardous waste "Treatment, Storage and Disposal Facility" under the emerging requirements of federal law. The technical, operational, monitoring, and record-keeping requirements of this classification would be administratively complex and highly expensive.

-- Reliance on the pond entailed the small but ever-present possibility of a leak, which could contaminate groundwater and expose Borden to high remedial costs and long-term liabilities.

Thus, Borden began exploring alternative means of handling the **phenolic** wastewater in 1980. By 1981, it decided to install an ozone treatment system, which was capable of destroying almost 99 percent of **phenols** (by chemically decomposing it with ozone) in wastewater containing 250 to 300 parts per million (ppm) of **phenols.** This system would reduce the **phenols** level in Borden's wastewater to one to five ppm, making it acceptable for discharge to the Union Sanitary District. Initially, Borden purchased a two-gallon-per-minute capacity ozone treatment unit to control the rising wastewater level in the evaporation pond.

Borden's first efforts at waste reduction sought to minimize the level of **phenol** and other organic materials in the wastewater, since high **phenol** concentrations impede effective operation of the ozone treatment unit.

A second reason why waste reduction measures were explored was the rapidly rising sludge level in the evaporation pond. Borden feared that the pond's 750,000 gallons of working capacity would soon be completely filled with sludge, rendering the pond unusable.

Additionally, federal and state regulation of all hazardous waste facilities had raised the cost of disposing of the hazardous sludge generated in the evaporation pond. Based on 1982 costs, Borden would have to spend more than $100,000 to dispose of sludge already accumulated in the pond.

One final influence on Borden's waste reduction efforts was the economic slump facing Borden (and the entire chemical industry) in the early 1980s, the worst since the Great Depression. The resulting slowdown in Borden's pace of business gave the plant manager time to identify and implement waste reduction practices.

In order for the evaporation pond to be phased out, the **phenol** content of the wastewater had to be consistently and dramatically reduced so that the final wastewater discharge from the ozone treatment unit would not exceed the limits established by USD. In order to do this, Borden concentrated its waste reduction efforts on the plant's resin operations, primarily because they were the source of all **phenol** wastes. The far-greater diversity and complexity of Borden's adhesives operations meant that, in contrast, they were the focus of only limited waste reduction efforts. Borden reported to INFORM that their waste reduction measures were successful, allowing the plant to discontinue use of the evaporation ponds in 1982.

The following waste reduction practices were reported by Borden:

RESIN OPERATIONS

Using two straightforward procedures, Borden began analyzing the ways in which **phenol** wastes from the resin operations were entering the plant's wastewater stream. In the first procedure, plant personnel monitored wastewater daily for organic loadings to quickly spot any sudden or unexpected surges in contaminants and then, in order to identify the source of these contaminants, reviewed staff activities and plant operations occurring before a surge. In the second procedure, plant personnel simply, in the manager's words, "walked through" the plant's operations — that is, they followed each step in the movement of materials, from arrival through processing to final off-site shipment — to identify all potential or actual sources of waste generation. Through these procedures, plant management identified and altered three practices.

Filter rinses

Before **phenolic** and other resins are loaded on trucks and shipped to customers, they are first passed through a filter that removes large particles of resinous materials which form in the reaction vessels and storage tanks. The filter canister, which holds about 25 gallons of resin, had to be drained for every truckload so that the filter could be rinsed. It had long been standard practice at the plant to empty the filter canisters directly into the plant's wastewater stream. Borden found that this practice, besides wasting a significant amount of product, was also the single most important source of contamination in the wastewater and subsequent sludge generation at the plant.

Borden implemented two changes to eliminate this waste source:

> 1) In late 1982, Borden installed a 250-gallon recovery tank next to the filters at the **phenolic** resin loading system and connected the tank to the filters with a pipe. When the filters are now cleaned, the loading operation is stopped, and the filter canister is pressurized with air, which forces the 25 gallons of resin into the tank. The filter screens are then removed for cleaning with minimal loss of product material.

2) Borden reported that the following year, an even better waste reduction system was adopted for the filter system at the urea resin loading area, made possible by the experience gained from the **phenolic** resins as well as the higher volumes of urea resins produced at the plant. When the filters are ready for cleaning, an air vent on the filter canister is opened and the filter pump is reversed, sucking the resin back into the product storage tank.

As a result of the above two modifications, Borden recovers virtually all material previously drained from the filters as waste. The plant manager told INFORM that this practice has brought about Borden's single largest reduction in waste.

Reactor vessel rinses

Rinsing the plant's reactor vessels -- 11,000 to 15,000-gallon closed, pressurized chambers -- between batches to remove residues requires enough rinsewater (2,000 to 3,000 gallons) to reach the agitator blades. In the past, Borden routinely cleaned the vessels by rinsing them with water, heating and stirring the water to remove residues, and draining the resulting material into the plant's wastewater.

Trial-and-error tests conducted in 1982 showed, however, that by using a two-rinse method, the plant could reduce its discharge of organic materials. In the current procedure, a small (approximately 100 gallon) first rinse removes most residual material and produces a rinsewater with a very high **phenol** concentration. Borden recovers this rinsewater and stores it for eventual reuse as a raw material in a later batch. A second, full-volume rinse is still required to fully clean the reactor vessel, but this rinse now has a greatly reduced **phenol** concentration. As a result, even though the overall quantity of wastewater remains the same, the quantity of organic materials lost as waste has dropped dramatically. This procedure is responsible for the second-largest amount of Borden's waste reduction, according to the plant manager, although the company did not supply estimates of the actual quantity of material saved.

Employees' practices

At Borden, small amounts of chemical waste can strongly influence the amount of overall waste generation. With direction and routine communication from the plant supervisors, employees have become increasingly aware of this potentially large impact, and management and workers alike have subsequently identified ways of reducing **phenolic** wastes. Plant staff learned, for example, that truck deliveries of **phenol** to the plant were generating unnecessary waste. On delivery, **phenol** is emptied into storage tanks through a hose; after the hose is disconnected, it always retains a small amount of the **phenol.** Formerly, the **phenol** was allowed to drip from the hose to the ground through a floor drain and into the wastewater stream. Plant personnel revised their unloading procedures and now flush the hose with a few gallons of water to unload the last amount of **phenol** into the storage tank. (Not all raw materials could readily tolerate this technique because the quality of the final product would suffer. However, the introduction of water has no effect on resin production, which is an aqueous process.)

ADHESIVES OPERATIONS

According to the plant manager, two operational changes have enabled Borden to reduce some of the waste generated by its adhesives operations. Samples of adhesives from each batch, which are taken and tested for quality, are now returned to the process when feasible instead of being automatically discarded; sample size, in addition, has been reduced from a quart to a pint. The second operational change introduced in 1983, was a new procedure for rinsing the adhesive mixing vessels, similar to the one procedure initiated for the resin reactor vessels, whereby vessel rinses are stored as a raw material. The detailed impact of the revised rinsing procedure on adhesive waste generation was not reported.

For several reasons, Borden has made only a limited attempt to reduce waste generated by its adhesives operations.

-- Adhesives operations are far more complex than those of resin. The manufacture of more than 250 adhesives products from five basic adhesives categories means that rinsewaters intended for reuse must be separated according to the type of adhesive produced and then stored before their return to the process.

-- Adhesives manufacture is more labor-intensive than resin manufacture and requires proportionally more effort to identify and remold workers' habits that are generating unnecessary wastes.

-- Unlike many resins, adhesives are not self-compatible. That is, waste from assorted end products cannot be mixed together and still be used as a raw material.

In addition to waste reduction at the source, the plant annually reclaims about 10,000 gallons of adhesives solvents and other solvents for recycling at a commercial facility, although such recycling opportunities are limited by the fact that many of the adhesives materials would clog up conventional solvent recovery equipment.

Fremont's manager told INFORM that future waste reduction changes at the plant might be able to reduce adhesives wastes by as much as 25 percent.

OBSERVATIONS

Plant management at Borden reported that a serious obstacle to implementing waste reduction practices at the Fremont plant was employees' resistance to the introduction of new procedures. Borden attributed this resistance to the additional time and effort on the part of the employees demanded by the new waste reduction measures, as well as an unwillingness to discard long-established procedures and learn new ones. In addition, Borden reported that employees often had difficulty grasping the fact that seemingly small spills of only a few gallons of **phenol** or other chemicals could result in millions of gallons of wastewater contaminated beyond acceptable discharge levels.

Continued attention by plant management to oversee employee practices and increase awareness about the importance of all losses of chemicals was eventually able to overcome most of the resistance to the revised procedures. On several occasions plant management found it necessary to threaten disciplinary action against some of its employees.

Other obstacles inhibited the exchange of waste reduction practices among the chemical plants owned by Borden. Although Fremont's waste reduction practices are both straightforward and cost-effective, the plant manager thought it unlikely that any of these changes have been adopted at other Borden plants, even those with operations similar to Fremont's. He cited four reasons for what he termed the lack of cross-pollination of waste reduction practices:

> 1) The autonomous nature of each Borden plant means that environmental problem-solving occurs primarily at the local level. There is no formal company-wide mechanism for disseminating information on waste reduction practices.
>
> 2) Waste reduction measures must be specially devised for each plant since equipment, layout, product mix and capacity to recycle or handle wastes differ greatly between locations and the measures may not prove effective at another plant.
>
> 3) Waste disposal costs -- and, as a result, economic incentives for waste reduction -- are

greater in California than elsewhere. Solid waste disposal costs in California since 1979 have increased from $50 to $150 per cubic yard according to Borden.

4) The Fremont plant faces more stringent regulatory pressures (including limits on **phenol** discharges) than do plants in other states.

SOURCES OF INFORMATION

SOURCE	LOCATION
1. Computer printout of air emissions from individual sources, 1984 (based on 1980 estimates)	Bay Area Quality Management District (BAAQMD) 939 Ellis Street San Francisco, CA 94109 (415) 771-6000
2. Computer printout of plant-wide air emissions, 1984 (based on 1983 estimates)	BAAQMD (see #1)
3. "Human Exposure to Atmospheric Concentrations of Selected Chemicals," prepared by Systems Applications, Inc. (Contract #68-02-3066), March 1983* (based on 1978 data)	U.S. Environmental Protection Agency (U.S. EPA) Office of Air Quality Planning and Standards Research Triangle Park, NC 27111 (919) 541-5315
4. Sewage Treatment Plant Survey, 1980	Union Sanitary District (USD) 4057 Baine Avenue Fremont, CA 94536 (415) 656-7584
5. Sewage Treatment Plant Wastewater Discharge Application, 1983	USD (see #4)

* This report gives 1978 estimates of air emissions from individual plants. The U.S. EPA considers the figures unreliable because of the estimation techniques used.

6. RCRA Part A Application, November 1980

U.S. EPA,
 Region IX
215 Fremont Street
San Francisco, CA 94105
(415) 974-8071

7. Information provided by Borden to INFORM

8. Environmental files, 1983

Department of Health
 Services
2151 Berkeley Way
Berkeley, CA 94704
(415) 540-2303

9. Environmental files, 1983

California Regional
 Water Quality Control
 Board
1111 Jackson Street
Oakland, CA 94607
(415) 464-1255

WASTE REDUCTION

BORDEN

Type of Waste	G-Generation D-Disposal Q-Quantity (year)	Waste Reduction Practice	Results	Sources of Information
HAZARDOUS Adhesives manufacturing wastes	Sent off-site for disposal 200,000 lbs/yr (1980)	Storage and reuse of adhesives samples taken for quality control OP (1982)	Small reduction in sludge	(7)
		Reuse of adhesive mixer rinses as raw material OP (1983)		(7)
Resin manufacturing wastes containing: **HAZARDOUS** Phenol	Sent off-site for disposal 500,000 lbs/yr (1980) G - ? D - ? Q - ? (1980)	Phenol resins drained from filters for reuse as raw material OP (1982)	Reduction of 25 gallons of phenolic resin per filter rinse. Total of 93% reduction of organics in wastewater for all these practices, leading to a reduction in solid wastes.	(7)
		Phenol rinsewater from reactor vessel reused as raw material OP (1982)		(7)
		Changes in employees' habits to reduce spills OP (1980-83)		(7)
NON-HAZARDOUS Urea	G - ? D - ? Q - ? (1984)	Urea resins drained from filters, pumped back into product tanks OP (1983)		(7)

Key to Waste Reduction Changes
PR – Product
PS – Process
EQ – Equipment
CH – Chemical Substitution
OP – Operational

BORDEN
WASTE GENERATION & DISPOSAL

Type of Waste ● Indicates Wastes Affected by Reduction	Generation or Disposal	Quantity of Waste	Sources of Information
Air Emissions			
HAZARDOUS			
Formaldehyde	Emissions from 5 process sources	800 lbs/yr (1980)	(1)
	Plant-wide emissions	170,000 lbs/yr* (1978)	(3)
Methanol	Emissions from 14 process sources	18,250 lbs/yr (1980)	(1)
Trichloroethane	Emissions from 4 process sources	11,600 lbs/yr (1980)	(1)
Trichloroethylene	Emissions from 1 process source	36 lbs/yr (1980)	(1)
Total organic emissions	Plant-wide emissions	116,580 lbs/yr (1983)	(2)
LIKELY TO CONTAIN HAZARDOUS CHEMICALS			
Adhesives emissions	Emissions from adhesives manufacturing process	53,000 lbs/yr (1980)	(1)
Fugitive organic emissions	Plant-wide vapor leaks from valves, pipes, pumps, etc.	22,300 lbs/yr (1980)	(1)
Wastewater Discharges			
Wastewater, including:	Discharged to the Union Sanitary District sewage treatment plant	36,385 gals/day (1983)	(5)
		41,844 gals/day (1980)	(4)
HAZARDOUS			
Formaldehyde	"	?** (1983)	(5)
Methyl ethyl ketone	"	?** (1983)	(5)
Methylene chloride	"	?** (1983)	(5)
● Phenol	"	?** (1980)	(5)

* The U.S. EPA considers this figure unreliable. Borden reported that formaldehyde emissions have significantly changed since it shut down part of its formaldehyde operations in December 1983, but did not provide updated emission figures.

** Listed as potentially present, but no information is available on actual quantities.

WASTE GENERATION & DISPOSAL

BORDEN

● Type of Waste Indicates Wastes Affected by Reduction	Generation or Disposal	Quantity of Waste	Sources of Information
Wastewater Discharges			
Wastewater (cont.) HAZARDOUS (cont.)	Discharged to the Union Sanitary District sewage treatment plant		
Toluene	"	?* (1983)	(5)
Trichloroethane	"	?* (1983)	(5)
Cadmium	"	1.6 gms/day (1980)	(4)
Copper	"	4.8 gms/day (1980)	(4)
Mercury	"	0.2 gms/day (1980)	(4)
Silver	"	?* (1983)	(5)
Zinc	"	53.9 gms/day (1980)	(4)
		?* (1983)	(5)
NON-HAZARDOUS			
Acetone	"	?* (1983)	(5)
Hexane	"	?* (1983)	(5)
Nalco 7326**	"	50-150 ppm (1980)	(4)
		?* (1983)	(5)
Nalco 7328**	"	40-150 ppm (1980)	(4)
	"	?* (1983)	(5)
● Urea	"	? (1984)	(7)
Barium	"	?* (1983)	(5)

* Listed as potentially present, but no information is available on actual quantities.

** These are commercial algicides.

WASTE GENERATION & DISPOSAL

●	Type of Waste Indicates Wastes Affected by Reduction	Generation or Disposal	Quantity of Waste	Sources of Information
	Solid Wastes			
	HAZARDOUS			
●	Adhesives manufacturing waste	Sent off-site for disposal	200,000 lbs/yr (1980)	(6)
	Asbestos waste	"	1,000 lbs/yr (1980)	(6)
●	Resin-manufacturing waste containing phenol and urea	"	500,000 lbs/yr (1980)	(6)
	Solvents, halogenated	"	500 lbs/yr (1980)	(6)

Carstab Division

Morton-Thiokol, Inc.
1560 West Street
Cincinnati, Ohio 45215
(513) 773-2195

SUMMARY

Carstab Corporation's plant (Cincinnati, Ohio) manufactures a wide variety of specialty chemicals, including additives for the petroleum, plastics, road paving, textiles and paper industries. The plant uses numerous organic and inorganic chemicals in its production processes, including **methanol**, numerous other organic solvents, and acids. The plant emits almost 400,000 pounds per year of organic chemicals to the air and generates 600,000 pounds of waste solvents and 1.6 million pounds of corrosive wastes. There is no indication from available information that Carstab has adopted any waste reduction techniques.

PLANT OPERATIONS

Carstab Corporation operates a 250-employee plant in Cincinnati, Ohio. This plant, established in 1949, was owned by Cincinnati Milacron, Inc. until 1980, when it was purchased by what is now Morton-Thiokol, Inc. (In 1982, Morton-Norwich, Inc. purchased Thiokol Corporation and became Morton-Thiokol, Inc., with headquarters in Chicago.)

Morton-Thiokol's three main business areas are specialty chemicals, aerospace products and salt. The Specialty Chemicals Division, to which the Carstab plant belongs,

NOTE: Chemicals in **boldface** type are hazardous, according to INFORM's criteria (see Appendix A).

manufactures adhesives, coatings and sealants; chemicals for the electrical and electronics industry; specialty dyes and colorants; reducing agents; plastics additives; and biocides at 23 U.S. manufacturing facilities. World sales for the division were $478 million in 1983; total corporate sales were over $1.5 billion in 1983.

The Carstab plant manufactures specialty chemicals for the plastics, petroleum, road paving, textile and paper industries. Its batch chemical operations, which comprise the bulk of its manufacturing, include: sulfurized oil and lard (petroleum additives), thioester compounds (antioxidants), sulfur-chlorinated styrene compounds*, tin mercaptide compounds, organo-sulfur compounds (plastics stabilizers), benzophenone compounds, lubricating stabilizers, **coumarin** compounds (optical brighteners)*, synthetic amide wax, butyltin carboxylates and phosphonium compounds. Continuous production processes are used to make organotin intermediates (plastics stabilizers) and lubricating stabilizers. A wide variety of raw materials including mineral spirits, **toluene, methanol,** isopropyl alcohol, heptane, chlorinated solvents* and numerous acids (hydrochloric, e.g.) are used in these production processes. Carstab operates research and development facilities at the Cincinnati site.

WASTE GENERATION AND WASTE REDUCTION

AIR EMISSIONS

Generation

Air permit file records indicate emissions from this plant of **toluene****, **methanol,** hydrochloric acid**,

---

* Carstab states that they do not currently use or manufacture these substances.
** Carstab reports that these chemicals are not currently used at the plant.

ammonia, and other chemical vapors, along with dust emissions from organotin and thioester manufacturing processes. The quantities of individual discharges were not available from air permit records, although total hydrocarbon emissions from the plant are reported as 391,720 pounds per year, according to the Southwest Ohio Air Pollution Control Agency (SWOAPCA).

SWOAPCA did not allow INFORM to fully review air permit files due to the confidential nature of some of the file information. Agency officials did comply with our request to provide summary data from the files, but this data did not include actual emissions quantities.

WASTEWATER DISCHARGES

Generation

Carstab's wastewater discharge to the Metropolitan Sewer District (MSD) of Greater Cincinnati includes low concentrations of **chloroethane, chloromethane** and a variety of heavy metals, according to its 1981 Discharge Permit Application. MSD has set limits on the plant's discharge for the following substances: amines, total halogenated organics, tin and the total concentration of organic vapors which can be emitted from the wastewater.

In a 1982 letter to the U.S. EPA, Carstab reported that it was discontinuing the use of a treatment tank which handled 59,000 gallons per day of wastewater. This tank had been used to separate **toluene** and other organic contaminants from the plant's wastewater prior to discharge to MSD, as reported in the 1982 RCRA Inspection Report. Carstab made this change so that it would not be regulated by the U.S. EPA as a hazardous waste treatment facility, thus avoiding the associated expense and administrative complexity of such a classification. The change was made possible, Carstab reported, by process changes made at the plant, but no information was provided as to the nature of these changes or whether they constituted a waste reduction practice. Carstab has since told INFORM that **toluene** is not currently used at the plant.

SOLID WASTES

Generation

Carstab annually generates substantial quantities of solvent and corrosive wastes, according to records of the Ohio Environmental Protection Agency (OEPA), including 342,000 pounds of **methanol;** 600,000 pounds of waste solvents and 1.6 million pounds of corrosive wastes. These wastes are temporarily stored on-site prior to off-site disposal at commercial waste handling facilities in Ohio, Indiana, Alabama and other states. Some solvent wastes were burned in an incinerator operated by the Metropolitan Sewer District of Greater Cincinnati, according to 1981 OEPA files, although Carstab reports that this incinerator is not currently in operation.

SOURCES OF INFORMATION

SOURCE	LOCATION
1. Air permit files, 1983	Southwestern Ohio Air Pollution Control Agency 2400 Beekman Street Cincinnati, OH 45214 (513) 251-8777
2. Emissions Inventory System Point Source for 1983, June 1984	Ohio Environmental Protection Agency (OEPA) Division of Air Pollution Control 361 East Broad Street Columbus, OH 43215 (614) 466-6116
3. Metropolitan Sewer District (MSD) Wastewater Discharge Permit Application, May 1981	Metropolitan Sewer District (MSD) 1600 Gest Street Cincinnati, OH 45204 (513) 352-4829
4. MSD metals analyses, July 30, 1980 and October 18, 1982	MSD (see #3)
5. RCRA Part A Application, November 1980	OEPA Division of Hazardous Waste Management Permits and Manifest Records Section 361 East Broad Street Columbus, OH 43215 (614) 466-1586
6. Facility Hazardous Waste Annual Report, 1981	OEPA (see #5)

7. Correspondence from
   Carstab to OEPA,
   September 1981

   OEPA
   Southwest District Office
   Division of Hazardous
    Waste Management
   7 East Fourth Street
   Dayton, OH 45402
   (614) 461-4670

8. Generator Annual
   Hazardous Waste
   Report, 1981

   OEPA
   (see #5)

9. Correspondence from
   Carstab to U.S. En-
   vironmental Protection
   Agency, September 1982

   OEPA
   (see #7)

10. RCRA Inspection
    Report, January 1982

    OEPA
    (see #7)

11. Information supplied
    by Carstab to INFORM

CARSTAB

WASTE GENERATION & DISPOSAL

Type of Waste ● Indicates Wastes Affected by Reduction	Generation or Disposal	Quantity of Waste	Sources of Information
Air Emissions			
HAZARDOUS			
Methanol	Emissions from 1 process source	? (1983)	(1)
Toluene*	"	? (1983)	(1)
LIKELY TO CONTAIN HAZARDOUS CHEMICALS			
Hydrocarbons	Emissions from 16 process sources	391,720 lbs/yr (1983)	(2)
MAY CONTAIN HAZARDOUS CHEMICALS			
Alcohols	Emissions from 1 process source	? (1983)	(1)
Dust from organotin intermediates manufacture	"	? (1983)	(1)
NON-HAZARDOUS			
Ammonia	Emissions from 2 process sources	? (1983)	(1)
Dust from thioester production	Emissions from 1 process source	? (1983)	(1)
Dust, miscellaneous	Emissions from 2 process sources	? (1983)	(1)
Hydrochloric acid**	"	? (1983)	(1)
Lime dust	Emissions from 1 process source	? (1983)	(1)
Sulfur monochloride**	"	? (1983)	(1)

* Carstab reports that toluene is not currently used at the plant.

** Carstab reports these processes have been discontinued.

CARSTAB

WASTE GENERATION & DISPOSAL

Type of Waste ● Indicates Wastes Affected by Reduction	Generation or Disposal	Quantity of Waste	Sources of Information
Wastewater Discharges			
Cooling, process and other wastewaters, including:	Discharged to the Metropolitan Sewer District of Greater Cincinnati	402,000 gals/day (of which 15,000 gals/day is process water) (1981)	(3)
HAZARDOUS			
Chloroethane	"	?* (1981)	(3)
Chloromethan	"	?* (1981)	(3)
Antimony	"	?* (1981)	(3)
Arsenic	"	?* (1981)	(3)
Cadmium	"	.003–.009 mg/l (1982)	(4)
Chromium	"	.008–.096 mg/l (1982)	(4)
Copper	"	.033–.282 mg/l (1982)	(4)
Lead	"	.014–.405 mg/l (1982)	(4)
Nickel	"	.072–.137 mg/l (1982)	(4)
Silver	"	?* (1981)	(3)
Zinc	"	.100–1.248 mg/l (1982)	(4)

* Suspected present in low concentrations.

CARSTAB

WASTE GENERATION & DISPOSAL

Type of Waste ● Indicates Wastes Affected by Reduction	Generation or Disposal	Quantity of Waste	Sources of Information
Solid Wastes			
HAZARDOUS			
Methanol	Stored on-site	150,000 lbs/yr (1980)	(5)
		342,000 lbs/yr (1981)	(6)
Solvents, halogenated*	"	200,000 lbs/yr (1980)	(5)
Solvents, non-halogenated	"	140,000 lbs/yr (1980)	(5)
		96,000 lbs/yr (1981)	(6)
Solvents, non-halogenated and unspecified corrosive wastes	"	20,000 lbs/yr (1980)	(5)
LIKELY TO CONTAIN HAZARDOUS CHEMICALS			
Waste solvents	Disposed of off-site by CECOS/CER Co., Williamsburg, OH; U.S. Ecology, Sheffield, IL; or Ohio Liquid Disposal Inc., Fremont, OH	600,000 lbs/yr** (1981)	(3)
	Incinerated at the Metropolitan Sewer District of Greater Cincinnati, Cincinnati, OH***	? (1981)	(7)
MAY CONTAIN HAZARDOUS CHEMICALS			
Acid layers, recovered	Disposed of off-site by ILWD Inc., Indianapolis, IN	372,000 lbs/yr (1981)	(8)
	Disposed of off-site by Ohio Liquid Disposal, Vickery, OH	506,000 lbs/yr (1981)	(8)

* Carstab reports that it no longer uses halogenated solvents in its operations.

** Probably included in other waste streams listed in this section.

*** Carstab reports that the Metropolitan Sewer District incinerator is not currently in operation.

210

CARSTAB

WASTE GENERATION & DISPOSAL

Type of Waste ● Indicates Wastes Affected by Reduction	Generation or Disposal	Quantity of Waste	Sources of Information
Solid Wastes			
MAY CONTAIN HAZARDOUS CHEMICALS (cont.)			
Acid residue, recovered	Disposed of off-site by Chemical Waste Management, Inc., Emelle, AL	2,000 lbs/yr (1981)	(8)
	Disposed of off-site by CECOS/CER Co., Williamsburg, OH	18,000 lbs/yr (1981)	(6)
Acid solvent residues	Stored on-site	136,000 lbs/yr (1981)	(6)
Acids and alkalies	Disposed of off-site by CECOS/CER Co., Williamsburg, OH; U.S. Ecology, Sheffield, IL; or Ohio Liquid Disposal Inc., Fremont, OH	1,600,000 lbs/yr (1981)	(3)
Corrosive wastes, unspecified	Stored on-site	1,600,000 lbs/yr (1980)	(5)
Filter press cakes	Disposed of off-site by CECOS/CER Co., Williamsburg, OH; U.S. Ecology, Sheffield, IL; or Ohio Liquid Disposal Inc., Fremont, OH	340,000 lbs/yr (1981)	(3)
Ignitable and corrosive wastes, unspecified	Stored on-site	550,000 lbs/yr (1980)	(5)
Pretreatment sludge	Disposed of off-site by CECOS/CER Co., Williamsburg, OH; U.S. Ecology, Sheffield, IL; or Ohio Liquid Disposal Inc., Fremont, OH	280,000 lbs/yr (1981)	(3)
Reactive wastes, unspecified	Stored on-site	80,000 lbs/yr (1980)	(5)
Scrubber solution	Disposed of off-site by Chemical Waste Management, Inc., Emelle, AL	400 lbs/yr (1981)	(8)
	Disposed of off-site by ILWD Inc., Indianapolis, IN	50,000 lbs/yr (1981)	(8)
Solvents, non-halogenated	Stored on-site	800,000 lbs/yr (1980)	(5)
Spent solvents and residues	Disposed of off-site by CECOS/CER Co., Williamsburg, OH	4,000 lbs/yr (1981)	(8)

Chevron Chemical Company

940 Hensley Street
Richmond, California 94804
(415) 231-8100

SUMMARY

Chevron Chemical Company's plant (Richmond, California) manufactures agricultural chemicals, including pesticides, herbicides, fungicides and fertilizers. This plant uses almost 10 million pounds per year of **trichloroethylene** as a raw material. Organic air emissions from the plant are 12,220 pounds per year. Wastewater is either treated in on-site evaporation ponds, discharged to surface water or discharged to the City of Richmond Water Pollution Control Plant. **Arsenic** and **Lindane** are present in the discharge to the sewage treatment plant. Solid wastes are either sent off-site for disposal, incinerated on-site or discharged to on-site evaporation ponds. Hazardous components of these solid wastes include **benzene, chloroform, dichloromethane, toluene** and **toxaphene**. A small quantity of **mercury** is recycled. Despite the company's position that the plant's waste reduction practices are documented in public records, no such records were identified by INFORM. There is no indication from available information that Chevron has adopted any waste reduction techniques.

PLANT OPERATIONS

Chevron Chemical Company operates a 755-employee agricultural chemical manufacturing plant in Richmond, California, ten miles northeast of San Francisco. This

> NOTE: Chemicals in **boldface** type are hazardous, according to INFORM's criteria (see Appendix A).

plant is part of a Chevron complex, which also includes a refinery and a research center. The Chevron Chemical Division of Chevron Corp., with 16 U.S. chemical plants, manufactures agricultural chemicals, fibers, resins, petrochemicals and fuel additives. U.S. sales for the division were $934 million in 1983, reflecting an operating loss for the second consecutive year. Pesticides, herbicides, fungicides and fertilizers are manufactured at the Richmond plant. Chevron uses 9,746,000 pounds per year of **trichloroethylene** as a raw material, according to its April 1983 City of Richmond Water Pollution Control Plant Application and Permit. Pilot plant formulation of new chemicals is also conducted on-site.

WASTE GENERATION AND WASTE REDUCTION

AIR EMISSIONS

Generation

Chevron discharges a total of 12,220 pounds per year of organic chemicals from its 161 air emissions sources (over one-third of which are tanks), according to Bay Area Air Quality Management District records for 1983.

WASTEWATER DISCHARGES

Generation

San Francisco Regional Water Quality Control Board files for 1980 indicate that Chevron generated 53,000,000 gallons of process wastewater and 79,258,029 gallons of other wastewater, including stormwater runoff, pollution control wastewater, cooling water and incinerator water, in 1978. Wastewaters are disposed of by four different methods:

> -- Potentially hazardous chemicals, primarily pesticide manufacturing wastes, are discharged to on-site evaporation ponds. Chevron told INFORM that through in-plant controls and incineration

of wastes, hazardous process water is not currently discharged to the evaporation ponds. However, it provided no details about modifications at the plant that allowed these discharges to be discontinued. Process wastewater is now incinerated, and the ponds receive only stormwater runoff (85 percent) and wastewater from incinerator operation (15 percent).

-- Wastewater from pollution control of the incinerator containing dissolved salts is neutralized and oxidized and then piped to the nearby Chevron refinery. It is combined with refinery wastewaters prior to discharge to Castro Creek, which feeds into San Pablo Bay.

-- Potentially hazardous stormwater runoff from the fertilizer and agricultural chemical plants is normally collected in the on-site ponds, but is discharged directly into Herman's Slough during periods of extra-heavy rainfall. The Slough eventually feeds into San Pablo Bay.

-- Surface runoff, pollution control equipment wastewater and cooling tower wastewater, containing the hazardous chemicals **arsenic** and **Lindane**, are discharged to the City of Richmond Water Pollution Control Plant.

In January 1983, Chevron was ordered by the Control Plant to begin monitoring its wastewater, on a monthly basis, for **arsenic, chromium** and total chlorinated hydrocarbons, because of Chevron's high discharge volumes. The concern was that even small fluctuations in the amount of toxic substances discharged by such a large-volume contributor could interfere with the City of Richmond's treatment plant processes. In March 1983, the City of Richmond also requested Chevron to analyze its wastewater for specific chlorinated hydrocarbons, such as **trichloroethylene,** and the pesticides **Lindane** and Sevin for the next three months. Chevron told INFORM that it is no longer required to analyze for **Lindane** and Sevin.

SOLID WASTES

Generation

Solid wastes produced at Chevron are either incinerated on-site, discharged to on-site evaporation ponds or sent off-site for disposal at a hazardous waste disposal facility. Chevron reported to INFORM that its use of an incinerator has lowered the amount of wastes it must send off-site for disposal. However, it provided no description of the plant's current solid waste disposal types and quantities.

According to the Water Pollution Control Plant's permit for Chevron, hazardous wastes disposed of by one or more of these methods include 1,838,400 pounds per year of **dichloromethane,** 1,634,100 pounds per year of **toluene,** and much smaller quantities of **benzene, chloroform, copper, 2,4-dichlorophenoxy acetic acid, hexachlorocyclohexane, toxaphene** and **zinc**; these chemicals are used as process raw materials and solvents and in laboratory research. **Mercury,** used in pressure indicators, is recycled. A 1978 State of California Industrial Waste Survey indicated that potentially hazardous wastes from pesticide manufacture are sent off-site for disposal.

OBSERVATIONS

There is no indication from available information sources that Chevron has adopted any waste reduction techniques. However, Chevron told INFORM that it "has an active waste management program which includes continual evaluation of the newest and best waste reduction technologies and has already proceeded a long way towards our ultimate goal of minimizing, if not eliminating, waste generation and disposal." In a letter to INFORM detailing recent changes in waste management at the plant, Chevron described incineration, the use of evaporation ponds, and rainwater runoff discharge practices at the Richmond plant. The only reference to waste reduction in the document refers to "effective in-plant source reduction programs" but lacks any details about the nature or impact of these practices.

Chevron also reported to INFORM that its reduction practices are documented in government files with "jurisdiction over health, safety and environmental matters." In addition to the normal review of state and federal environmental agency files responsible for Chevron (conducted for each study plant), INFORM also wrote a letter to the director of each of these agencies, specifically requesting any information available in public files about waste reduction efforts at the plant. These letters were sent to the directors of the Bay Area Air Quality Management District; the San Francisco Regional Water Quality Control Board; the U.S. Environmental Protection Agency, Region IX; the City of Richmond Water Pollution Control Plant; the Toxic Substances Control Division, North Coast California Section, Department of Health Services; and the California Occupational Safety and Health Administration.

None of these offices provided us with any waste reduction information. The Water Pollution Control Plant failed to respond to INFORM'S request, while the Air Quality Management District, the Department of Health Services and the Occupational Safety and Health Administration responded, but had no information to provide about waste reduction at Chevron. The latter two agencies explained that they simply do not retain records on waste reduction. The U.S. EPA told INFORM that it found "no information...pertaining to waste reduction practices at the plant" in its air and water records. Finally, the Regional Water Quality Control Board told INFORM that its files "contain descriptions of some of the waste treatment and reduction methods used at Chevron." However, it explained that "Chevron has claimed that innovative techniques used for waste reduction are 'trade secrets' and therefore not available for public review."

SOURCES OF INFORMATION

SOURCE	LOCATION
1. Bay Area Air Quality Management District Emissions Summary printout, June 1983	Bay Area Air Quality Management District 939 Ellis Street San Francisco, CA 94109
2. Regional Water Quality Control Board - San Francisco Bay Region, Internal Memo, May 1980	Regional Water Quality Control Board 1111 Jackson Street, Room 6040 Oakland, CA 94607 (415) 464-1255
3. State of California, Industrial Waste Survey, September 1978	Department of Health Services Toxic Substances Control Division 2151 Berkeley Way Berkeley, CA 94704 (415) 540-2043
4. City of Richmond Water Pollution Control Plant, Application and Permit, April 1983	City of Richmond Water Pollution Control Plant (RWPCP) 601 Canal Boulevard Richmond, CA 94804 (415) 231-2145
5. Chevron letter to the City of Richmond Water Pollution Control Plant, February 1983	RWPCP (see #4)
6. Chevron letter to the City of Richmond Water Pollution Control Plant, April 1983	RWPCP (see #4)

7. City of Richmond Water   RWPCP
   Pollution Control Plant   (see #4)
   letter to Chevron,
   January 1983

8. City of Richmond Water   RWPCP
   Pollution Control Plant   (see #4)
   letter to Chevron,
   March 1983

9. Information provided
   by Chevron to INFORM

CHEVRON

WASTE GENERATION & DISPOSAL

Type of Waste ● Indicates Wastes Affected by Reduction	Generation or Disposal	Quantity of Waste	Sources of Information
Air Emissions			
LIKELY TO CONTAIN HAZARDOUS CHEMICALS Organics	Plant-wide emissions	12,220 lbs/yr (1983)	(1)
Wastewater Discharges			
Wastewaters, including: LIKELY TO CONTAIN HAZARDOUS CHEMICALS	Treated in on-site evaporation ponds		
Process waste from Difolatan manufacture*	"	22 million gals/yr (1978)	(3)
Process waste from pesticides formulation and packaging*	"	31 million gals/yr (1978)	(3)
Stormwater runoff from Difolatan manufacture	"	2.7 million gals/yr (1978)	(3)
Stormwater runoff including Orthene	"	9.5 million gals/yr (1978)	(3)
Wastewater from pollution control of incinerator	"	? (1984)	(9)
MAY CONTAIN HAZARDOUS CHEMICALS Stormwater runoff	Discharged to Herman's Slough during periods of heavy rainfall	? (1980)	(2,3)
NON-HAZARDOUS Wastewater from pollution control of incinerator containing dissolved salts	Treated and combined with Chevron Refinery wastewater prior to discharge to San Pablo Bay	40 million gals/yr (1978)	(2,3)

* Chevron reported that these wastes are not currently discharged to the evaporation ponds.

CHEVRON

WASTE GENERATION & DISPOSAL

Type of Waste ● Indicates Wastes Affected by Reduction	Generation or Disposal	Quantity of Waste	Sources of Information
Wastewater Discharges			
Surface runoff, pollution control equipment wastewater and cooling tower wastewater, including:	Discharged to the City of Richmond Water Pollution Control Plant	27,058,029 gals/yr (1983)	(4,5)
HAZARDOUS			
Arsenic	"	.73 ppm (1983)	(6)
Lindane	"	.001 ppm (1983)	(6)
NON-HAZARDOUS			
Sevin	"	.05 ppm (1983)	(6)
Solid Wastes			
HAZARDOUS			
Benzene	Sent off-site for disposal or incinerated on-site	140 lbs/yr* (1983)	(4)
Chloroform	Incinerated on-site	14.4 lbs/yr* (1983)	(4)
Dichloromethane	Sent off-site for disposal or incinerated on-site	1,838,400 lbs/yr* (1983)	(4)
2,4-Dichlorophenoxy acetic acid	Incinerated on-site	19,560 lbs/yr* (1983)	(4)
Hexachlorocyclohexane	Sent off-site for disposal, incinerated on-site or discharged to an on-site evaporation pond	456 lbs/yr* (1983)	(4)
Toluene	"	1,634,100 lbs/yr* (1983)	(4)
Toxaphene	"	28,656 lbs/yr* (1983)	(4)

* The Richmond Industrial Survey (4) lists these figures as "quantity used or discharged."

CHEVRON

WASTE GENERATION & DISPOSAL

Type of Waste ● Indicates Wastes Affected by Reduction	Generation or Disposal	Quantity of Waste	Sources of Information
Solid Wastes			
HAZARDOUS (cont.)			
Copper compounds	Sent off-site for disposal, incinerated on-site or discharged to an on-site evaporation pond	5,244 lbs/yr* (1983)	(4)
Mercury	Recycled	12 lbs/yr* (1983)	(4)
Zinc compounds	Sent off-site for disposal, incinerated on-site or discharged to an on-site evaporation pond	13,992 lbs/yr* (1983)	(4)
LIKELY TO CONTAIN HAZARDOUS CHEMICALS			
Difolatan manufacturing wastes:			
Off-test material and contaminated raw materials	Sent off-site for disposal	60,000 gals/yr (1978)	(3)
Off-test product	"	140 cubic yds/yr (1978)	(3)
Incinerator residue (salts and oxides)	"	350 cubic ft/yr (1978)	(3)
Pesticides formulation and packaging wastes:			
Drum liners	"	120 cubic yds/yr (1978)	(3)
Off-test products	"	380 cubic yds/yr (1978)	(3)
Organophosphate contaminated items, pesticides, etc.	"	360 cubic yds/yr (1978)	(3)
Solvents, wastewater from research operations	"	22,000 gals/yr (1978)	(3)

* The Richmond Industrial Survey (4) lists these figures as "quantity used or discharged."

CHEVRON

WASTE GENERATION & DISPOSAL

Type of Waste ● Indicates Wastes Affected by Reduction	Generation or Disposal	Quantity of Waste	Sources of Information
Solid Wastes			
LIKELY TO CONTAIN <u>HAZARDOUS CHEMICALS</u> (cont.) Wastewater solvents, waste hydrocarbons and sludge from Orthene manufacture	Sent off-site for disposal	14,000 gals/yr (1978)	(3)

222

CIBA-GEIGY Corporation

Toms River Plant
Route 37 West
P.O. Box 71
Toms River, New Jersey 08753
(201) 349-5200

SUMMARY

CIBA-GEIGY's plant (Toms River, New Jersey) manufactures a wide variety of dyes and epoxy resins from chemicals which include **chromium, nitrobenzene, toluene, epichlorohydrin,** and numerous other metals and chlorinated organics. Fundamental changes in process chemistry and manufacturing conditions have reduced **chromium** wastes by 25 percent and completely eliminated **mercury** wastes. On the other hand, the plant deemed it simpler to discontinue the small-volume manufacture of dyes that relied on **zinc** rather than search for possible waste reduction options for this metal. Process changes also resulted in reduced discharges of **nitrobenzene** and sulfuric acid. A second category of waste reduction used chemical substitutions for lime (used during on-site wastewater treatment) in order to minimize the quantity of sludge produced. The need to increase yields and reduce waste management costs, combined with regulatory pressures, were the principal motivations for waste reduction. The intense scrutiny that Toms River is currently receiving from environmental groups and regulatory authorities due to concern over possible contamination from the plant may prompt additional waste reduction measures.

NOTE: Chemicals in **boldface** type are hazardous, according to INFORM's criteria (see Appendix A).

PLANT OPERATIONS

The Toms River Plant is located near the New Jersey coast, about 50 miles south of Newark. The plant is owned by CIBA-GEIGY Corporation, the U.S. operations of CIBA-GEIGY Limited, a Swiss-based chemical company that manufactures agricultural chemicals, pharmaceuticals, dyes, plastics and photographic supplies. World-wide sales in 1983 were $7 billion; sales of CIBA-GEIGY Corp. in the U.S. were $1.3 billion.

The Toms River plant, built in 1952 and employing 1,050 people, has been fully owned by CIBA-GEIGY since 1981.* The plant, part of CIBA-GEIGY's Dyestuffs and Chemicals Division, produces a great many dyes and epoxy resins on a highly variable production schedule; about 500 different products are manufactured annually, as many as 100 on any given day. Production capacities are 220,000 pounds per day for dyes and 105,000 pounds per day for epoxy resins. Smaller quantities of some specialty chemicals, such as water repellents and fluorescent whiteners, are also produced at the plant. All production is from batch operations. The plant's sales in 1979 were $90 million.

WASTE GENERATION AND WASTE REDUCTION

AIR EMISSIONS

Generation

Air permit files of the New Jersey Department of Environmental Protection (NJDEP) list more than 100 permits for air emissions from tanks, processing

---

* Ciba and Geigy were separate companies and jointly owned Toms River (along with Sandoz, Inc.) when the plant was first built. They merged in 1970 to form CIBA-GEIGY and in 1981 bought the remaining share of the Toms River plant from Sandoz.

equipment and other sources at the Toms River plant. Annual air emissions include 68,100 pounds of **toluene**, 21,900 to 33,440 pounds of **epichlorohydrin**, 14,371 pounds of **nitrobenzene**, 8,865 pounds of **ethyl benzene** (present as a contaminant in **toluene**), 7,820 pounds of **dichlorobenzene**, 3,100 pounds of **trichloroethane**, 2,341 pounds of **trichlorobenzene**, and 1,010 pounds of **anthracene**. In some cases, these emissions are a substantial fraction of the overall quantity of the chemical in use. For instance, **dichlorobenzene** emissions in 1978 (the most recent year for which NJDEP data are available) were 12.7 percent of the total annual quantity (61,640 pounds) in use at Toms River.

Reduction

Waste reduction practices affecting air emissions have involved changes aimed at specific chemicals that have eliminated or reduced not only air emissions, but all wastestreams associated with these chemicals.

In one case reported to INFORM by CIBA-GEIGY, a process change allowed the plant to completely eliminate its reliance on a hazardous chemical, **mercury,** thereby eliminating associated wastes.

Mercury had been used at Toms River in the manufacture of anthraquinone (AQ) dyes, a common category of materials widely used for dyeing cotton. The classical process for manufacturing these dyes involved reacting AQ through a number of intermediate stages to produce aminoanthraquinone. A sulfonation reaction during this process requires the use of **mercury** as a catalyst, and results in the generation of mercuric wastes.

Of the 2,280 pounds of **mercury** used at Toms River in 1978, most of it ended up as waste material as follows: 10 pounds of air emissions, 58 pounds discharged to wastewater, 325 pounds of solid waste buried in an on-site landfill, and the remainder sent off-site as a component of 39,500 pounds of contaminated material trapped during a process filtering operation.

CIBA-GEIGY's corporate research group in Switzerland developed a novel chemical pathway for manufacturing aminoanthraquinone that circumvented the sulfonation step, thus eliminating the need for **mercury** as a catalyst. This process was introduced at Toms River in 1983 with the subsequent elimination of all wastes containing **mercury.**

The primary motivation for developing a mercury-free process, according to the Director of Production and Environmental Affairs at Toms River, was the "profound effect" on all industry of serious environmental contamination and poisoning problems involving **mercury.** Particularly worrisome was the "element of surprise" involved in any discharge of wastes containing **mercury** no matter how secure they were deemed to be.*

The company also reported to INFORM that considerable efforts have been directed toward reducing **nitrobenzene** wastes at their source. Although specific measures for reducing air emissions were not reported, CIBA-GEIGY estimates that the overall quantity of **nitrobenzene** wastes from all wastestreams has been reduced tenfold since 1972.

WASTEWATER DISCHARGES

Generation

Discharge of process wastewater from Toms River -- unique in that the plant has the only permit in New

---

*   Widespread **mercury** contamination has occurred in Sweden, Japan and Iraq and is responsible for some of the most serious environmental health effects known -- thousands of illnesses and hundreds of deaths have been caused by mercury poisoning. In some cases, industrial discharges of mercury that were thought to present no dangers to health, were unexpectedly introduced into the food-chain through chemical alterations in the environment and subsequently poisoned people eating mercury-contaminated foods.

Jersey for direct discharge to the Atlantic Ocean -- averages four to six million gallons per day. According to NJDEP and U.S. EPA records, the final discharge, which receives on-site biological treatment at the company's wastewater treatment plant, contains over 50 hazardous chemicals in quantities ranging from .03 to almost 200 pounds per day.

Although the NJDEP discharge permit for the plant restricts the quantity of hazardous organics and metals in the wastewater, the Toms River plant has exceeded the imposed limits on numerous occasions. For example, discharges of **nitrobenzene** have ranged as high as 432 pounds per day (despite a permit limitation of 79 pounds per day) and have been of particular concern to environmental officials, according to permit files at U.S. EPA and NJDEP.

A second untreated discharge of 11.4 million gallons per day of non-contact cooling water and storm water is discharged to Toms River and is monitored for four metals -- **copper, zinc, chromium** and aluminum -- present in generally trace quantities.

Reduction

As explained above, a process change to eliminate the use of **mercury** as a catalyst eliminated discharges to all wastestreams, including 58 pounds per year discharged in wastewater.

Nitrobenzene, used as a reactant and solvent in a number of dye manufacturing processes at the plant, has been reduced by recirculating wastewater from a steam distillation unit used to purify the chemical, according to information provided to INFORM by plant management. No information was available to document the impact of this practice on reducing **nitrobenzene** wastewater discharges, however.

An example of a waste reduction change that minimized both wastewater discharges and solid waste generation is CIBA-GEIGY's development of a new dye manufacturing process. The classic method of manufacturing dyes containing **chromium** is to use a water-based process that relies on excess quantities of this metal to insure the maximum yield of dyestuff from the reaction. As a result, large quantities of unreacted **chromium** in the process water became part of the plant's wastewater and solid waste streams.

In 1978, the most recent year that data is available from the NJDEP, Toms River used 35,461 pounds of **chromium** in dye manufacturing; 2,946 pounds (8 percent) were discharged as wastewater and another 12,460 pounds (35 percent) as solid waste sludges. In other words, almost half of all **chromium** in use at the plant in 1978 became waste material.

CIBA-GEIGY reported to INFORM that a new solvent-based process, developed by the corporate research group in Switzerland, is now in use at Toms River that uses **chromium** far more efficiently than the old water-based process. Total loss of **chromium** as a waste material is conservatively estimated to have been reduced by 25 percent. Despite the reduced waste generation however, the cost of producing dyes is greater with the new process than with the old, chiefly because of the added cost of using solvent instead of water. However, the company views the advantages of reducing its long-term liability associated with **chromium** wastes, particularly solid wastes that are landfilled, as outweighing the additional costs of the new dye-manufacturing process.

A third metal used in dye-manufacturing, **zinc,** is similar to **chromium** in that substantial quantities -- 50 percent of the metal -- ended up as wastewater and solid waste discharges, rather than in the finished dye product. Processes using **zinc** were examined for ways in which waste discharges could be reduced. However, Toms River opted to discontinue the manufacture of all dyes containing **zinc,** rather than pursue possible waste reduction initiatives, because the volume of dyes produced with this chemical and the overall profit to the plant were too small to justify either the additional cost of waste reduction research or the continued disposal of contaminated wastes.

A final example of waste reduction of wastewater discharges are process changes at Toms River that have reduced the generation of acid wastewaters. Traditionally, large excess quantities of sulfuric acid, an inexpensive industrial chemical, have been used to insure complete reaction of the more valuable dye-forming raw materials during sulfonation reactions, a very common step in dye manufacturing. After the reaction occurs and the solid dye product is filtered out of solution, the remaining acid is discharged to the wastewater stream.

Through laboratory and plant-scale experimentation, Toms River identified the minimum amounts of acid needed to insure completion of its major sulfonation reactions (each reaction in different dye manufacturing operations had to be independently evaluated -- a process that is still continuing at the plant) and estimates that total sulfuric acid waste discharges have been reduced 10 to 40 percent as a result. Additional research is aimed at developing a non-acidic sulfonation process which will eliminate the need for sulfuric acid altogether and, as the company reported to INFORM, is nearing the point where the process can soon be adopted on a commercial scale at Toms River.

SOLID WASTES

Generation

Waste generation information from U.S. EPA and NJDEP files list six inorganic and more than 40 organic hazardous chemical wastestreams generated at the Toms River plant in quantities ranging from 100 pounds per year (**methyl chlorocarbonate**) to more than 80,000 pounds per year (**toluene**). In addition, 7.3 million pounds per year of solid waste sludge likely to contain hazardous chemicals are generated from the operation of an on-site sewage treatment plant.

The plant has an on-site landfill for disposal of sewage treatment plant sludge and other solid wastes. Ignitable wastes, including **aniline, epichlorohydrin, methanol, phenol, perchloroethylene,** and **toluene** are sent to the CIBA-GEIGY plant in McIntosh, Alabama, for incineration. Other hazardous wastes disposed of off-site are sent to commercial waste disposal firms and are landfilled.

Reduction

All of the process changes mentioned under the sections on air emissions and wastewater discharges also affect the quantity of solid wastes generated. Thus, solid wastes containing **mercury** have been eliminated, and **nitrobenzene** and **chromium** solid wastes have been reduced, by the changes already described.

The decision to discontinue the manufacture of dyes requiring **zinc,** although not a waste reduction practice by INFORM's criteria, has nonetheless eliminated the solid wastes associated with these processes.

The changes which lowered the discharge of acid wastewaters also reduce solid waste generation. Although the immediate impact of minimizing sulfuric acid use is the reduction of acid discharges to the plant's wastewater, the ultimate goal is to prevent the generation of large quantities of solid waste sludges during the plant's wastewater treatment process. Acid wastewaters must be treated with caustic materials in order to reduce the acidity of the water prior to discharge. Toms River relies on one of the most common treatment methods -- the addition of lime to the wastewater -- to neutralize acid levels, but this procedure results in the generation of large quantities of solid waste sludge during the wastewater treatment process. The steps taken at Toms River to reduce acid discharges in the wastewater have the eventual aim of reducing the quantity of lime that is required to treat the acid and, hence, reduce the quantity of sewage sludge generated during the treatment process.

Two other practices at Toms River are also geared towards reducing the volume of wastewater treatment sludge. A high-quality lime has been substituted for a lower-quality material, with the result that more acid can be neutralized -- and less sludge generated -- per pound of lime added to the wastewater. Although the higher quality material is more expensive than the original grade of lime used, the plant considers the advantage of reduced sludge generation to outweigh the additional cost of wastewater treatment.

Additionally, Toms River accepts caustic wastewaters from other industrial plants if the waters can be used to neutralize their own acidic wastewater. The caustic liquids offer the advantage of effective neutralization at less cost and with far less solid sludge generation than the use of lime. The caustic wastewaters have come from diverse industrial operations including a Coca Cola plant manufacturing fruit drinks; a Nestles plant manufacturing instant coffee; rinsewaters from plants manufacturing vitamins, pharmaceuticals, and dyes; wastewaters from a photographic lab and bilge water from a leaking ship. (NJDEP is currently investigating the composition of these wastewaters out of concern that the advantages of reduced sludge generation may be offset by introducing additional toxic components to the wastestream by accepting these off-site wastes.)

Both practices -- use of high-quality lime and replacement of lime with caustic wastewaters -- are examples of waste reduction at the source, but of a different nature than most other waste reduction practices covered in this study since the source of the waste, in this case, is the wastewater treatment plant rather than the manufacturing process. Only one of the practices reducing sludge wastes had its origin in the manufacturing process -- the reduction in the quantity of acid used.

OBSERVATIONS

The waste reduction changes reported to INFORM by the company and recorded in public files have chiefly been process changes to eliminate or reduce particular chemicals which, in turn, had the effect of reducing discharges to all wastestreams -- air, water and solid wastes. All reported process changes have occurred at the dye manufacturing operations, as described to INFORM during meetings with the manager of these operations. The management of the epoxy operations at Toms River declined to meet with INFORM and we have not learned of any waste reduction practices at the plant's epoxy operations.

At the time of this writing, CIBA-GEIGY's plant at Toms River was in the midst of intense public, regulatory and media scrutiny regarding the plant's current and past waste disposal practices, and the role that waste reduction should be playing in the overall strategy of waste management. Although the impact of this scrutiny cannot be documented with numerical precision, it doubtless adds momentum to the plant's search for waste reduction alternatives.

In July 1984, the Toms River plant was the target of action by members of the environmental group Greenpeace, who scaled a water tower at the plant to display the sign "Reduce It, Don't Produce It," and simultaneously had scuba divers locate the plant's pipeline in the Atlantic Ocean and attempt to plug it up to prevent the discharge of wastewaters. The action received widespread media attention.

NJDEP, which had been conducting its own investigation of waste management at Toms River, expanded the scope of its efforts after Greenpeace's actions. Both NJDEP and U.S. EPA officials have expressed concerns about some waste management practices at Toms River due to initial indications that past disposal practices at the plant's on-site landfills may have resulted in the disposal of enormous quantities of previously unreported hazardous chemical wastes. In addition to state and federal oversight, county health and legal authorities are also undertaking their own investigations into possible contamination of local groundwater by waste disposal at the plant.

While a high level of governmental scrutiny is not uncommon for large industrial plants, the combination at Toms River of regulatory oversight and highly visible action by a nationally-based environmental group creates an enormous amount of pressure for this plant to undertake actions to revise their waste management practices (even if the company feels that current waste management practices are already sufficient). Grass-roots environmental groups in the community of Toms River have held public meetings attended by hundreds of citizens demanding that the plant "do something" about pollution.

As this report was being written, the pace of government action regarding Toms River was increasing. Groundwater contamination was detected at the site as a result of past land disposal practices, and Toms River was declared by the U.S. EPA as a federal Superfund site. NJDEP levied the largest environmental fine ever imposed in the state -- $1.45 million -- against the plant. Toms River's renewal permit to discharge wastewater to the Atlantic Ocean is considered the most stringent set of environmental restrictions ever imposed in New Jersey. The state is considering bringing criminal charges against the company.

The impact that this public scrutiny will have on long-term implementation of waste reduction at the plant cannot be fully anticipated, but is likely to be substantial.

Other factors also act to influence the nature and extent of waste reduction at Toms River. Although, in theory, wastewater generation could be minimized by repeatedly producing back-to-back batches of the same or very similar dyes (thus eliminating the need to rinse equipment in between batches) the company reported to INFORM that this is not the practice at

Toms River or anywhere else in the dye industry. The demands on production — influenced by customer demand, seasonal changes in dye use, availability and price of raw materials, and the capacity to store excess dye production -- are so highly variable as to preclude the possibility of scheduling production strictly on the basis of minimizing waste generation.

On the other hand, CIBA-GEIGY reported to INFORM that the trend among larger dye manufacturers in recent years has been to consolidate product lines. That is, companies are reducing the total number of dyes manufactured at a given plant while striving for increased yields (and more favorable economics) of the remaining processes. The primary motivation for this change is to increase the profitability of dye manufacturing operations (a necessity, rather than a luxury, in the face of increasingly stiff competition from foreign manufacturers). However, the management at Toms River speculated that there is a trend towards waste reduction as well, since consolidated production reduces the number of rinses required (hence the quantity of wastewater generated) and increased process efficiencies means less materials lost to the wastestream.

SOURCES OF INFORMATION

SOURCE	LOCATION
1. Computer printout of air emissions, 1984	New Jersey Department of Environmental Protection (NJDEP) Bureau of Air Pollution Control CN027 Trenton, NJ 08625 (609) 633-7994
2. New Jersey Industrial Survey, 1978 data	NJDEP Office of Science and Research CN402 Trenton, NJ 08625 (609) 292-6714
3. Wastewater Discharge Permit Application (NJPDES), June 1981	NJDEP Division of Water Resources CN029 Trenton, NJ 08625 (609) 292-5602
4. Compliance Monitoring Report, March 1984	NJDEP (see #3)
5. Chemical and Mutagenicity Analysis Report, October 1981	NJDEP (see #3)
6. Chemical and Mutagenicity Analysis Report, July 1982	NJDEP (see #3)
7. Report of the Office of Science and Research, March 1984	NJDEP (see #3)

8. Compliance Monitoring
   Report, March 1981

   NJDEP
   (see #3)

9. Effluent Limitations
   Violation Summary
   report, February 1982

   NJDEP
   (see #3)

10. Information provided by
    CIBA-GEIGY to INFORM

11. Wastewater Discharge
    Factsheet submitted to
    the Division of Water
    Resources, April 1984

    NJDEP
    Division of Waste
    Management
    32 East Hanover St.
    Trenton, NJ 08625
    (609) 292-9879

12. "Human Exposure to
    Atmospheric Concentra-
    tions of Selected
    Chemicals," prepared
    by Systems Applica-
    tions, Inc. (Contract
    #68-02-3066) March
    1983*

    U.S. Environmental
    Protection Agency
    Office of Air Quality
    Planning and Standards
    Research Triangle Park,
    NC 27111
    (919) 541-5315

13. Internal New Jersey
    Department of Environ-
    mental Protection
    memo, February 26, 1982

    NJDEP
    (see #3)

14. RCRA Part A Applica-
    tion, November 1980

    U.S. EPA, Region II
    Permits Administration
     Branch
    26 Federal Plaza
    New York, NY 10278
    (212) 264-5602

---

* This report gives 1978 estimates of air emissions for individual plants. The U.S. EPA considers these figures unreliable because of the estimation techniques used.

15. Hazardous Waste Annual   NJDEP
    Report, 1982   (see #11)

16. Hazardous Waste Annual   NJDEP
    Report, 1981   (see #11)

17. Hazardous Waste   NJDEP
    Manifest Records,   Bureau of Hazardous
    1980   Wastes
                                CN027
                                Trenton, NJ 08625
                                (609) 984-2302

18. Correspondence from   NJDEP
    CIBA-GEIGY to NJDEP   (see #3)
    June 30, 1983

WASTE REDUCTION

CIBA-GEIGY

Type of Waste	G-Generation D-Disposal Q-Quantity (year)	Waste Reduction Practice	Results	Sources of Information
HAZARDOUS				
Mercury	Plant-wide air emissions 10 lbs/yr (1978)	Use of a new process chemistry for manufacturing dyes to eliminate mercury as a catalyst PS (1983)	All mercury wastes from this process were eliminated	(10)
	Discharged in wastewaters to Atlantic Ocean 0.62 lbs/day (1981)			
	Landfilled on-site 325 lbs/yr (1978)			
	Off-site landfill 1,185 lbs/yr (max) (1978)			
Mercury wastes:				(10)
Concentrated	Stored on-site 600 lbs/yr (1980)			
Sludges	Off-site landfill 3,520 gals/yr (1982)			
Other	Stored on-site 86,000 lbs/yr (1980)	↓	↓	
Nitrobenzene	Plant-wide air emissions 14,371 lbs/yr (1978)	Type of practice not reported ? (year unknown)	Total nitrobenzene wastes reduced 90%	(10)
	Discharged in wastewaters to Atlantic Ocean 8 lbs/day (1981)			
Nitrobenzene tars	Stored on-site 670 gals/yr (1982)	↓	↓	(10)

Key to Waste Reduction Changes
PR – Product
PS – Process
EQ – Equipment
CH – Chemical Substitution
OP – Operational

WASTE REDUCTION

CIBA-GEIGY

Type of Waste	G-Generation D-Disposal Q-Quantity (year)	Waste Reduction Practice	Results	Sources of Information
HAZARDOUS (cont.)				
Nitrobenzene wastes	Stored on-site 500 lbs/yr (1980)	Type of practice not reported ? (year unknown)	Total nitrobenzene wastes reduced 90%	(10)
Chromium	Discharged in wastewaters to Atlantic Ocean 14.74 lbs/day (1981)	New solvent-based process increases retention of chromium PS (early 1980s)	Total chromium loss reduced by 25 percent	(10)
	Discharged in cooling and storm waters to Toms River .45 lbs/day (1981)			
	Landfilled on-site 12,460 lbs/yr (1978)			
LIKELY TO CONTAIN HAZARDOUS CHEMICALS				
Sludge from operation of sewage treatment plant	Landfilled on-site 20,000 lbs/day (1981)	Optimized acid additions to process through research and development tests PS (1983)	?	(10)
	Off-site disposal 38,340 lbs/yr (1980)	↓		
NON-HAZARDOUS				
Sulfuric acid	Discharged in wastewaters to Atlantic Ocean Q - ? (1978)		Reduced waste acid 10 to 40 percent	(10)
		Use of high-quality lime to minimize sludge generation CH (early 1980s)	?	

Key to Waste Reduction Changes
- PR – Product
- PS – Process
- EQ – Equipment
- CH – Chemical Substitution
- OP – Operational

CIBA-GEIGY

WASTE GENERATION & DISPOSAL

Type of Waste ● Indicates Wastes Affected by Reduction	Generation or Disposal	Quantity of Waste	Sources of Information
Air Emissions			
HAZARDOUS			
Aniline	Plant-wide emissions	5.0 lbs/yr (1978)	(2)
Anthracene	"	1,010 lbs/yr (1978)	(2)
Chlorobenzene	"	319 lbs/yr (1978)	(2)
Chlorophenol	Emissions from 1 source	1.0 lbs/hour (1984)	(1)
Cresol	Emissions from 2 sources	.19 lbs/hour (1984)	(1)
1,2 Dichlorobenzene	Plant-wide emissions	7,820 lbs/yr (1978)	(2)
1,3 Dichlorobenzene	Emissions from 1 source	.1 lbs/hour (1978)	(2)
Dimethyl sulfate	Emissions from 4 sources	.03 lbs/hour (1984)	(1)
Epichlorhydrin	Plant-wide emissions	21,900 lbs/yr (1978)	(2)
	Plant-wide emissions	33,440 lbs/yr (1978)	(12)
	Emissions from 10 sources	14.3 lbs/hour (1984)	(1)
Ethyl benzene	Plant-wide emissions	8,865 lbs/yr (1978)	(2)
Formaldehyde	Plant-wide emissions	161 lbs/yr (1978)	(2)
	Emissions from 3 sources	.36 lbs/hour (1984)	(1)
Hydrogen sulfide	Emissions from 8 sources	1.3 lbs/hour (1984)	(1)
● Mercury	Plant-wide emissions	10 lbs/yr (1978)	(2)

CIBA-GEIGY

WASTE GENERATION & DISPOSAL

Type of Waste ● Indicates Wastes Affected by Reduction	Generation or Disposal	Quantity of Waste	Sources of Information
<td colspan="3">**Air Emissions**</td>		
HAZARDOUS (cont.)			
Mercury (cont.)	Emissions from 1 source	.001 lbs/hour (1984)	(1)
Methanol	Emissions from 12 sources	9.0 lbs/hour (1984)	(1)
Naphthalene	Plant-wide emissions	100 lbs/yr (1978)	(2)
	Emissions from 1 source	1.0 lbs/hour (1984)	(1)
● Nitrobenzene	Plant-wide emissions	14,371 lbs/yr (1978)	(2)
	Emissions from 9 sources	6.4 lbs/hour (1984)	(1)
Phenol	Emissions from 3 sources	.2 lbs/hour (1984)	(1)
Phosgene	Plant-wide emissions	136 lbs/yr (1978)	(2)
Phthalic anhydride	Emissions from 1 source	.1 lbs/hour (1984)	(1)
Pyridine	Emissions from 1 source	.02 lbs/hour (1984)	(1)
Toluene	Plant-wide emissions	68,100 lbs/yr (1978)	(2)
	Emissions from 8 sources	9.4 lbs/hour (1984)	(1)
1,2,4-Trichlorobenzene	Plant-wide emissions	2,341 lbs/yr (1978)	(2)
1,1,1-Trichloroethane	Plant-wide emissions	3,100 lbs/yr (1978)	(2)
LIKELY TO CONTAIN HAZARDOUS CHEMICALS			
Dyestuff intermediate chemicals	Emissions from 25 sources	21.3 lbs/hour (1984)	(1)

CIBA-GEIGY

WASTE GENERATION & DISPOSAL

Type of Waste ● Indicates Wastes Affected by Reduction	Generation or Disposal	Quantity of Waste	Sources of Information
Air Emissions			
NON-HAZARDOUS			
Ammonia	Emissions from 19 sources	4.8 lbs/hour (1984)	(1)
Sulfuric acid	Emissions from 41 sources	21.9 lbs/hour (1984)	(1)
Chlorotoluene	Emissions from 3 sources	.2 lbs/hour (1984)	(1)
Wastewater Discharges			
Process wastewaters containing the following chemicals:	Discharged to the Atlantic Ocean	4-6 million gals/day (1981, 1983)	(3, 18)
HAZARDOUS			
Acenaphthene	"	4.2 lbs/day (max.) (1981)	(3)
Aniline	"	5.8 lbs/day (1984)	(7)
Anthracene	"	200 lbs/yr (1978)	(2)
		8.9 lbs/day (max.) (1981)	(3)
Anthracene (deuterated)	"	7.0 lbs/day (1984)	(7)
Benzo fluoranthene	"	.7 lbs/day (1981)	(5)
Bis(2-chloroethoxy) methane	"	1.6 lbs/day (max.) (1981)	(3)
Bis(2-ethylhexyl) phthalate	"	0.5-51.5 lbs/day (1981, 1982)	(5,6)
Chlorobenzene	"	31.1 lbs/yr (1978)	(2)
p-Chloro-m-cresol	"	130 lbs/day (max.) (1981)	(3)
		1-2.07 lbs/day (1981, 1982)	(5,6)

242

CIBA-GEIGY

WASTE GENERATION & DISPOSAL

Type of Waste ● Indicates Wastes Affected by Reduction	Generation or Disposal	Quantity of Waste	Sources of Information
Wastewater Discharges			
Process wastewaters ...chemicals (cont.) HAZARDOUS (cont.)	Discharged to the Atlantic Ocean		
● Nitrobenzene	"	1,150 lbs/yr (1978)	(2)
		8 lbs/day (1981)	(3)
		95-432 lbs/day* (1982)	(9)
2-Nitrophenol	"	1.41 lbs/day (max.) (1981)	(3)
4-Nitrophenol	"	3.98 lbs/day (max.) (1981)	(3)
N-Nitrosodiphenylamine	"	1.08 lbs/day (1981)	(3)
Phenol	"	150 lbs/yr (1978)	(2)
		2.82 lbs/day (max.) (1981)	(3)
Tetrachloroethylene	"	57.33 lbs/day (max.) (1981)	(3)
Toluene	"	1,200 lbs/yr (1978)	(2)
		74 lbs/day (max.) (1981)	(3)
		.02 lbs/day (1981)	(5)
1,2,4-Trichlorobenzene	"	694 lbs/yr (1978)	(2)
		.70-7.85 lbs/day (1981, 1982, 1984)	(3,4, 5,6, 7,8)

* Reported violations of the plant's wastewater permit discharge limitations.

CIBA-GEIGY

WASTE GENERATION & DISPOSAL

Type of Waste ● Indicates Wastes Affected by Reduction	Generation or Disposal	Quantity of Waste	Sources of Information
Wastewater Discharges			
Process wastewaters ...chemicals (cont.) HAZARDOUS (cont.)	Discharged to the Atlantic Ocean		
1,1,1-Trichloroethane	"	200 lbs/yr (1978)	(2)
		.12 lbs/day (1984)	(4)
		2.33 lbs/day (max.) (1981)	(3)
Trichlorofluoromethane	"	59.8 lbs/day (max.) (1981)	(3)
2,4,6-Trichlorophenol	"	6.29 lbs/day (max.) (1981)	(3)
m-Xylene	"	.92 lbs/day (1984)	(4)
o-Xylene	"	.03 lbs/day (1981)	(5)
p-Xylene	"	.53 lbs/day (1984)	(7)
		.03 lbs/day (1981)	(5)
Arsenic	"	.55 lbs/day (max.) (1981)	(3)
Barium	"	116.6 lbs/day (1981)	(3)
Beryllium	"	.29 lbs/day (max.) (1981)	(3)
Cadmium	"	.29 lbs/day (max.) (1981)	(3)
● Chromium	"	2,946 lbs/yr (1978)	(2)
		14.74 lbs/day (1981)	(3)

WASTE GENERATION & DISPOSAL

CIBA-GEIGY

Type of Waste ● Indicates Wastes Affected by Reduction	Generation or Disposal	Quantity of Waste	Sources of Information
Wastewater Discharges			
Process wastewaters ...chemicals (cont.)	Discharged to the Atlantic Ocean		
HAZARDOUS (cont.)			
Copper	"	4,817 lbs/yr (1978)	(2)
		34.54 lbs/day (1981)	(3)
Lead	"	.90 lbs/day (1981)	(3)
● Mercury	"	58 lbs/yr (1978)	(2)
		.62 lbs/day (1981)	(3)
Nickel	"	9.46 lbs/day (1981)	(3)
Selenium	"	1.01 lbs/day (max.) (1981)	(3)
Silver	"	.40 lbs/day (max.) (1981)	(3)
Zinc	"	957 lbs/yr (1978)	(2)
		21.34 lbs/day (1981)	(3)
NON-HAZARDOUS			
Bromochloromethane	"	20.25 lbs/day (1984)	(7)
2-Bromo-1-chloro-propane	"	21.05 lbs/day (1984)	(7)
N,N-dimethylaniline	"	38.97 lbs/day (1981)	(5)
Dimethylphenol	"	.80 lbs/day (1984)	(7)
● Sulfuric acid	"	? (1984)	(10)

CIBA-GEIGY

WASTE GENERATION & DISPOSAL

Type of Waste ● Indicates Wastes Affected by Reduction	Generation or Disposal	Quantity of Waste	Sources of Information
Wastewater Discharges			
Cooling water and storm water containing:	Discharged to Toms River	11.4 million gals/day (1981)	(3)
HAZARDOUS			
● Chromium	"	.45 lbs/day (1981)	(8)
Copper	"	.59 lbs/day (1981)	(8)
Zinc	"	2.20 lbs/day (1981)	(3)
		3.06 lbs/day (1981)	(8)
NON-HAZARDOUS			
Aluminum	"	25.96 lbs/day (1981)	(3)
		13.67 lbs/day (1981)	(8)
Solid Wastes			
HAZARDOUS			
Aniline wastes	Stored on-site	24,000 lbs/yr (1980)	(14)
	Incinerated at company plant in McIntosh, AL	2,145 gals/yr (1981)	(16)
Anthraquinone wastes	Stored on-site	1,000 lbs/yr (1980)	(14)
Benzotrichloride wastes	"	1,000 lbs/yr (1980)	(14)
Chloral wastes	"	700 lbs/yr (1980)	(14)
p-Chloroaniline wastes	"	100 lbs/yr (1980)	(14)
Chlorinated solvent wastes	"	1,595 gals/yr (1982)	(15)
		5,555 gals/yr (1981)	(16)

WASTE GENERATION & DISPOSAL

CIBA-GEIGY

Type of Waste ● Indicates Wastes Affected by Reduction	Generation or Disposal	Quantity of Waste	Sources of Information
Solid Wastes			
HAZARDOUS (cont.)			
Chlorobenzene sludges	Stored on-site	75,000 lbs/yr (1980)	(14)
Chlorobenzene wastes	"	13,400 lbs/yr (1980)	(14)
Cresol wastes	"	10,000 lbs/yr (1980)	(14)
Di-n-butyl phthalate wastes	"	100 lbs/yr (1980)	(14)
1,2-Dichlorobenzene	Landfilled on-site	570 lbs/yr (1978)	(2)
1,2-Dichlorobenzene wastes	Stored on-site	1,000 lbs/yr (1980)	(14)
Diethyl phthalate wastes	"	100 lbs/yr (1980)	(14)
3,3-Dimethoxybenzidine wastes	"	1,000 lbs/yr (1980)	(14)
3,3-Dimethylbenzidine wastes	"	500 lbs/yr (1980)	(14)
Dimethyl sulfate wastes	"	10,000 lbs/yr (1980)	(14)
Epichlorohydrin wastes	"	13,400 lbs/yr (1980)	(14)
	Incinerated at company plant in McIntosh, AL	1,485 gals/yr (1981)	(16)
Formaldehyde wastes	Stored on-site	5,000 lbs/yr (1980)	(14)
Isobutanol wastes	"	700 lbs/yr (1980)	(14)
Methanol wastes	"	13,100 lbs/yr (1980)	(14)
		220 gals/yr (1982)	(15)
	Incinerated at company plant in McIntosh, AL	275 gals/yr (1981)	(16)

CIBA-GEIGY
WASTE GENERATION & DISPOSAL

Type of Waste ● Indicates Wastes Affected by Reduction	Generation or Disposal	Quantity of Waste	Sources of Information
Solid Wastes			
HAZARDOUS (cont.)			
Methyl chloride wastes	Stored on-site	100 lbs/yr (1980)	(14)
Methyl chlorocarbonate wastes	"	100 lbs/yr (1980)	(14)
Methyl ethyl ketone wastes	"	60,300 lbs/yr (1980)	(14)
Methyl isobutyl ketone wastes	"	10,000 lbs/yr (1980)	(14)
Naphthalene	Landfilled on-site	570 lbs/yr (1978)	(2)
Naphthalene wastes	Stored on-site	1,000 lbs/yr (1980)	(14)
p-Nitroaniline	"	3,000 lbs/yr (1980)	(14)
● Nitrobenzene tars	"	15,000 lbs/yr (1980)	(14)
		670 gals/yr (1982)	(15)
● Nitrobenze wastes	"	500 lbs/yr (1980)	(14)
Perchloroethylene wastes	Incinerated at company plant in McIntosh, AL	110 gals/yr (1982)	(15)
Phenacetin	Stored on-site	100 lbs/yr (1980)	(14)
Phenol wastes	"	10,000 lbs/yr (1980)	(14)
		4,015 gals/yr (1982)	(15)
		6,090 gals/yr (1981)	(16)
	Incinerated at company plant in McIntosh, AL	4,490 gals/yr (1982)	(15)

WASTE GENERATION & DISPOSAL

CIBA-GEIGY

Type of Waste ● Indicates Wastes Affected by Reduction	Generation or Disposal	Quantity of Waste	Sources of Information
Solid Wastes			
HAZARDOUS (cont.)			
Phosgene wastes	Stored on-site	2,000 lbs/yr (1980)	(14)
Phthalic anhydride wastes	"	3,000 lbs/yr (1980)	(14)
Polychlorinated biphenyl wastes	Disposed of an in off-site landfill	550 gals/yr (1982)	(15)
	Sent off-site for disposal	1,980 gals/yr (1980)	(17)
Pyridine wastes	Stored on-site	1,000 lbs/yr (1980)	(14)
Resorcinol wastes	"	300 lbs/yr (1980)	(14)
Toluene wastes	"	80,000 lbs/yr (1980)	(14)
		3,825 gals/yr (1982)	(15)
		11,790 gals/yr (1981)	(16)
	Incinerated at company plant in McIntosh, AL	5,812 gals/yr (1982)	(15)
		11,100 gals/yr (1981)	(16)
Toluene diamine wastes	Stored on-site	300 lbs/yr (1980)	(14)
o-Toluidine wastes	"	300 lbs/yr (1980)	(14)
1,2,4-Trichlorobenzene wastes	Disposed of in an on-site landfill	2,125 lbs/yr (1978)	(2)
Trichloroethane wastes	Stored on-site	300 lbs/yr (1980)	(14)
Trypan blue wastes	"	100 lbs/yr (1980)	(14)
Xylene wastes	"	10,000 lbs/yr (1980)	(14)

CIBA-GEIGY

WASTE GENERATION & DISPOSAL

Type of Waste ● Indicates Wastes Affected by Reduction	Generation or Disposal	Quantity of Waste	Sources of Information
Solid Wastes			
HAZARDOUS (cont.)			
Solvent sludges and tars	Stored on-site	79,000 lbs/yr (1980)	(14)
Ammonium meta vanadate wastes	"	100 lbs/yr (1980)	(14)
Asbestos wastes	"	5,000 lbs/yr (1980)	(14)
	Disposed of in an on-site landfill	1,540 gals/yr (1982)	(15)
● Chromium	"	12,460 lbs/yr (1978)	(2)
Copper	"	6,036 lbs/yr (1978)	(2)
	Sent off-site to commercial recycler	26,158 lbs/yr (1978)	(2)
Copper cyanide waste	Stored on-site	100 lbs/yr (1980)	(14)
Hydrazine waste	"	1,000 lbs/yr (1980)	(14)
Hydrofluoric acid waste	"	100 lbs/yr (1980)	(14)
● Mercury	Disposed of in an on-site landfill	325 lbs/yr (1978)	(2)
	"	1,185 lbs/yr* (1978)	(2)
● Mercury wastes:			
Concentrated	Stored on-site	600 lbs/yr (1980)	(14)
Sludges	Disposed of in an off-site landfill	3,520 gals/yr (1982)	(15)
		3,630 gals/yr (1981)	(16)
Other wastes containing mercury	Stored on-site	86,000 lbs/yr (1980)	(14)

* Maximum quantity of mercury (3%) contained in 39,500 pounds per year of filter cake wastes.

252

WASTE GENERATION & DISPOSAL

CIBA-GEIGY

Type of Waste ● Indicates Wastes Affected by Reduction	Generation or Disposal	Quantity of Waste	Sources of Information
Solid Wastes			
HAZARDOUS (cont.)			
Vanadium pentoxide wastes	Stored on-site	100 lbs/yr (1980)	(14)
Zinc	Disposed of in an on-site landfill	3,830 lbs/yr (1978)	(2)
LIKELY TO CONTAIN HAZARDOUS CHEMICALS			
Organic and metal residues	Disposed of off-site	137,235 lbs/yr (1980)	(17)
Organic still bottoms	"	724,505 lbs/yr (1980)	(17)
Paint and pigment wastes	"	38,900 lbs/yr (1980)	(17)
● Sludge from operation of wastewater treatment plant	Disposed of in an on-site landfill	20,000 lbs/day (1981)	(3)
	Disposed of off-site	38,340 lbs/yr (1980)	(17)
Solvent wastes:			
Halogenated	"	23,265 lbs/yr (1980)	(17)
Non-halogenated	"	319,131 lbs/yr (1980)	(17)
Laboratory solvents	Incinerated at company plant in McIntosh, AL	3,410 gals/yr (1982)	(15)
Miscellaneous	Stored on-site	4,730 gals/yr (1981)	(16)
Waste liquids from dye operations	Disposed of off-site	45,980 lbs/yr (1980)	(17)

253

WASTE GENERATION & DISPOSAL

CIBA-GEIGY

● Type of Waste Indicates Wastes Affected by Reduction	Generation or Disposal	Quantity of Waste	Sources of Information
Solid Wastes			
NON-HAZARDOUS			
Acetone wastes	Stored on-site	10,000 lbs/yr (1980)	(14)
	Incinerated at company plant in McIntosh, AL	2,145 gals/yr (1981)	(16)
Cyclohexane wastes	Stored on-site	10,000 lbs/yr (1980)	(14)
Isopropanol wastes	Incinerated at company plant in McIntosh, AL	220 gals/yr (1982)	(15)
		1,705 gals/yr (1981)	(16)
Polypropylene wastes	Disposed of off-site	32,400 lbs/yr (1980)	(17)

Colloids of California

Colloids, Inc.
350 Freethy Boulevard
Richmond, California 94804
(415) 235-2225

SUMMARY

Colloids of California (Richmond, California) is a small manufacturer of antifoam chemicals used primarily in the adhesives industry. This plant intermittently discharges cooling water and dilute tank washings to the West Contra Costa Sanitary District. There is no indication from available information that Colloids has adopted any waste reduction techniques.

PLANT OPERATIONS

Colloids of California is a five-employee plant located in Richmond, California, ten miles northeast of San Francisco. This plant, established in 1967, has sales of $250,000 to $500,000 per year. The parent company, Colloids, Inc., is located in Newark, New Jersey.

The Richmond plant manufactures antifoam compounds (used to minimize detergent foam, for example) using light petroleum oils, glycol esters, metal soap, surfactants and various other raw materials. These antifoams are produced in batches, by blending and heating compounds in two 1,500-gallon blending tanks. The compounds, many of which are received by Colloids already in a semifinal state, are custom-manufactured for specific uses; 70 percent of these products are used by the adhesives industry.

WASTE GENERATION AND WASTE REDUCTION

Cooling water, plant laboratory wastewater and highly diluted tank washings are discharged intermittently to the West Contra Costa Sanitary District, Richmond, California. The wastewater flow, comprised mainly of cooling water, varies daily from 0 to 1,000 gallons. No information was available in public files on air emissions or solid waste generation.

SOURCES OF INFORMATION

SOURCE	LOCATION
1. West Contra Costa Sanitary District files, 1983	West Contra Costa Sanitary District 2910 Hilltop Drive Richmond, CA 94806 (415) 222-6700

COLLOIDS

WASTE GENERATION & DISPOSAL

Type of Waste ● Indicates Wastes Affected by Reduction	Generation or Disposal	Quantity of Waste	Sources of Information
Wastewater Discharges			
NON-HAZARDOUS Cooling water	Discharged to the West Contra Costa County Sanitary District, Richmond, CA	0-1,000 gals/day (1983)	(1)

Dow Chemical U.S.A.

Loveridge Road
Pittsburg, California 94565
(415) 432-5000

SUMMARY

Dow Chemical's facility (Pittsburg, California), one of the largest chemical manufacturing operations in the western U.S., manufactures chlorine and sodium hydroxide as starting materials for the production of chlorinated organic chemicals such as **carbon tetrachloride, perchloroethylene,** bacteriocides, and fungicides. Other chemicals used at this plant include **picoline, dichloropropene, trichloroethylene, xylene, trichloroethane,** and **dichloroethylene.** Chlorinated organic chemical by-products such as **polychlorinated biphenyls** appear in some process wastestreams and are destroyed by high-temperature incineration. Dow has replaced a gas pressurizing technique with a pumping mechanism that eliminates air emissions from a material transfer step. Hydrochloric acid discharges in the plant's wastewater have been reduced by recovering this chemical for use as a raw material. In addition to waste reduction, Dow has implemented water recycling practices for its cooling waters and the use of evaporation ponds for process waters, and has, as a result, achieved plant-level wastewater discharge reductions of 95 percent in the past ten years.

NOTE: Chemicals in **boldface** type are hazardous, according to INFORM's criteria (see Appendix A).

PLANT OPERATIONS

Dow Chemical's Pittsburg, California, plant is situated in a highly industrialized area 35 miles northeast of San Francisco. One of the largest chemical plants in the western United States, it employs 650 people and annually manufactures over $100 million worth of products. The plant, built in 1916, is one of five major U.S. manufacturing facilities operated by the Dow Chemical Company, the second largest chemical company in the U.S. (1983 sales: nearly $11 billion). Dow operates a total of 32 manufacturing plants in the U.S. and 82 plants in 29 foreign countries.

The Pittsburg facility is a complex of several interrelated chemical plants that manufacture both organic and inorganic chemicals. At the heart of the complex is Dow's Chlor-Alkali Plant, an inorganic chemical manufacturing operation that annually produces about 260 million pounds each of sodium hydroxide (also known as caustic or lye) and chlorine and seven million pounds per year of a secondary product, hydrogen, which it sells or burns on site as supplemental fuel. The caustic, chlorine and hydrogen are sold as products or used as raw materials in other manufacturing operations at the complex. In particular, chlorine is used in the manufacture of a wide variety of chlorinated organic chemicals and other products at five separate Pittsburg operations:

-- The Per-Tet Plant manufactures two main products: 22 to 38 million pounds per year of **perchloroethylene** and 27 million pounds per year of **carbon tetrachloride,** and the by-product hydrogen chloride, all three by a process in which an organic raw material (chiefly methane, from on-site natural gas wells) reacts with chlorine from the Chlor-Alkali Plant. **Perchloroethylene** is a common solvent widely used for dry cleaning and industrial degreasing. **Carbon tetrachloride,** sold to a nearby Du Pont plant, is used as a raw material in the manufacture of chlorofluorocarbon refrigerants (Freons).

-- The Sym-Tet Plant (an abbreviation for symmetrical tetrachloropyridine) annually manufactures 18 million pounds of N-SERVE nitrogen stabilizer, Dow's brand name for a material whose active ingredient is 2-chloro-6-trichloromethyl pyridine. N-SERVE is a soil bacteriostat manufactured by a process that reacts **picoline** with chlorine and caustic and uses **xylene** as a process solvent.

-- The DOWICIL plant produces 2.8 million pounds per year of DOWICIL, Dow's brand name for a bacteriocide produced by the reaction of chloropropylene with hexamethylenetetraamine.

-- The Latex Plant produces 24 million pounds per year of latexes that have a wide variety of applications in the manufacture of water-based paints, coated papers, textiles, and carpets. Raw materials used in this operation include styrenes, butadienes and **1,1 dichloroethylene.**

-- The VIKANE plant manufactures two million pounds per year of VIKANE, Dow's brand name for a sulfuryl fluoride compound used as a space fumigant. VIKANE and its raw materials -- including chlorine, sulfur dioxide, and hydrogen fluoride -- are inorganic compounds.

The Pittsburg facility is also the site of a large Research and Pilot Plant operation, which refines chemical production methods and develops new products for test marketing.

In addition to carrying out chemical manufacturing, the Dow complex stores and packages **dichloropropene** for sale as a soil fumigant under the company's brand name, TELONE. Other chemicals handled at the plant include **trichloroethylene** and **trichloroethane.**

WASTE GENERATION AND WASTE REDUCTION

AIR EMISSIONS

Generation

A report by the California Air Resources Board (ARB) has estimated emissions of two chlorinated organic chemicals from the Pittsburg plant: **carbon tetrachloride,** 320,000 pounds per year (one-third of all **carbon tetrachloride** emissions in the entire state) and **perchloroethylene,** 44,000 pounds per year. However, air permit files of the Bay Area Air Quality Management District (BAAQMD) give substantially lower annual figures for emissions of these two chemicals -- **carbon tetrachloride,** 58,360 pounds and **perchloroethylene** (from nine process and storage sources), 3,800 pounds. However, BAAQMD files also report large quantities of unspecified emissions that could include **carbon tetrachloride** and **perchloroethylene** (or many of the other organic chemicals used at Dow). For instance, the files list 77,380 pounds per year of chemical emissions from a single process tank as "general organics" -- without identifying the substance emitted. The ARB report notes that **carbon tetrachloride** storage tank emissions are "currently essentially uncontrolled," but says that "plans for future control are under way."*

Total organic chemical emissions from the plant's process sources, storage tanks, and loading areas are 598,000 pounds per year, according to BAAQMD estimates.

Reduction

INFORM has learned of one waste reduction practice affecting air emissions at the Pittsburg complex. Dow explains that waste is produced during the transfer

---

* See page 100 of the ARB report ("Inventory of Carcinogenic Substances Released into the Ambient Air of California," prepared by Science Applications, Inc., November 1982).

of a raw material from a tank car to a storage tank, and subsequently, from the storage tank into the reactor vessels. Raw material vapors were lost as waste when they became mixed with nitrogen gas that was used to pressurize the tanks in order to push the raw material from the tank car into the storage tank and from the storage tank into the reactor. These gases then had to be vented to a pollution-control scrubber for disposal.*

In order to eliminate air emissions from the storage tank-to-reactor vessel transfer step, Dow replaced the gas pressurizing technique with a pumping mechanism, thereby eliminating the tank pressurizing step and its associated raw material losses. Dow also mentioned future plans to make a similar change in the tank car-to-storage tank transfer step.

Dow explained that this reduction technique was both originated and implemented by the plant superintendent and that source reduction was the sole reason for the change.

WASTEWATER DISCHARGES

Generation

The Pittsburg plant discharges from 0.9 to 1.5 million gallons per day of treated wastewater -- non-contact cooling water and other non-process waters -- to the New York Slough (an estuary leading into San Francisco Bay), according to the Regional Water Quality Control Board.

---

*Dow considers this process to be proprietary, so it did not provide the chemical name of the waste being reduced. However, it did note that this waste is legally classified as "extremely hazardous" by California.

According to state and federal records, the facility also discharges highly contaminated process wastewaters to six on-site evaporation ponds, where the volume of water is reduced by evaporation and the contaminants are concentrated in the ponds (some may also be decomposed by sunlight). The San Francisco region's sunny and dry climate, with a net evaporation rate of more than 50 inches of water per year, greatly facilitates use of the evaporation ponds. This wastewater typically contains 85 to 97 percent water, three to 15 percent inorganic sodium salts, and trace levels to one percent of organic chemicals. Of these discharges, the largest single stream is caustic wastewater containing trace organic chemicals from the air pollution scrubber used at Dow's Research and Pilot Plant.

These concentrated contaminants will not be removed until Dow ceases discharge to the ponds, which it must do within 12 years, according to California law. Dow explains that its long-term plans are to decontaminate and then recycle the sodium salts which are concentrated in the ponds.

Reduction

INFORM has learned of one waste reduction practice affecting wastewater discharges. Whereas hydrochloric acid wastewater produced by manufacturing processes and air pollution control devices at the N-SERVE, Per-Tet and other plants was formerly discharged to the New York Slough, it is now directly reused in several on-site processes as a raw material or sold as a commercial product. Dow explains that changes at the plant made possible the recovery of hydrochloric acid beginning in the late 1960s. These changes included installation of incinerators, whose by-products contain hydrochloric acid; installation of expanded storage capacity for hydrochloric acid; and installation of equipment to concentrate hydrochloric acid and to make it into a product of saleable quality. In addition, this reduction option required Dow to develop markets in the late 1960s and early 1970s for sale of its excess hydrochloric acid. However, the details of these changes and the impact of this waste reduction practice on hydrochloric acid waste generation were not disclosed by the company.

During this period, both corporate and local management personnel at the company decided not to renew the plant's permit to discharge hydrochloric acid to New York Slough, and by the early 1970s, Dow had completely ceased this discharge. This reduction option offered economic advantages, since, as Dow explained to INFORM: "We receive value for the (hydrochloric acid) now, whereas it had no value or possibly a negative value when we discharged it as waste to the river."

Dow officials told INFORM that, in addition to waste reduction, increased internal recycling of its cooling waters and the use of evaporation ponds to dispose of process waters has resulted in lower wastewater discharges. Since the early 1970s, these measures have lowered wastewater discharges to the New York Slough from 30 million gallons per day to 1.5 million gallons -- an overall reduction of 95 percent. For example, the majority of the facility's process wastewaters, which contain dissolved salts and various organic chemicals, are no longer discharged but are instead now sent to the plant's evaporation ponds.

SOLID WASTES

Generation

The Pittsburg facility's solid waste generation includes 10.2 million pounds of chlorinated organic chemicals, which are incinerated or recycled on site. Dow refers to these incinerators as thermal oxidation units, since it considers them to be part of their production processes rather than their waste treatment operations. Consequently, the plant does not consider the materials sent to these units to be wastes, as INFORM refers to them. A large portion of the incinerated waste stream -- six million pounds -- originates from the

Per-Tet plant and contains from 50 to 100 parts per million of **polychlorinated biphenyls** (PCBs). In October 1982, Dow (which has been incinerating this wastestream since 1973) received formal approval to continue burning PCBs under the provisions of what is called a "Trial Burn" permit issued by the U.S. EPA.

The plant ships an additional 6.9 million pounds of organic and inorganic chemical wastes off site for disposal. Wastes that are burnable but incompatible with the plant's incinerator are sent to a Dow plant in Louisiana for incineration.

Reduction

INFORM has not learned of any waste reduction practices affecting solid wastes at the Pittsburg complex. However, company officials told INFORM that plant management has imposed a $215 per drum surcharge on wastes being landfilled as an incentive for increasing incineration. The company felt it likely that this large surcharge may also have encouraged process managers to reduce the generation of hazardous solid waste. Dow has not documented what impact this surcharge may have had on waste generation and reduction.

OBSERVATIONS

Impact of Company Environmental Policy:

Dow's internal document, "Environmental Policy and Guidelines," influences the entire spectrum of waste generation and waste management practices at the company's facilities, including the Pittsburg complex. Wherever possible, company policy is to handle hazardous wastes internally, that is, either at the site of generation or at another Dow facility. Waste disposal at a non-company site, such as a commercial waste handler, requires approval of the site by state and federal environmental authorities as well as by Dow's Environmental and General managers.

This official statement also spells out several policies, in order of preference, for managing hazardous wastes:

-- Minimization or elimination of waste through research efforts, process design, and revised plant operations;

-- Reuse and recycling of materials whenever feasible;

-- Adoption of a hierarchy of waste disposal options, in the following order of preference: 1) incineration of burnable hazardous wastes, 2) land disposal at a facility owned by Dow, and 3) land disposal at a site not owned by Dow.

The policy also explicitly rejects deep-well injection, which it calls "not a continuing solution for hazardous waste disposal."

Dow views its commitment to high-temperature incineration as the best possible means of disposing of their chlorinated organic wastestreams. The chief rationale is that incineration permanently destroys and converts most of the wastestreams into water vapor, carbon dioxide, and hydrogen chloride. The latter is recovered as hydrochloric acid, a step that is part of the facility's efforts to recycle and reduce wastewater discharges. Moreover, Dow is acutely aware of the potential long-term liability associated with land burial and burns even those waste streams that can be legally discharged to landfills, a method whose cost is only one-third that of incineration.

Dow's policies do not ban landfilling outright. Rather, the company uses economic measures and management direction to encourage incineration and discourage landfilling. Processes that generate hazardous solid wastes are assessed a surcharge, above the cost of disposal, of $215 per drum for hazardous wastes going to a landfill. The fee, imposed by the plant manager on each of the plant's process managers, equalizes the costs of the two disposal methods, thereby removing the economic advantage of landfill disposal.

Dow reported that both the high costs of incinerating wastes (particularly those sent off-site, which have added shipping expenses) and Dow's self-imposed surcharge on solid wastes discharged to landfills may play a role in waste reduction. Large waste disposal costs charged directly to the process that generates wastes reduce the overall profitability of the process, which is a key influence on a process manager's professional advancement. Dow added that while they have not developed waste disposal cost ratios for their production processes, these costs are significant relative to other production costs, such as energy, raw materials and labor. Thus, such costs may encourage the generation of less waste. However, Dow did not have documentation available showing the impact of the surcharge since they have not set up the ongoing program of materials and economic accounting that would make such an assessment possible.

Waste Reduction Not Always Seen
As The Best Option:

Dow officials reported to INFORM that waste reduction at the source is not always a feasible option and, even if it is, is not always the most desirable means of waste management. They cited three examples at Pittsburg where emphasis was placed on waste management practices other than waste reduction.

Per-Tet Operations: Dow reported that because of the chemical equilibrium of the reaction that manufactures **carbon tetrachloride** and **perchloroethylene,** the formation of small amounts of other heavier chlorinated organic compounds is unavoidable, based on current knowledge about this reaction's chemistry. This process was already upgraded in the 1940s and 1950s to maximize the product yield and minimize the by-product yield. There is currently no known way to alter the manufacturing process to prevent formation of these waste materials -- in other words, waste reduction at the source is not a viable option. This wastestream -- which contains **hexachlorobenzene, hexachlorobutadiene, PCBs,** and other chlorinated organic chemicals — is instead incinerated on-site at high temperatures.

Dow admits that "it is conceivable that through a research project we could yet find a method by which we could produce **perchloroethylene** and **carbon tetrachloride** and reduce the amount of by-product production." However, in light of the fact that incineration yields a commercial product, hydrochloric acid, Dow does not consider source reduction to be a better option, especially given the commitment of research time that would be required to explore new reaction mechanisms. As Dow reported, "we feel our limited resources for research are better spent in other areas."

Therefore, source reduction is not feasible here for a combination of reasons: the absence of an existing production process alternative, the attractiveness of the hydrochloric acid-yielding incineration option, and the absence of any incentive to undertake a costly research project to explore reduction methods based on Dow's feeling that the existing process is entirely appropriate both economically and environmentally.

Laboratory research wastes: A second example where Dow finds waste reduction at the Pittsburg facility to be less desirable than other handling techniques involves the plant's research on new products. Here, Dow's Environmental Manager reports that extensive waste reduction at the source could "severely curb innovation." Wastes generated during laboratory synthesis of new chemicals are quantitatively small but complex in nature. Dow feels that it would be best to incinerate the diverse but small-volume organic wastes produced by these operations.

Pilot Plant Operations: Finally, Dow also explained why waste reduction is a limited option at the plant's research and process development pilot plant operations, which generate about 50 percent of the wastewater entering the evaporation ponds. The company is currently looking into ways to reduce water and salt discharges (99+ percent of the total discharge) since the ponds must be closed within 12 years. However, Dow believes that reduction of the small quantities of organic wastes entering the ponds may not be the best option and might even be counterproductive. This part of the plant handles materials and uses chemical reactions and processes that are constantly changing -- meaning that equipment requires frequent cleaning in between processes. As a result, its operations are sometimes highly waste-intensive, and there may be few opportunities for modifying processes at this stage to minimize waste generation.

Dow explains that waste reduction is not appropriate here because "it is in this pilot plant stage where processes are developed to minimize or eliminate waste generation prior to the construction of larger production units." Thus, waste reduction efforts are not applicable to these early stages of process development.

SOURCES OF INFORMATION

SOURCE	LOCATION
1. "Human Exposure to Atmospheric Concentrations of Selected Chemicals," prepared by Systems Applications Inc. (Contract #68-02-3066), March 1983* (based on 1978 data)	U.S. Environmental Protection Agency (U.S. EPA) Office of Air Quality Planning and Standards Research Triangle Park, NC 27111 (919) 541-5315
2. "Inventory of Carcinogenic Substances Released into the Ambient Air of California," California Air Resources Board, prepared by Science Applications, Inc., November 1982	California Air Resources Board (ARB) 1102 Q Street Sacramento, CA 95812 (916) 322-2990
3. Report of Task #2 of Inventory of Carcinogenic Substances Released into the Ambient Air of California" (see #2)	ARB (see #2)
4. Computer printout of plant-wide air emissions at Dow, 1983	Bay Area Air Quality Management District 939 Ellis Street San Francisco, CA 94109 (415) 771-6000

* This report gives 1978 estimates of air emissions for individual plants. The U.S. EPA considers these figures unreliable because of the estimation techniques used.

5. INFORM's estimates based on knowledge of processes and chemicals in use

6. Water use flow diagram

    Regional Water Quality
    Control Board (RWQCB)
    1111 Jackson Street
    Room 6040
    Oakland, CA 94607
    (415) 464-1255

7. Information provided by Dow to INFORM

8. Wastewater Discharge Permit Application file - renewal application, November 1979; revisions to renewal, March 1981

    RWQCB
    (see #6)

9. Surveillance and Compliance Inspection Report, February, 1982*

    RWQCB
    (see #6)

10. Hazardous Waste Operating Plan, June 1983

    U.S. EPA, Region IX
    215 Fremont Street
    San Francisco, CA 94105
    (415) 974-8119

11. RCRA Part A Application, November 1980

    U.S. EPA, Region IX
    (see #10)

12. Trial Burn Permit (issued by U.S. EPA) October 1982

    RWQCB
    (see #6)

13. Correspondence from Dow to Regional Water Quality Board, July 1979

    RWQCB
    (see #6)

* Also on file with Citizens for a Better Environment, 88 First Street, S. Francisco, CA 94105, (415) 777-1984.

WASTE REDUCTION

DOW

Type of Waste	G-Generation D-Disposal Q-Quantity (year)	Waste Reduction Practice	Results	Sources of Information
LIKELY TO CONTAIN HAZARDOUS CHEMICALS Unspecified hazardous materials	Air emissions from a transfer operation Q – ? (1984)	Pressurizing technique transferring raw material from storage tank to reactor vessel replaced by pumping mechanism EQ/OP (year unknown)	Raw material losses from this transfer step were eliminated	(7)
NON-HAZARDOUS Hydrochloric acid	G – ? D – ? Q – ? (1984)	Isolation from wastewaters of hydrochloric acid for reuse as a process raw material or for sale as a commercial product PS/OP (year unknown)	?	(7)

Key to Waste Reduction Changes
PR – Product
PS – Process
EQ – Equipment
CH – Chemical Substitution
OP – Operational

WASTE GENERATION & DISPOSAL

DOW

Type of Waste ● Indicates Wastes Affected by Reduction	Generation or Disposal	Quantity of Waste	Sources of Information
\multicolumn{3}{c}{**Air Emissions**}	
HAZARDOUS			
Carbon tetrachloride	Process emissions	58,360 lbs/yr (1983)	(4)
	Plant-wide emissions	91,070 lbs/yr (1978)	(1)
		35,724-58,883 lbs/yr (1982)	(2)
		320,000 lbs/yr (1982)	(3)
Perchloroethylene	Emissions from 9 process and storage sources	3,800 lbs/yr (1983)	(4)
	Plant-wide emissions	44,000 lbs/yr (1978, 1982)	(1,3)
		8,309-10,530 lbs/yr (1982)	(2)
Trichloroethane	Process emissions	14,490 lbs/yr (1983)	(4)
Trichloroethylene	"	4,270 lbs/yr (1983)	(4)
Xylene	"	5,180 lbs/yr (1983)	(4)
Organic chemicals	Plant-wide emissions	598,000 lbs/yr (1983)	(4)
	Emissions from 1 process tank	77,380 lbs/yr (1983)	(4)
	Emissions from 6 on-site evaporation ponds	?* (1984)	(5)

* Likely to be present. (Assumption based on INFORM's knowledge of the types of discharges to the ponds.)

DOW
WASTE GENERATION & DISPOSAL

Type of Waste ● Indicates Wastes Affected by Reduction	Generation or Disposal	Quantity of Waste	Sources of Information
Air Emissions			
LIKELY TO CONTAIN HAZARDOUS CHEMICALS			
Halogenated organic chemicals	Process emissions from latex operations	2,230 lbs/yr (1983)	(4)
	Process emissions, non-latex	5,500 lbs/yr (1983)	(4)
● Unspecified hazardous material	Emissions from a transfer operation	? (1984)	(7)
NON-HAZARDOUS			
Acetone	Process emissions	11,570 lbs/yr (1983)	(4)
Naphtha	"	9,855 lbs/yr (1983)	(4)
Wastewater Discharges			
HAZARDOUS			
Process wastewater containing heavy concentrations of salt and a wide variety of chlorinated organic chemicals	Discharged to 6 on-site evaporation ponds	15,000-89,000 gals/day (1979, 1982)	(6,9)
Non-contact cooling water and other waters, including:	Discharged to New York Slough	.90-1.5 million gals/day (1981)	(8)
HAZARDOUS			
Antimony	"	.20 lbs/day (max.) (1981)	(8)
Chromium	"	.29 lbs/day (1981)	(8)
Copper	"	64 lbs/yr (1981)	(5,8)
Lead	"	.24 lbs/day (1982)	(9)
Mercury	"	3.2 lbs/day (1981)	(5,8)
Nickel	"	.07 lbs/day (1981)	(8)
Zinc	"	.07 lbs/day (1981)	(8)

WASTE GENERATION & DISPOSAL

DOW

Type of Waste ● Indicates Wastes Affected by Reduction	Generation or Disposal	Quantity of Waste	Sources of Information
Wastewater Discharges			
NON-HAZARDOUS ● Hydrochloric acid	?	? (1984)	(7)
Solid Wastes			
HAZARDOUS			
Chlorinated organic chemical wastes from DOWICIL plant maintenance and cleanup	?	35 drums/yr (1982)	(9)
Chlorinated organic wastes from chlorine production	Incinerated on-site*	3,000,000 lbs/yr (1983)	(10)
Chlorinated organic wastes from chloropyridine production	Incinerated on-site*	1,200,000 lbs/yr (1983)	(10, 11)
Dichloropropene (TELONE soil fumigant) and other chlorinated propenes	Stored and recycled on-site	? (1982)	(9)
Phenol wastes	?	124,800 lbs/yr (1980)	(11)
Still bottoms and other chlorinated organic wastes from carbon tetrachloride and perchloroethylene production**	Incinerated on-site*	6,000,000 lbs/yr (1983)	(10, 11, 12)
Asbestos waste from chlor-alkali process diaphragm cells	?	10 drums/yr (1982)	(9)
Asbestos-containing baghouse filter waste from chlorine production	Sent off-site for disposal	500 lbs/yr (1983)	(10)

* Dow refers to these incinerators as thermal oxidation units, since it considers them to be part of their production processes rather than their waste treatment operations.

** Portions of this waste stream contain from 50 to 100 parts per million of PCBs.

WASTE GENERATION & DISPOSAL

Type of Waste ● Indicates Wastes Affected by Reduction	Generation or Disposal	Quantity of Waste	Sources of Information
Solid Wastes			
LIKELY TO CONTAIN HAZARDOUS CHEMICALS			
Chlorinated organic chemicals from chloralkali plant chlorine purification	?	40 drums/yr (1982)	(9)
Chlorinated pyridine wastes	Incinerated off-site	250,000 lbs/yr (1983)	(10)
Compressor oil contaminated with chlorinated organic chemicals from entire complex	?	200 drums/yr (1982)	(9)
Process cleaning wastes from entire complex	?	1,500 drums/yr (1982)	(9)
Sediments, filters and other residue contaminated by organic chemicals from chlorination process	Incinerated off-site	40,000 lbs/yr (1983)	(10)
MAY CONTAIN HAZARDOUS CHEMICALS			
Butadiene, styrene and other latex sludges from latex production	Coagulated in on-site evaporation ponds and disposed of in an on-site landfill	? (1982)	(9, 13)
Filters contaminated by paint and solvents	Sent off-site for disposal	6,000 lbs/yr (1983)	(10)
Muds from salt processing operations (organic chemicals and caustics)	Disposed of in an on-site landfill	540,000–1,000,000 lbs/yr (1982)	(9)

WASTE GENERATION & DISPOSAL

DOW

Type of Waste ● Indicates Wastes Affected by Reduction	Generation or Disposal	Quantity of Waste	Sources of Information
Solid Wastes			
NON-HAZARDOUS			
Assorted laboratory wastes from research operations	Sent off-site for disposal	14,000 lbs/yr (1983)	(10)
Chlorine from chlorination processes	Incinerated on-site	500,000 lbs/yr (1983)	(10)
Hexamethylene tetramine	Stored and recycled on-site	? (1982)	(9)
Latex production wastes containing acrylic and butyl acrylic acids	?	10 drums/yr (1982)	(9)
Sodium fluoride wastes	Sent off-site for disposal	10,000 gals/month (1982)	(9)
		6,000,000 lbs/yr (1983)	(10)

278

E.I. Du Pont de Nemours & Company

Chambers Works
Deepwater, New Jersey 08023
(609) 299-5000

SUMMARY

Du Pont's Chambers Works plant (Deepwater, New Jersey) is one of the oldest and largest chemical plants in New Jersey, occupying 619 acres and employing 4,100 people. Its more than 700 products include general organic intermediates, petroleum additives, general industrial chemicals, prescription drugs, flame-retardant nylon, textile fibers, chemical components of Teflon, fluorinated hydrocarbons (Freons), elastomers, specialty chemicals, aromatics and sulfuric acid. Raw materials used in the Chambers Works' production processes include **benzene, carbon tetrachloride, chloroform, chlorobenzene, o-dichlorobenzene, ethylene dibromide, ethylene dichloride, nitrobenzene, phosgene, toluene** and **trichloroethylene.** Du Pont was the first plant in New Jersey to apply the "Bubble" policy to its air emissions control efforts. This option designates a company as a single air pollution source enclosed by an imaginary bubble, thereby allowing it to control individual emissions sources within this bubble as it sees fit. Wastewater generated by Du Pont and hundreds of outside companies is treated in the Chambers Works' 40 million gallons per day wastewater treatment plant which makes use of Du Pont's unique Powdered Activated Carbon Treatment technology. The plant generates 80 billion pounds per year of hazardous solid wastes, and is the largest waste generator in the country. These wastes are landfilled, stored, treated

NOTE: Chemicals in **boldface** type are hazardous, according to INFORM's criteria (see Appendix A).

or incinerated on- and off-site. There are numerous references in public files to steps Du Pont has taken to reduce the generation of wastewater it discharges to its treatment plant or to groundwater. However, a lack of details on these efforts precluded INFORM from determining whether or not these are waste reduction practices.

PLANT OPERATIONS

Du Pont operates a large, 4,100-employee chemical complex called the Chambers Works in Deepwater, a city on the Delaware River in the southwestern corner of New Jersey. The Chambers Works, established in 1917, is one of the oldest and largest chemical manufacturing plants in the state, and is the largest of Du Pont's U.S. plants. It is spread across 619 acres, contains several hundred buildings, 200 miles of paved roads, tracks for 750 rail cars, vast pipeline networks, and manufactures more than 700 products through 2,000 different chemical processes.

Du Pont, the largest U.S. chemical company, has five major business areas in chemical manufacturing: Fibers (such as Orlon and Dacron); Polymer Products (such as plastic resins, elastomers and films); Industrial and Consumer Products (such as photographic and electronic products, finishes, non-stick coatings, explosives and analytical instruments); Agricultural and Industrial Chemicals (such as fungicides, herbicides, insecticides and commodity and specialty chemicals); and Biomedical Products. The Chambers Works manufactures products for all five of these business areas. Products from the chemical divisions are sold to the textile, transportation, electronics, food packaging, construction, paper, automotive, petroleum, agriculture and numerous other industries. Sales for these divisions in 1983 were as follows: Fibers, $4.8 billion; Polymer Products, $3.4 billion; Industrial and Consumer Products, $2.5 billion; Agricultural and Industrial Chemicals, $3.5 billion; and Biomedical Products, $1.1 billion. Total corporate sales from both chemical operations and petroleum and coal production were $35.4 billion in 1983.

The Deepwater plant manufactures general organic intermediates, petroleum additives (organo-lead anti-knocks, e.g.), prescription drugs, flame-retardant nylon, textile fibers (Kevlar, e.g.), chemical building blocks for the manufacture of Teflon (Du Pont's well-known non-stick coating which was invented at Deepwater), fluorinated hydrocarbons (Freons), **isocyanates**, elastomers (Neoprene rubber, e.g.), specialty chemicals (organic titanates, e.g.), aromatics (**phenylene diamine**, e.g.), sulfuric acid and a variety of other chemical products.

Most of Du Pont's chemical manufacturing is carried out in batch operations, except for some large continuous production processes such as **tetraethyl lead** and **tetramethyl lead** manufacturing. Production quantities for 1978, as estimated by Systems Applications, Inc. (see [13] in "Sources of Information"), were: **trichlorofluoromethane** (Freon 11), 36.8 million pounds; **dichlorodifluoromethane** (Freon 12), 52.3 million pounds; monochlorodifluoromethane (Freon 22), 48.8 million pounds; and **phosgene** (used for the production of other chemicals at Chambers Works), 102.6 million pounds.

Raw materials used annually by the Chambers Works include: 112.6 million pounds of **carbon tetrachloride** and 71.9 million pounds of **chloroform,** both used in fluorocarbon manufacture; 34 million pounds of **chlorobenzene**, used in **nitrochlorobenzene** manufacture; 782,610 pounds of **o-dichlorobenzene,** used as a solvent in **toluene diisocyanate** production and as an intermediate in dichloroaniline production; **nitrobenzene,** used as an intermediate in 3,4-dichloroaniline production; 70.8 million pounds of **phosgene,** used in **toluene diisocyanate** manufacture; and 38 million pounds of **toluene,** used in **toluene diisocyanate** manufacture, according to SAI's report. Information from New Jersey's Air Pollution Emissions Data System also lists **ethylene dichloride, ethylene**

dibromide, **benzene** and **trichloroethylene** as raw materials used at Deepwater.

Du Pont explained in its 1981 "Bubble Proposal" to U.S. EPA that the scope and complexity of the facility necessitated development of state-of-the-art computerized data management systems for such tasks as emissions inventory, budget management, wastewater treatment plant operation and groundwater monitoring. One such data base is Du Pont's Pollution Standards System, which contains detailed data on air, liquid and solid wastes generated at each process at the plant.

Du Pont's sale of its dyestuffs business in 1981, and a number of other factors, such as the chemical industry's economic hardships in the early 1980s and regulatory constraints on **lead** gasoline additives, left the Chambers Works with excess service and resource capacity as plant operations were trimmed to make it more competitive (by paring down the work force and demolishing unneeded buildings, for example), according to Chemical Week [12,14]. This newly-available capacity encouraged the Deepwater plant to offer its excess resources and services for sale to other companies. Closure of the Chambers Works' large dye production facilities, for example, created excess capacity at the wastewater treatment plant, which Du Pont now operates as a commercial enterprise to treat the wastewaters from more than 200 outside customers.

WASTE GENERATION AND WASTE REDUCTION

AIR EMISSIONS

Generation

The Chambers Works emits a wide variety of chemicals from 1,500 stacks and vents and other sources, according to its 1981 "Bubble Proposal" [5]. Among Du Pont's largest organic chemical air emissions are: **dichlorodifluoromethane**, 110,376 pounds per year; **methyl chloride**, 1,466,300 pounds per year; **tetra ethyl lead**,

153,300 pounds per year; and **trichlorobenzene**, 153,300 pounds per year.

Of the 200 to 300 volatile organic substances (VOS) emissions sources at Chambers Works, air emissions from 119 sources, including the seven largest, are controlled under a regulatory strategy known as the "Bubble Concept."* Under this strategy, a plant is considered a single pollution source, instead of an aggregate of individual pollution sources. By putting an imaginary "bubble" around the plant, the company is then free to control its individual air emissions sources as it sees fit, as long as total emissions from the bubble do not exceed regulatory limits. Du Pont was the first plant in New Jersey to apply the Bubble Concept to its air emissions control efforts.

Du Pont's 1981 "Bubble Proposal" [5] details the specific control techniques employed by the plant under this strategy. Du Pont proposed to feed emissions from three continuous VOS sources to four furnaces normally used to recover untreated **lead** in the **tetra ethyl lead** and **tetra methyl lead** manufacturing process. These emissions, including **methyl chloride**, would be destroyed while providing the furnaces with half of their fuel needs. The other four major VOS sources, which are of a less continuous nature, would be incinerated in a specially designed incinerator. The remaining 112 sources would not be controlled beyond existing levels.

Total VOS emissions under the Bubble Proposal are reduced from 9,500,000 pounds to 876,000 pounds per year. This reduction is 4,600,000 pounds per year more than conventional (individual source) controls

---
* Du Pont is still awaiting approval of this strategy from the New Jersey Department of Environmental Protection (NJDEP), according to the NJDEP's New Source Review Section.

would yield. In addition to the benefits of smaller VOS emissions, less testing and monitoring, simpler enforcement and earlier regulatory compliance, Chemecology [16] added that the bubble strategy would save Du Pont $12 million in control costs.

WASTEWATER DISCHARGES

Generation

Du Pont discharges 80 million gallons per day of process water, non-contact cooling water, wastewater it receives from outside sources and other waters to the Delaware River, according to its 1980 wastewater discharge permit application. Du Pont treats 36 million gallons per day of this wastewater prior to discharge, according to U.S. EPA records. The treated effluent is then pumped to an on-site holding pond and combined with untreated non-contact cooling water prior to discharge to the Delaware River. This discharge contains a wide variety of hazardous and non-hazardous chemicals including: **1,4-dichlorobenzene**, 12,000 pounds per year; **2,4-dinitrotoluene**, 16,000 pounds per year; **2,6-dinitrotoluene**, 19,000 pounds per year; **arsenic**, 28,400 pounds per year; and **zinc**, 22,000 pounds per year, according to the 1978 New Jersey Industrial Survey.

In addition to this major discharge, Du Pont also discharges untreated wastewaters, such as non-contact cooling water, from nine smaller outfalls to the Delaware River. According to NJDEP wastewater permit files, Du Pont also discharges an unspecified quantity of process and other wastewaters to groundwater from on-site ditches and lagoons used in wastewater treatment plant and **lead** recovery operations.

Du Pont's 40 million gallon-per-day capacity state-of-the-art wastewater treatment plant, built in 1975, uses the company's patented Powdered Activated Carbon Treatment (PACT) technology in its treatment operations. The PACT process supplements conventional activated sludge treatment with powdered activated carbon additions. PACT yields a better removal efficiency for toxic organic compounds than conventional treatment systems do, especially for wastewater containing highly variable organic loads (from batch operations, like the Chambers Works, or spills), according to a 1981 U.S. EPA report. Removal efficiency rates for 40 organic chemicals are greater than 90 percent. PACT yields greater than 95 percent removal of biodegradable organics and 80 percent removal of total organics from the Chambers Works' strongly acidic and highly variable wastewater. U.S. EPA's report noted, however, that the PACT system is not designed for treatment of metals, and removal rates are generally 20 to 50 percent.

In 1981, when Du Pont's large dye production facilities were shut down, the treatment plant was left with excess capacity which Du Pont put to use by operating the plant as a commercial treatment facility for wastewaters from other Du Pont plants and over 200 outside customers. The Chambers Works only accounts for 60 percent of the wastewater now being treated at Deepwater, the other 40 percent being generated by outside sources. In addition to treating wastewater, Du Pont also offers other services to its outside customers, including equipment decontamination, sludge disposal (in an on-site landfill) and a PACT licensing program, which allows municipalities and industries to use this Du Pont technology at their treatment plants. Average annual operating costs for the $10 million treatment plant are $16.5 million, of which $10.6 million are spent on PACT (secondary treatment) and $5.9 million are spent on primary treatment (neutralization), according to the U.S. EPA report.

There are numerous references in wastewater discharge permit files to steps Du Pont has taken to lower the quantity of wastewater it discharges to its treatment plant or to groundwater. However, a lack of information about these efforts precluded INFORM from determining whether or not these steps constituted waste reduction at the source. For example, in the "Notification of Non-Compliance - January 1984" [19], Du Pont informed the NJDEP of remedial measures it had taken to prevent future dissolved organic carbon (DOC) discharge limitation violations. These included a review of Chambers Works' DOC sources "to further reduce the DOC load" to the treatment plant. Du Pont also initiated a "maximum technical program...to study the various methods needed to reduce DOC load" to the treatment plant. However, no details were provided as to the specific nature of these DOC-lowering efforts, particularly whether or not they reduced wastes at the source.

Du Pont also discussed steps it had taken "to reduce and/or eliminate both present and past sources of groundwater contamination" in a January 1984 letter to the NJDEP [20]. The company reported that: "As a result of certain business decisions and installation of source treatment, (Chambers Works) reduced the amount of waste material to the contaminated ditch system by approximately 75 percent." As an example, Du Pont referred to a decrease in the amount of **lead** wastes going to the open ditches as a result of improvements in the **lead** recovery system. Yet, as in the case of the DOC-lowering efforts, the "business decisions" and "source treatments" which Du Pont mentions are not elaborated in NJDEP files, making it impossible to tell whether or not these steps constitute waste reduction.

SOLID WASTES

Generation

The Chambers Works is the largest solid hazardous waste producing plant in the nation, according to a U.S. EPA survey in 1983. NJDEP records list a halogenated solvent wastestream of 1.9 billion pounds. Other large hazardous solid waste streams are: **carbon tetrachloride** manufacturing wastes, 51.9 million pounds per year; **ethyl chloride** manufacturing wastes, 16.4 million pounds per year; **phenol**, 1.0 million pounds per year; **chromium** wastes, 2.0 million pounds per year; and **lead** wastes, 11.2 million pounds per year.

In August 1983, U.S. EPA released the results of a survey which found that the Chambers Works generated 40 million tons (80 billion pounds) of hazardous wastes in 1981, as much hazardous waste as U.S. EPA formerly estimated was produced nationwide, according to a New York Times article [21]. However, U.S. EPA officials qualified this figure by explaining that 99 percent of this hazardous waste is actually water used for dilution. (Even after dilution, hazardous wastes are still considered hazardous, according to federal regulations.) The agency further added that "while the plant was a major producer of hazardous waste, it had not created pollution problems," according to the New York Times.

Solid wastes generated at Chambers Works or received from outside sources are incinerated or landfilled on-site, or sent to off-site facilities in Louisiana, Texas and Nevada (either another Du Pont facility or a commercial waste management company) for disposal.

SOURCES OF INFORMATION

SOURCE	LOCATION
1. Computer printout of toxic air emissions in New Jersey, from Air Pollution Emissions Data System (APEDS), 1982	New Jersey Department of Environmental Protection (NJDEP) Bureau of Air Pollution Control CN027 Trenton, NJ 08625 (609) 633-7994
2. New Jersey Industrial Survey, 1978 data	NJDEP Office of Science and Research 190 West State Street Trenton, NJ 08625 (609) 292-6714
3. Estimates by Systems Applications, Inc. as reported in "Major Industrial Sources of Potential Toxic Air Pollutants," 1982*	National Clean Air Coalition 530 7th Street, SE Washington, DC 20003 (202) 543-8200
4. Chambers Works Bubble Proposal, 1980 data	U.S. Environmental Protection Agency (U.S. EPA), Region II Air Facilities Branch 26 Federal Plaza New York, NY 10278 (212) 264-9627

---

* This report gives 1978 estimates of air emissions from individual plants. U.S. EPA considers these figures unreliable because of the estimation techniques used.

5. Application for Approval of Alternative Pollution Control Strategy: Bubble Proposal, 1981 revision

    U.S. EPA, Region II
    (see #4)

6. Information provided by Du Pont as reported in "Major Industrial Sources of Potential Toxic Air Pollutants," 1982*

    National Clean Air Coalition
    (see #3)

7. Wastewater discharge permit application, 1980

    U.S. EPA, Region II
    Permits Administration Branch
    26 Federal Plaza
    New York, NY 10278
    (212) 264-9881

8. Draft "Administrative Order, Du Pont Chambers Works, Deepwater, Salem County," November 1983

    NJDEP
    Division of Water Resources
    CN029
    Trenton, NJ 08625
    (609) 292-5602

9. Revised RCRA Part A Application, 1980

    NJDEP
    Division of Waste Management
    32 East Hanover Street
    Trenton, NJ 08625
    (609) 292-9879

10. Computer printout of solid waste streams, 1983

    NJDEP
    (see #9)

---

* This report gives 1978 estimates of air emissions from individual plants. U.S. EPA considers these figures unreliable because of the estimation techiques used.

11. Hazardous Waste Manifest Records, 1980

    NJDEP
    Bureau of Hazardous
      Waste
    CN027
    Trenton, NJ 08625
    (609) 984-2302

12. "Pumping Vigor Into a Venerable Plant"

    Chemical Week, January 9, 1985; p.56.

13. "Human Exposure to Atmospheric Concentrations of Selected Chemicals, Volume II," prepared by Systems Applications, Inc. (Contract #68-02-3066), March 1983*

    U.S. EPA
    Office of Air Quality
      Planning and Standards
    Research Triangle Park,
    NC 27111
    (919) 541-5315

14. "Du Pont's Wastewater Venture"

    Chemical Week, June 13, 1984; pp 11-12.

15. "Putting 'Bubble' Control of Pollution to the Test"

    Chemical Week, September 9, 1981; pp 22-24.

16. "Agency Approves 'Bubble' Policy"

    Chemecology, May 1982; p. 11.

17. "Full-Scale Demonstration of Industrial Wastewater Treatment Utilizing Du Pont's PACT Process," U.S. EPA Project Summary (EPA-600/S2-81-159), December 1981

    U.S. EPA
    Kerr Research
      Laboratory
    P.O. Box 1198
    Ada, OK 74820
    (405) 332-8800

---

* This report gives 1978 estimates of air emissions from individual plants. U.S. EPA considers these figures unreliable due to the estimation techniques used.

18. "'PACT': A Step Beyond   Du Pont Magazine,
    the Conventional"   January/February 1982;
                                 p. 18.

19. Notification of Non-   NJDEP
    Compliance, January 1984   (see #8)

20. Correspondence from Du   NJDEP
    Pont to the New Jersey   (see #8)
    Department of Environ-
    mental Protection,
    January 1984

21. "E.P.A. Study Has   New York Times,
    Jerseyans Puzzled"   September 1, 1983;
                                 p. B2.

DU PONT

WASTE GENERATION & DISPOSAL

Type of Waste ● Indicates Wastes Affected by Reduction	Generation or Disposal	Quantity of Waste	Sources of Information
Air Emissions			
HAZARDOUS			
Benzene	Emissions from 9 sources	5.28 lbs/hour (1982)	(1)
	Plant-wide emissions	38,100 lbs/yr (1978)	(2)
Carbon tetrachloride	"	75,420 lbs/yr* (1978)	(3)
	Emissions from 13 sources	188.02 lbs/hour (1982)	(1)
	Process emissions	8,585 lbs/yr (1980)	(4)
	Plant-wide emissions	7,100 lbs/yr (1978)	(2)
Chlorobenzene (Mono)	Plant-wide emissions	54,400 lbs/yr* (1978)	(3)
	Process emissions	58,867 lbs/yr (1980)	(4)
	Plant-wide emissions	124,000 lbs/yr (1978)	(2)
Chloroform**	"	148,800 lbs/yr* (1978)	(3)
1,2-Dichlorobenzene**	"	796,860 lbs/yr* (1978)	(3)
	Process emissions	19,009 lbs/yr (1980)	(4)
	Plant-wide emissions	44,152 lbs/yr (1978)	(2)
Dichlorodifluoromethane	Process emissions	110,376 lbs/yr (1980)	(4)

* The U.S. EPA considers these figures unreliable because of the estimation techniques used.

** Chambers Works has stopped manufacturing these chemicals.

DU PONT ## WASTE GENERATION & DISPOSAL

Type of Waste • Indicates Wastes Affected by Reduction	Generation or Disposal	Quantity of Waste	Sources of Information
Air Emissions			
HAZARDOUS (cont.)			
1,2-Dichloroethane	Emissions from 53 sources	35.16 lbs/hour (1982)	(1)
	Process emissions	47,216 lbs/yr (1980)	(4)
	Plant-wide emissions	64,800 lbs/yr (1978)	(2)
Dichloroethane, 1-chloro-1,1-difluoroethane and vinylidene fluoride	Emissions from 1 source	240,000 lbs/yr (1981)	(5)
Dinitrotoluene*	Process emissions	8,585 lbs/yr (1980)	(4)
2,4-Dinitrotoluene	Plant-wide emissions	11,400 lbs/yr (1978)	(2)
Ethylene dibromide	Emissions from 49 sources	5.16 lbs/hour (1982)	(1)
	Plant-wide emissions	3,950 lbs/yr (1978)	(2)
Ethylene oxide	Emissions from 2 sources	9,858 lbs/yr (1981)	(5)
Ethylene oxide and propylene oxide	Emissions from 1 source	5,520 lbs/yr (1981)	(5)
Formaldehyde	Plant-wide emissions	19,350 lbs/yr (1978)	(2)
Methyl alcohol	Emissions from 4 sources	76,815 lbs/yr (1981)	(5)
	Process emissions	58,254 lbs/yr (1980)	(4)
Methyl alcohol, methyl chloride, ethyl alcohol, Freons and nitromethane	Emissions from 1 source	18,666 lbs/yr (1981)	(5)
Methyl alcohol, methyl isobutyl ketone and isopropyl alcohol	"	2,116 lbs/yr (1981)	(5)

* This manufacturing operation was substantially reduced or eliminated in 1980.

WASTE GENERATION & DISPOSAL

DU PONT

Type of Waste ● Indicates Wastes Affected by Reduction	Generation or Disposal	Quantity of Waste	Sources of Information
Air Emissions			
HAZARDOUS (cont.)			
Methyl alcohol and methyl methacrylate	Emissions from 1 source	1,277 lbs/yr (1981)	(5)
Methyl chloride	Emissions from 3 sources	1,466,300 lbs/yr (1981)	(5)
	Plant-wide emissions	2,606,000 lbs/yr (1978)	(2)
Methyl chloride, tetra methyl lead and ethyl chloride	Emissions from 1 source	197,000 lbs/yr (1981)	(5)
Methyl isobutyl ketone and isopropyl alcohol	Emissions from 2 sources	9,816 lbs/yr (1981)	(5)
Methylene chloride	Process emissions	56,414 lbs/yr (1980)	(4)
	Plant-wide emissions	23,170 lbs/yr (1978)	(2)
Nitrobenzene	"	1,275 lbs/yr* (1978)	(3)
	"	1,560 lbs/yr (1978)	(2)
Phenol	"	2,210 lbs/yr (1978)	(2)
Phosgene	"	18,934 lbs/yr* (1978)	(3)
	Process emissions	31,886 lbs/yr (1980)	(4)
	Plant-wide emissions	47,978 lbs/yr (1978)	(2)
Tetrachloroethylene	Emissions from 4 sources	2.1 lbs/hour (1982)	(1)
Tetra ethyl lead	Process emissions	153,300 lbs/yr (1980)	(4)
Tetra methyl lead	"	36,792 lbs/yr (1980)	(4)

* The U.S. EPA considers these figures unreliable because of the estimation techniques used.

DU PONT
WASTE GENERATION & DISPOSAL

Type of Waste ● Indicates Wastes Affected by Reduction	Generation or Disposal	Quantity of Waste	Sources of Information
Air Emissions			
HAZARDOUS (cont.)			
Toluene	Plant-wide emissions	48,640 lbs/yr* (1978)	(3)
	"	17,654 lbs/yr (1978)	(6)
	Emissions from 1 source	45,550 lbs/yr (1981)	(5)
	Plant-wide emissions	3,980 lbs/yr (1978)	(2)
Toluene diisocyanate**	Process emissions	33,726 lbs/yr (1980)	(4)
Trichlorobenzene	"	153,300 lbs/yr (1980)	(4)
	Plant-wide emissions	6,760 lbs/yr (1978)	(2)
Trichloroethylene	Emissions from 4 sources	16.26 lbs/hour (1982)	(1)
Trichlorofluoromethane	Process emissions	110,376 lbs/yr (1980)	(4)
	Plant-wide emissions	1,031,000 lbs/yr (1978)	(2)
Xylenes, mixed	Process emissions	9,811 lbs/yr (1980)	(4)
Volatile organic substances***	"	195,611 lbs/yr (1980)	(4)
Lead	Plant-wide emissions	111,700 lbs/yr (1978)	(2)

* The U.S. EPA considers this figure unreliable because of the estimation techniques used.

** This manufacturing operation was substantially reduced or eliminated in 1980.

*** Emissions of 42 chemicals whose emissions rates are less than 1 pound per hour.

DU PONT

WASTE GENERATION & DISPOSAL

Type of Waste ● Indicates Wastes Affected by Reduction	Generation or Disposal	Quantity of Waste	Sources of Information
Wastewater Discharges			
Process water, non-contact cooling water, wastewater from outside sources and other waters including:	Treated and discharged to the Delaware Estuary	80 million gals/day (of which 36 million gals/day are treated) (1980, 1981)	(7, 17)
HAZARDOUS			
Bromoform	"	1,420 lbs/yr (1978)	(2)
Chlorobenzene	"	22 lbs/day (max.) (1980)	(7)
	"	2,100 lbs/yr (1978)	(2)
Chloroethane	"	70.4 lbs/day (max.) (1980)	(7)
1,2-Dichlorobenzene	"	623 lbs/day (max.) (1980)	(7)
		5,300 lbs/yr (1978)	(2)
1,4-Dichlorobenzene	"	6.6 lbs/day (max.) (1980)	(7)
		12,000 lbs/yr (1978)	(2)
1,2-Dichloroethane*	"	11 lbs/day (max.) (1980)	(7)
		2,500 lbs/yr (1978)	(2)
Diethyl phthalate	"	1,500 lbs/yr (1978)	(2)
4,6-Dinitro-o-cresol	"	1,500 lbs/yr (1978)	(2)
2,4-Dinitrotoluene	"	19.8 lbs/day (max.) (1980)	(7)
		16,000 lbs/yr (1978)	(2)

* Also referred to in company files as ethylene dichloride.

WASTE GENERATION & DISPOSAL

DU PONT

●	Type of Waste Indicates Wastes Affected by Reduction	Generation or Disposal	Quantity of Waste	Sources of Information
	Wastewater Discharges			
	Wastewater (cont.) HAZARDOUS (cont.)	Treated and discharged to Delaware River Estuary		
	2,6-Dinitrotoluene	"	246 lbs/day (max.) (1980)	(7)
			19,000 lbs/yr (1978)	(2)
	Ethyl benzene	"	11 lbs/day (max.) (1980)	(7)
			90 lbs/yr (1978)	(2)
	Methylene chloride	"	46.2 lbs/day (max.) (1980)	(7)
			9,100 lbs/yr (1978)	(2)
	Nitrobenzene	"	17.6 lbs/day (max.) (1980)	(7)
			260 lbs/yr (1978)	(2)
	2-Nitrophenol	"	1,500 lbs/yr (1978)	(2)
	4-Nitrophenol	"	1,687 lbs/day (max.) (1980)	(7)
			1,800 lbs/yr (1978)	(2)
	Phenols, total	"	90.2 lbs/day (1980)	(7)
	Tetrachloroethylene	"	79.2 lbs/day (max.) (1980)	(7)
			320 lbs/yr (1978)	(2)
	Toluene	"	33 lbs/day (max.) (1980)	(7)
			440 lbs/yr (1978)	(2)

WASTE GENERATION & DISPOSAL

DU PONT

Type of Waste ● Indicates Wastes Affected by Reduction	Generation or Disposal	Quantity of Waste	Sources of Information
Wastewater Discharges			
Wastewater (cont.) HAZARDOUS (cont.)	Treated and discharged to the Delaware River Estuary		
1,2,4-Trichlorobenzene	"	2,200 lbs/yr (1978)	(2)
1,1,1-Trichloroethane	"	11 lbs/day (max.) (1980)	(7)
		55 lbs/yr (1978)	(2)
Trichlorofluoro-methane	"	33 lbs/day (max.) (1980)	(7)
		2,200 lbs/yr (1978)	(2)
Antimony	"	8.8 lbs/day (1980)	(7)
		2,000 lbs/yr (1978)	(2)
Arsenic	"	20.9 lbs/day (1980)	(7)
		28,400 lbs/yr (1978)	(2)
Barium	"	48.4 lbs/day (1980)	(7)
Chromium	"	8.8 lbs/day (1980)	(7)
		2,000 lbs/yr (1978)	(2)
Copper	"	18.7 lbs/day (1980)	(7)
		2,900 lbs/yr (1978)	(2)
Cyanide	"	35.2 lbs/day (1980)	(7)
		7,600 lbs/yr (1978)	(2)
Lead	"	39.6 lbs/day (1980)	(7)
		6,200 lbs/yr (1978)	(2)

DU PONT

WASTE GENERATION & DISPOSAL

Type of Waste ● Indicates Wastes Affected by Reduction	Generation or Disposal	Quantity of Waste	Sources of Information
Wastewater Discharges			
Wastewater (cont.) HAZARDOUS (cont.)	Treated and discharged to the Delaware River Estuary		
Nickel	"	22 lbs/day (1980)	(7)
		4,500 lbs/yr (1978)	(2)
Selenium	"	8.8 lbs/day (1980)	(7)
		1,800 lbs/yr (1978)	(2)
Zinc	"	110 lbs/day (1980)	(7)
		22,000 lbs/yr (1978)	(2)
LIKELY TO CONTAIN HAZARDOUS CHEMICALS			
Process and other wastewaters	Discharged to groundwater from on-site ditches and lagoons used in wastewater treatment and lead recovery operations	? (1983)	(8)

299

DU PONT

WASTE GENERATION & DISPOSAL

Type of Waste ● Indicates Wastes Affected by Reduction	Generation or Disposal	Quantity of Waste	Sources of Information
Solid Wastes			
HAZARDOUS			
RCRA Regulated "Hazardous Waste"*	---*	88 billion lbs/yr* (1981)	(21)
Acrylonitrile	Incinerated on-site or disposed of in an on-site landfill	2,000 lbs/yr (1980)	(9)
Aniline	"	20,000 lbs/yr (1980)	(9)
	Incinerated on-site	153,300 lbs/yr (1978)	(2)
	?	665,400 lbs/yr (1978)	(2)
	Incinerated at Rollins Environmental Services, Baton Rouge, LA	46,500 lbs/yr (1978)	(2)
Benzal chloride	Incinerated on-site or disposed of in on-site landfill	2,000 lbs/yr (1980)	(9)
Benzene	"	20,000 lbs/yr (1980)	(9)
Bis(2-ethyl hexyl) phthalate	Incinerated on-site	340,000 lbs/yr (1978)	(2)
Carbon disulfide	Incinerated on-site or disposed of in on-site landfill	2,000 lbs/yr (1980)	(9)
Carbon tetrachloride manufacturing wastes	Treated on-site or disposed of in on-site landfill	51,891,840 lbs/yr (1983)	(10)
Chloral	Incinerated on-site or disposed of in on-site landfill	2,000 lbs/yr (1980)	(9)
p-Chloroaniline	"	2,000 lbs/yr (1980)	(9)

* Total hazardous waste generation includes most chemical wastes listed below.

DU PONT
WASTE GENERATION & DISPOSAL

Type of Waste ● Indicates Wastes Affected by Reduction	Generation or Disposal	Quantity of Waste	Sources of Information
Solid Wastes			
HAZARDOUS (cont.)			
Chlorobenzene	Incinerated on-site or disposed of in on-site landfill	20,000 lbs/yr (1980)	(9)
	Treated on-site	651,600 lbs/yr (1978)	(2)
	Incinerated on-site	7,000 lbs/yr (1978)	(2)
	Incinerated at Rollins Environmental Services, Baton Rouge, LA	46,800 lbs/yr (1978)	(2)
Chloroform	Disposed of in on-site landfill	1,000 lbs/yr (1980)	(9)
2,4-Diaminotoluene	?	3,000 lbs/yr (1978)	(2)
1,2-Dichlorobenzene	Incinerated on-site or disposed of in on-site landfill	20,000 lbs/yr (1980)	(9)
	Incinerated on-site	16,300 lbs/yr (1978)	(2)
1,3-Dichlorobenzene	Incinerated on-site or disposed of in on-site landfill	20,000 lbs/yr (1980)	(9)
1,4-Dichlorobenzene	"	20,000 lbs/yr (1980)	(9)
Dichlorodifluoromethane	"	20,000 lbs/yr (1980)	(9)
1,2-Dichloroethane	"	20,000 lbs/yr (1980)	(9)
Dichloromethane	"	20,000 lbs/yr (1980)	(9)
	Incinerated on-site	16,320 lbs/yr (1978)	(2)
Dimethyl phthalate	Incinerated on-site or disposed of in on-site landfill	2,000 lbs/yr (1980)	(9)

WASTE GENERATION & DISPOSAL

Type of Waste ● Indicates Wastes Affected by Reduction	Generation or Disposal	Quantity of Waste	Sources of Information
Solid Wastes			
HAZARDOUS (cont.)			
Dimethyl sulfate	Incinerated on-site or disposed of in on-site landfill	2,000 lbs/yr (1980)	(9)
2,4-Dinitrotoluene	"	2,000 lbs/yr (1980)	(9)
	Sent off-site for disposal	6,000 lbs/yr (1980)	(11)
	Incinerated at Rollins Environmental Services, Baton Rouge, LA	160,000 lbs/yr (1978)	(2)
	Incinerated at Rollins Environmental Services, Deer Park, TX	20,400 lbs/yr (1978)	(2)
	Incinerated at Du Pont Ponchartrain Plant, LaPlace, LA	20,000 lbs/yr (1978)	(2)
	?	6,840 lbs/yr (1978)	(2)
2,6-Dinitrotoluene	Incinerated on-site or disposed of in on-site landfill	2,000 lbs/yr (1980)	(9)
	Incinerated at Rollins Environmental Services, Baton Rouge, LA	109,000 lbs/yr (1978)	(2)
	Incinerated at Du Pont Ponchartrain Plant, LaPlace, LA	13,400 lbs/yr (1978)	(2)
	Incinerated at Rollins Environmental Services, Bridgeport, NJ	13,400 lbs/yr (1978)	(2)
Di-n-octylphthalate	Incinerated on-site or disposed of in on-site landfill	2,000 lbs/yr (1980)	(9)
Ethyl chloride manufacturing wastes	Treated or incinerated on-site or disposed of in on-site landfill	16,365,888 lbs/yr (1983)	(10)
Ethylene diamine	Incinerated on-site or disposed of in on-site landfill	2,000 lbs/yr (1980)	(9)

WASTE GENERATION & DISPOSAL

DU PONT

Type of Waste ● Indicates Wastes Affected by Reduction	Generation or Disposal	Quantity of Waste	Sources of Information
Solid Wastes			
HAZARDOUS (cont.)			
Ethylene dibromide	Incinerated on-site or disposed of in on-site landfill	2,000 lbs/yr (1980)	(9)
Ethylene oxide	"	2,000 lbs/yr (1980)	(9)
Ethylene thiourea	"	2,000 lbs/yr (1980)	(9)
Formaldehyde	"	2,000 lbs/yr (1980)	(9)
Hydrazine	"	2,000 lbs/yr (1980)	(9)
Isobutyl alcohol	"	3,991 lbs/yr (1983)	(10)
Maleic anhydride	"	2,000 lbs/yr (1980)	(9)
Methyl alcohol	"	173,837 lbs/yr (1983)	(10)
Methyl chloride	Incinerated on-site	7,000 lbs/yr (1978)	(2)
Methylene chloride	"	16,320 lbs/yr (1978)	(2)
Methyl ethyl ketone	Incinerated on-site or disposed of in on-site landfill	40,000 lbs/yr (1980)	(9)
Methyl isobutyl ketone	"	2,000 lbs/yr (1980)	(9)
Methyl methacrylate	"	2,000 lbs/yr (1980)	(9)
Naphthalene	"	20,000 lbs/yr (1980)	(9)

DU PONT

WASTE GENERATION & DISPOSAL

●	Type of Waste Indicates Wastes Affected by Reduction	Generation or Disposal	Quantity of Waste	Sources of Information
	Solid Wastes			
	HAZARDOUS (cont.)			
	1-Naphthylamine	Incinerated on site or disposed of in on-site landfill	2,000 lbs/yr (1980)	(9)
	p-Nitroaniline	"	20,000 lbs/yr (1980)	(9)
	Nitrobenzene	"	20,000 lbs/yr (1980)	(9)
		?	899,660 lbs/yr (1978)	(2)
	Paraldehyde	Incinerated on-site or disposed of in on-site landfill	2,000 lbs/yr (1980)	(9)
	Phenol	Incinerated at Rollins Environmental Services, Deer Park, TX	1,047,600 lbs/yr (1978)	(2)
	Polychlorinated biphenyl and polybrominated biphenyl-contaminated material	Sent off-site for disposal	36,000 lbs/yr (1980)	(11)
	Pyridine	Incinerated on-site or disposed of in on-site landfill	2,000 lbs/yr (1980)	(9)
	Tetrachloroethylene	Incinerated on-site	3,800 lbs/yr (1978)	(2)
	Tetrachloromethane	Incinerated on-site or disposed of in on-site landfill	40,000 lbs/yr (1980)	(9)
	Tetra ethyl lead	Treated on-site	20,000 lbs/yr (1980)	(9)
	Thiourea	Incinerated on-site or disposed of in on-site landfill	2,000 lbs/yr (1980)	(9)

WASTE GENERATION & DISPOSAL
DU PONT

Type of Waste ● Indicates Wastes Affected by Reduction	Generation or Disposal	Quantity of Waste	Sources of Information
Solid Wastes			
HAZARDOUS (cont.)			
Toluene	Incinerated on-site or disposed of in on-site landfill	85,820 lbs/yr (1983)	(10)
	Stored on-site	269,280 lbs/yr (1978)	(2)
	Incinerated on-site	15,500 lbs/yr (1978)	(2)
Toluene diisocyanate	Incinerated on-site or disposed of in on-site landfill	20,000 lbs/yr (1980)	(9)
1,1,1-Trichloroethane	"	2,000 lbs/yr (1980)	(9)
Trichloroethylene	Incinerated on-site	1,940 lbs/yr (1978)	(2)
	?	3,400 lbs/yr (1978)	(2)
Trichlorofluoromethane	Incinerated on-site or disposed of in on-site landfill	2,000 lbs/yr (1980)	(9)
2,4,5-Trichlorophenol	"	2,000 lbs/yr (1980)	(9)
2,4,6-Trichlorophenol	"	2,000 lbs/yr (1980)	(9)
Xylene	"	20,000 lbs/yr (1980)	(9)
Solvents, halogenated	Treated or incinerated on-site or disposed of in on-site landfill	1,898,672,985 lbs/yr (1983)	(10)
Solvents, non-halogenated	"	5,388,766 lbs/yr (1983)	(10)
Wastewater treatment sludges from electroplating operations	Treated on-site or disposed of in on-site landfill	47,900,160 lbs/yr (1983)	(10)

DU PONT

WASTE GENERATION & DISPOSAL

Type of Waste ● Indicates Wastes Affected by Reduction	Generation or Disposal	Quantity of Waste	Sources of Information
Solid Wastes			
HAZARDOUS (cont.)			
Antimony	?	2,350 lbs/yr (1978)	(2)
	Disposed of in off-site sanitary landfill at Nuclear Engineering, Beatty, NV	156,000 lbs/yr (1978)	(2)
Arsenic wastes	Treated on-site or disposed of in on-site landfill	766,401 lbs/yr (1983)	(10)
Asbestos	Sent off-site for disposal	39,417 lbs/yr (1980)	(11)
	Disposed of in an on-site landfill	12,600 lbs/yr (1978)	(2)
Chromium	?	47,580 lbs/yr (1978)	(2)
Chromium wastes	Treated or incinerated on-site or disposed of in on-site landfill	1,999,831 lbs/yr (1983)	(10)
Copper	?	27,600 lbs/yr (1978)	(2)
Copper cyanide	Disposed of in on-site landfill	2,000 lbs/yr (1980)	(9)
Hydrofluoric acid	"	1,995 lbs/yr (1983)	(10)
Lead	?	2,460 lbs/yr (1978)	(2)
Lead wastes	Treated on-site or disposed of in on-site landfill	11,176,704 lbs/yr (1983)	(10)
Nickel	?	4,470 lbs/yr (1978)	(2)
Potassium cyanide	Disposed of in on-site landfill	2,000 lbs/yr (1980)	(9)
Sodium cyanide	"	2,000 lbs/yr (1980)	(9)
Zinc	?	80,250 lbs/yr (1978)	(2)

Additional Wastes Reported by Du Pont Not Included in the Chart

AIR EMISSIONS

HAZARDOUS*

2,6-Dinitrotoluene

MAY CONTAIN
HAZARDOUS CHEMICALS

Halon 1301**

NON-HAZARDOUS

Acetone
Butane
Butyl Alcohol
1-Chloro-1,1-difluoro-
  ethane
1,2-Dichloro-1,
  1-difluoroethane

NON-HAZARDOUS (cont.)

Difluoroethane
Ethane
Ethyl chloride
Ethylene
Fluoroethane
Freons, unspecified
Heptane
Hexafluoropropylene
Hexane
Isobutylene
Isopropyl alcohol
Monochlorodifluoro-
  methane
alpha-Naphthylamine
Nonene
Octene
Pentane
Propane
Propylene oxide
Tetrafluoroethylene
Vinylidene fluoride
Zepel Fluoromonomer**

 * All hazardous wastes listed here are generated in quantities of less than 1,000 pounds per year. Hazardous wastes above 1,000 pounds per year are listed on the waste generation table.

** Du Pont's tradename for this unspecified organic chemical.

Additional Wastes Reported by Du Pont Not Included in the Chart

WASTEWATER DISCHARGES

HAZARDOUS*

Aniline
Benzene
Benzyl chloride
Beryllium
Bis(2-ethyl hexyl) phthalate
Cadmium
Carbon disulfide
Chloroform
2-Chlorophenol
Cresol
1,3-Dichlorobenzene
2,4-Dichlorophenol
2,4-Dimethylphenol
Dinitrobenzene
Epichlorohydrin
Ethylene diamine
Ethylene dibromide
Formaldehyde
Mercury
Methyl methacrylate
Naphthalene
Phenol
Phosgene

HAZARDOUS (cont.)

Silver
Thallium
Trichloroethylene
Xylene

NON-HAZARDOUS

Aluminum
Cobalt
Cyclohexane
Dimethyl amine
Iron
Magnesium
Manganese
Methyl mercaptan
Molybdenum
Nitrotoluene
Propylene oxide
Quinolene
Styrene
Titanium
Triethanolamine
Trimethyl amine
Xylenol

* All hazardous wastes listed here are generated in quantities of less than 1,000 pounds per year. Hazardous wastes above 1,000 pounds per year are listed on the waste generation table.

Additional Wastes Reported by Du Pont Not Included in the Chart

SOLID WASTES

HAZARDOUS*

Formic acid
Mercury
Research wastes**
Resorcinol

MAY CONTAIN HAZARDOUS CHEMICALS

Alkaline solution
Chlorinated (dioxin furan) resin
Dye wastes
Explosive residue
Fluorocarbon
Groundwater, contaminated
Halogenated solvents, unspecified
Lacramators, amines and mercaptan
Non-halogenated solvents, unspecified
Oil and oil sludges
Packed lab chemicals
Plastics, plasticizers, resins and elastomers

NON-HAZARDOUS

Acetone
n-Butyl alcohol
Corrosive wastes, unspecified
Cyanates
Cyclohexane
Dimethyl amine
Ethyl acetate
Ignitable wastes, unspecified
alpha-Naphthylamine
Non-halogenated solvents, unspecified
Phosphorus
Reactive wastes, unspecified
Research wastes***
Rhodamine B
Tetrahydrofuran

---

* All hazardous wastes listed here are generated in quantities of less than 1,000 pounds per year. Hazardous wastes above 1,000 pounds per year are listed on the waste generation table.
** Contains 69 chemicals used in research.
*** Contains 5 chemicals used in research.

Exxon Chemical Americas

Bayway Chemical Plant
P.O. Box 23
Linden, New Jersey 07036
(201) 474-0100

SUMMARY

Exxon Chemical Americas' Bayway plant (Linden, New Jersey), adjacent to a larger refinery at the same site, manufactures oil and fuel additives, synthetic lubricants, solvents, and specialty chemicals. Hazardous chemicals handled include the products **methyl ethyl ketone** and **methyl isobutyl ketone** and the raw materials **maleic anhydride** and **phenol**. Installation of floating roofs (required for some emission sources by New Jersey air pollution regulations) has reduced emissions of several chemicals by over 90 percent, for an overall savings since 1974 of more than five million pounds of organic chemicals (annual savings in 1983 of 680,000 pounds valued at over $200,000). Use of conservation vents (also a requirement of state regulations) has reduced air emissions by unspecified amounts. Exxon credits its Stewardship Program -- designed to monitor and control wastewater discharges -- with a 75 percent decrease, since the mid-1970s, in the level of organic chemicals discharged in the plant's wastewater. The company also reports that a variety of measures have reduced the generation of hazardous solid wastes, but did not provide the detailed information that would allow an evaluation of their impact.

NOTE: Chemicals in **boldface** type are hazardous, according to INFORM's criteria (see Appendix A).

PLANT OPERATIONS

Exxon Chemical Americas' 550-employee Bayway plant, next to a larger petroleum refining operation at the same site, is located in a heavily industrialized area of northern New Jersey. The refinery has operated since 1907; the chemical plant, the first commercial petrochemical facility in the United States, has operated since 1921. Exxon Chemical Americas, a division of the Exxon Chemical Company (which is wholly owned by Exxon Corporation) is the fourth largest chemical company in the United States. Exxon Corporation is the largest corporation in the world.

The plant manufactures four major product lines:

-- ORGANIC SOLVENTS, including such commonly used industrial chemicals as **methyl ethyl ketone** (MEK), secondary butyl alcohol, **methyl isobutyl ketone** (MIBK), and acetone.

-- OLEFINS, a category of organic chemicals that includes propylene and butylenes, used as raw materials in the manufacture of other industrial chemicals and plastics.

-- "PARAMINS" ADDITIVES, Exxon's trade name for its line of chemicals that, when added to fuels and lubricating oils, inhibit corrosion, improve flow properties, stabilize viscosity, and otherwise enhance their quality. The Paramins operations manufacture dispersant additives (with the raw material **maleic anhydride**), detergent inhibitor additives (with the raw material **phenol**), and synthetic lubricating oils.

-- SPECIALTY CHEMICALS, a minor product line of the Bayway plant, includes such isobutylene polymers as "LM VISTANEX," Exxon's trade name for an ingredient in chewing gum and surgical adhesives.

Both the solvent and the olefin operations, which use hydrocarbons from the Exxon refinery as their main source of raw materials, are carried out at the area of the Bayway complex known as the East Side Chemical Plant. The manufacture of Paramins additives and specialty chemicals, which takes place at the Bayway complex's West Side Chemical Plant, requires catalysts and a wide variety of organic and inorganic raw materials.

WASTE GENERATION AND WASTE REDUCTION

AIR EMISSIONS

Generation

The Bayway plant has more than 250 active air permits covering emissions from over 200 storage tanks as well as from process equipment and transfer (loading and unloading) operations. Chemicals emitted in the largest quantities are **MEK,** 27,140 pounds per year; acetone, 14,580 pounds per year; **MIBK,** 12,860 pounds per year; dodecyl phenol, 4,860 pounds per year; and **phenol,** 2,230 pounds per year.

Reduction

Two kinds of equipment changes have had major effects on air emissions at the Bayway plant, preventing the loss of over five million pounds of organic chemicals since 1974. Emissions reductions in 1983 totalled 680,000 pounds per year, worth more than $200,000.

Exxon's major air emissions reductions resulted from the installation of floating roofs on the 16 tanks storing the most volatile (easily evaporated) chemicals used at the plant, including acetone, **MEK, MIBK,** alcohols, and other industrial solvents. A floating roof -- the purpose of which is to reduce vapor losses from large tanks -- is a cover that rests on the surface of the liquid being stored in the tank. The roof rises and falls as the level of the liquid changes, thus minimizing the formation of chemical vapors which are then subsequently vented from the tank. The plant installed floating roofs on 11 storage tanks in 1975 and later added five more to new or existing tanks. According to Exxon's air permit files, the efficiency of the floating roofs varied little among the tanks and averaged 90.4 percent.

SAVINGS FROM FLOATING ROOF INSTALLATION, 1975-83

Chemical	Total Emissions Reduction Since Installation (pounds)	1983 Material Savings (pounds)	1983 Cost Savings (dollars)
Acetone	2,550,320	237,980	$ 52,355
MEK	2,318,350	239,220	$ 86,120
Isopropyl alcohol	274,720	24,240	$ 6,545
MIBK	6,640	3,320	$ 1,630
Secondary butyl ether	5,260	2,630	$ 1,095
Assorted industrial solvents*	174,420	174,420	$ 57,560
Other ketones and alcohols*	43,120	--	--
TOTAL SAVINGS	5,372,830 pounds	681,810 pounds	$205,305

Current dollar savings are based on 1983 prices per pound as follows: acetone, 22¢; MEK, 36¢; MIBK, 49¢, and isopropyl alcohol, 27¢. Values for the other organic chemicals were based on an average price of 33¢ per pound.

---

* These categories contain a wide variety of organic chemicals, which are not specifically identified in air permit files.

The floating roofs have been effective in reducing emissions of chemicals that are handled in large quantities at the plant. A large **MEK** storage tank, for example, annually handles more than 68 million pounds (10.2 million gallons) of the chemical. Whereas annual emissions from this tank without a floating roof were 48,120 pounds (0.07 percent of total annual **MEK** production), use of a floating roof reduced annual emissions to 4,620 pounds -- 0.007 percent of the amount stored.

The capital costs of the floating roofs ranged from $5,200 to $13,000. Many of the floating roofs (even those with maximum installation and operating costs) saved enough material to pay for themselves in one year.

Exxon's second waste reduction practice is the addition, since 1976, of seven conservation vents to storage tanks and process sources emitting dodecyl phenol, **maleic anhydride,** nonyl phenol, and other hydrocarbons. Conservation vents -- which release a tank's accumulating vapors less freely and continuously than do standard vents that are routinely installed as part of large storage tanks and process equipment -- reduce air emissions by a range of 30 to 75 percent. Detailed information on the types and quantities of materials saved were not reported in air permit files and were not made available by Exxon to INFORM.

New Jersey air pollution control regulations, which became effective in 1976, require floating roofs and conservation vents for many sources of organic emissions. The type and degree of control depend on the size of the emissions source and vapor pressure of the chemical (an indication of the ease with which a substance can evaporate). Although Exxon installed floating roofs in anticipation of these state regulations, a growing appreciation of the cost-effectiveness of floating roofs was also a motivating factor. Consequently, Exxon reported to INFORM that they have installed more floating roofs than environmental regulations require.

WASTEWATER DISCHARGES

Generation

The Bayway plant discharges process waters to nearby Morses Creek after treating them on-site; the plant also discharges untreated non-contact cooling waters. Information sources reviewed by INFORM do not distinguish between discharges from Exxon's chemical operations and its refinery located at the same site, and thus give no detailed information about the discharges from the chemical plant alone.

Reduction

Plant management told INFORM of two changes affecting its wastewater discharges. One of these, which Exxon calls its Stewardship Program, has reduced by 75 percent the level of organic wastes entering the Bayway facility's wastewater. The second change, while not constituting hazardous waste reduction as INFORM uses the term, caused the declassification of the plant's acidic wastewater stream as hazardous.

Exxon implemented its two-part Stewardship Program as a result of its concern about the large fluctuations in the quantity and quality of its wastewater discharges, which were interfering with the operation of its wastewater treatment plant. Exxon first established a series of wastewater sampling stations to monitor concentrations of organic chemicals. Plant operators are, as a result, quickly made aware of unusual changes in these concentrations in the wastewater flowing to the treatment plant and can trace them back to their source and quickly remedy the problem. The program also established for the first time the levels of organic waste being generated by each of Exxon Chemical's manufacturing processes. The company uses this information to charge to each process a portion (based on waste generation) of the multi-million dollar cost of operating the wastewater treatment system and to set targets for waste reduction for each process. Plant officials told INFORM that the combination of minimizing the impact of process upsets, charging process managers for the actual cost of wastewater treatment, and setting waste reduction goals for each process have led to an overall reduction of 75 percent of the quantity of organic chemicals in the plant's wastewater. Exxon did not provide INFORM with more detailed information of how these wastes were reduced.

Exxon also implemented a change which, although not constituting waste reduction as INFORM uses the term, did affect the plant's generation of 21.6 million pounds per year (2.7 million gallons) of wastewater containing acids which are discharged to Exxon's on-site wastewater treatment facilities. This wastestream would ordinarily not have been legally classified as a hazardous waste, because the acid levels were within the limits established by law.* However, because of variability in Exxon's processes, the acid levels of the wastewater occasionally became so great that the entire wastestream had to be considered a hazardous waste. Federal and state hazardous waste laws would have required the plant to file as a "Treatment, Storage, or Disposal" (TSD) facility if the acidity continued to exceed the hazardous level. Rather than undergo the costly and complex TSD application process, Exxon instituted procedures that equalized and neutralized the generation of acid waste and prevented excess acidity — meaning that the wastewaters were no longer considered legally hazardous. Since no other hazardous wastes were treated, stored or disposed of on-site (all other hazardous wastes were shipped off-site for disposal), it was not necessary for the plant to file as a TSD facility.

----
* Acidity is measured by a chemical index known as pH. Under federal and New Jersey State law, a waste is legally hazardous if the pH is below 2.0.

SOLID WASTES

Generation

The Bayway plant generates solid wastes containing **phenol, maleic anhydride,** and other process chemicals in use at the plant. All solid hazardous waste is transported off-site for disposal at commercial waste management facilities in New Jersey, New York and Ohio.

Reduction

According to Exxon, aggregate data for its two New Jersey chemical operations -- the Bayway plant and one in Bayonne -- show a 55 percent reduction in hazardous waste generated (per ton of product manufactured) and a 45 percent reduction in total (hazardous and non-hazardous) solid waste. However, these waste reduction data do not distinguish between the two plants. Although Exxon told INFORM of several waste reduction measures -- process changes and technological innovations, chemical substitutions, the prevention of spills and leaks through improved procedures, and the elimination of contaminants through better quality control -- it neither identified the plant where these were implemented, gave details of these practices nor documented their specific impact on waste reduction.

SOURCES OF INFORMATION

SOURCE	LOCATION
1. Air permit files, 1984	New Jersey Department of Environmental Protection (NJDEP) Division of Environmental Quality Newark Field Office 1100 Raymond Boulevard Newark, NJ 07102 (201) 648-2075
2. New Jersey Industrial Survey, 1978 data	NJDEP Office of Science and Research 190 West State Street Trenton, NJ 08625 (609) 292-6714
3. Wastewater discharge files, 1984	U.S. Environmental Protection Agency (U.S. EPA), Region II Permits Administration Branch 26 Federal Plaza New York, NY 10278 (212) 264-9881
4. Wastewater discharge files, 1984	NJDEP Division of Water Resources 1474 Prospect Street Trenton, NJ 08625 (609) 292-5602
5. RCRA Hazardous Waste Notification, 1981	U.S. EPA, Region II (see #3)

6. Hazardous Waste Annual   NJDEP
   Report, 1981   Division of Waste
                                 Management
                               32 East Hanover Street
                               Trenton, NJ 08625
                               (609) 292-9879

7. Hazardous Waste Annual   NJDEP
   Report, 1982   (see #6)

8. Information provided by
   Exxon to INFORM

WASTE REDUCTION

EXXON

Type of Waste	G–Generation D–Disposal Q–Quantity (year)	Waste Reduction Practice	Results	Sources of Information
HAZARDOUS				
Maleic anhydride	Air emissions from 2 storage tanks 820 lbs/yr (1984)	Addition of conservation vent to 1 tank EQ (1980)	?	(1)
Methyl ethyl ketone (MEK)	Air emissions from 8 storage tanks 27,140 lbs/yr (1984)	Addition of floating roofs to all tanks EQ (1975, 1976, 1977)	Current emissions reductions of 239,220 lbs/yr (89.8 percent reduction)	(1)
Methyl isobutyl ketone (MIBK)	Air emissions from 4 storage tanks 12,860 lbs/yr (1984)	Addition of floating roof to 1 tank EQ (1975)	Current emissions reductions of 3,320 lbs/yr (94.9 percent reduction)	(1)
Hazardous solid wastes	G – ? D – ? Q – ? (1984)	General program of waste reduction; details of practices not reported ? (year unknown)	55 percent reduction of hazardous solid wastes from two Exxon plants. Impact at individual plants was not reported.	(8)
LIKELY TO CONTAIN HAZARDOUS CHEMICALS				
Organic chemical process wastes	Wastewaters treated and discharged to Morses Creek Q – ? (1984)	Plant-wide stewardship program to monitor concentrations of organic chemicals in wastestreams OP (early 1970s)	Reduction of organic loading in wastewater by 75 percent	(8)

NOTE: Exxon has installed floating roofs on 16 storage tanks which are used for storing many different chemicals. The chemical listed in the first column is the chemical currently stored in the tank. The dates in the third column indicate the year the floating roof was installed on the tank.

Key to Waste Reduction Changes
PR– Product
PS– Process
EQ– Equipment
CH– Chemical Substitution
OP– Operational

WASTE REDUCTION

EXXON

Type of Waste	G-Generation D-Disposal Q-Quantity (year)	Waste Reduction Practice	Results	Sources of Information
Hydrocarbons	Air emissions from 6 storage tanks 13,760 lbs/yr (1984)	Addition of floating roofs to 3 tanks EQ (1975, 1976)	Current emissions reductions of 174,420 lbs/yr (93.2 percent reduction)	(1)
		Addition of conservation vents to 3 tanks EQ (1983)	Current emissions reductions of 2,400 lbs/yr	(1)
NON-HAZARDOUS				
Acetone	Air emissions from 5 storage tanks 14,580 lbs/yr (1984)	Addition of floating roofs to all tanks EQ (1975, 1976, 1979)	Current emissions reductions of 237,980 lbs/yr (94.2 percent reduction)	(1)
Dodecyl phenol	Air emissions from 8 process sources 680 lbs/yr (1984)	Addition of conservation vents to 4 process sources EQ (1976)	?	(1)
	Air emissions from 1 storage tank 478 lbs/yr (1984)	Addition of conservation vent EQ (1976)	Current emissions reductions of 236 lbs/yr (33 percent reduction)	(1)
	Air emissions from 1 storage tank 1.8 lbs/yr (1984)	Addition of conservation vent EQ (1981)	?	(1)
Nonyl phenol	Air emissions from 1 storage tank 8.96 lbs/yr (1984)	Addition of conservation vent EQ (1980)	30 percent reduction of emissions	(1)
Secondary butyl ether	Air emissions from 1 storage tank 350 lbs/yr (1984)	Addition of floating roof EQ (1982)	Current emissions reductions of 2,630 lbs/yr	(1)

Key to Waste Reduction Changes
PR- Product
PS- Process
EQ- Equipment
CH- Chemical Substitution
OP- Operational

WASTE GENERATION & DISPOSAL

EXXON

Type of Waste ● Indicates Wastes Affected by Reduction	Generation or Disposal	Quantity of Waste	Sources of Information
Air Emissions			
HAZARDOUS			
● Maleic anhydride	Emissions from 2 storage tanks	820 lbs/yr (1984)	(1)
	Plant-wide emissions	1,300 lbs/yr (1978)	(2)
● Methyl ethyl ketone (MEK)	Emissions from 8 storage tanks	27,140 lbs/yr (1984)	(1)
● Methyl isobutyl ketone (MIBK)	Emissions from 4 storage tanks*	12,860 lbs/yr (1984)	(1)
Phenol	Emissions from 5 process sources	3.6 lbs/yr (1984)	(1)
	Emissions from 3 storage tanks	700-800 lbs/yr (1984)	(1)
	Plant-wide emissions	2,230 lbs/yr (1978)	(2)
MAY CONTAIN HAZARDOUS CHEMICALS			
● Hydrocarbons	Emissions from 6 storage tanks	13,760 lbs/yr (1984)	(1)
	Emissions from 1 process source	17,520 lbs/yr (1984)	(1)
NON-HAZARDOUS			
● Acetone	Emissions from 5 storage tanks	14,580 lbs/yr (1984)	(1)
● Dodecyl phenol	Emissions from 8 process sources	680 lbs/yr (1984)	(1)
	Emissions from 3 storage tanks	4,389 lbs/yr (1984)	(1)
●	Emissions from 1 storage tank	478 lbs/yr (1984)	(1)
●	"	1.8 lbs/yr (1984)	(1)

* Emissions from a fifth tank are exempt from reporting requirements because the tank has been in continuous use since before 1968, when the requirements became effective. No information is available on emissions from this tank.

EXXON
WASTE GENERATION & DISPOSAL

Type of Waste ● Indicates Wastes Affected by Reduction	Generation or Disposal	Quantity of Waste	Sources of Information
Air Emissions			
NON-HAZARDOUS (cont.)			
● Nonyl phenol	Emissions from 1 storage tank	8.96 lbs/yr (1984)	(1)
● Secondary butyl ether	"	350 lbs/yr (1984)	(1)
	Emissions from 1 process source	116 lbs/yr (1984)	(1)
Wastewater Discharges			
● Organic chemical process wastes, including: HAZARDOUS	Treated and discharged to Morses Creek	?* (1984)	(3,4)
Phenol	"	2,200 lbs/yr (1978)	(2)
MAY CONTAIN HAZARDOUS CHEMICALS Cooling water	Discharged to Morses Creek	?* (1984)	(3,4)
NON-HAZARDOUS Acidic wastewaters	Discharged to an on-site surface impoundment and treated	21.6 million lbs/yr (1981)	(5)

* Since all wastewater discharge reporting for the Bayway operations combines information from its refinery and its chemical plant, amounts and types of discharges for the chemical plant alone are not specified.

EXXON
WASTE GENERATION & DISPOSAL

●	Type of Waste Indicates Wastes Affected by Reduction	Generation or Disposal	Quantity of Waste	Sources of Information
	Solid Wastes			
	HAZARDOUS			
	Corrosive liquids containing Plastigon	Incinerated at Rollins Environmental Services, Bridgeport, NJ	17,740 lbs/yr (1981)	(6)
●	Hazardous wastes*	?	? (1984)	(8)
	Maleic anhydride	Disposed of in an off-site landfill at Newco, Inc., Niagara, NY	28,000 lbs/yr (1978)	(2)
	Maleic anhydride and phenol wastes	Sent off-site for disposal at Browning Ferris Industries, Elizabeth, NJ and CECOS/CER Co., Williamsburg, OH	18,920 lbs/yr (1981)	(6)
	Phenol wastes	Sent off-site for disposal at Browning Ferris Industries, Elizabeth, NJ and SCA Chemicals, Newark, NJ	32,780 lbs/yr (1982)	(7)
	Solvents, halogenated	Incinerated at Rollins Environmental Services, Bridgeport, NJ	12,100 lbs/yr (1982)	(7)
	NON-HAZARDOUS			
	Aluminum chloride, phosphorus pentasulfide, sulfuric acid coke and polymerized acrylic acid wastes	Sent off-site for disposal at CECOS International, Niagara Falls, NY; CECOS/CER Co., Williamsburg, OH; and Browning Ferris Industries, Elizabeth, NJ	253,460 lbs/yr (1981)	(6)
	Cyclodiene dimer, dicyclopentadiene and methylcyclopentadiene wastes	Incinerated at Rollins Environmental Services, Bridgeport, NJ	2,352 gals/yr (1982)	(7)
	Isophorone wastes	Sent off-site for disposal at Solvents Recovery Service of New Jersey, Linden, NJ	422 lbs/yr (1981)	(6)

* Includes the solid wastes listed here which are considered hazardous by INFORM's criteria, as well as other wastes legally classified as hazardous by the State of New Jersey.

EXXON

WASTE GENERATION & DISPOSAL

Type of Waste ● Indicates Wastes Affected by Reduction	Generation or Disposal	Quantity of Waste	Sources of Information
Solid Wastes			
NON-HAZARDOUS (cont.)			
Phosphorus pentasulfide waste and other corrosive solids	Sent off-site for disposal at CECOS International, Niagara Falls, NY	149,080 lbs/yr (1982)	(7)
Vinyl acetate wastes	Incinerated at Rollins Environmental Services, Bridgeport, NJ	42,200 lbs/yr (1982)	(7)

Fibrec, Inc.

1154 Howard Street
San Francisco, California 94103
(415) 431-8214

Fibrec, Inc. is a small dyestuffs distributor in downtown San Francisco. A total absence of information on this plant in public records precluded analysis of its waste generation and handling methods. The company wrote INFORM that it receives drums of dyes and then repacks them in smaller containers for sale and that it handles no toxic or hazardous chemicals. There is no indication from available information that Fibrec has adopted any waste reduction techniques.

Fisher Scientific Company

Allied Corporation
1 Reagent Lane
Fair Lawn, New Jersey 07410
(201) 796-7100

SUMMARY

Fisher Scientific's plant (Fair Lawn, New Jersey) manufactures reagent chemicals through batch processes and then packages and distributes these chemicals to clinical and industrial laboratories. The plant receives damaged, expired or discontinued products from other Fisher facilities to determine if they are reusable and either resells or disposes of these materials. Numerous solvents are routinely used at the plant, including **benzene, chloroform** and **trichloroethylene.** Wastewater containing heavy metals is treated on-site and then discharged to the Passaic Valley Sewerage Commission. Solid wastes are stored and treated on-site and sent off-site for disposal. Flammable solvents are burned off-site as a supplemental fuel. Fisher has not adopted any waste reduction techniques because the small volumes and constant variety of materials handled at the plant have not readily lent themselves to waste reduction practices. However, a changing focus at the plant, including the hiring of a materials manager, may lead to waste reduction in the future.

PLANT OPERATIONS

Fisher Scientific Company, owned by Allied Corporation, operates a plant in Fair Lawn, a city in northeastern

NOTE: Chemicals in **boldface** type are hazardous, according to INFORM's criteria (see Appendix A).

New Jersey, less than fifteen miles from New York City and Newark. Established in 1955, this 325-employee facility is Fisher's main manufacturing plant in the U.S.

Fisher was purchased by the Allied Corporation in late 1981 and has since been merged into the health and scientific products unit of the company's Industrial and Technology sector. Allied's five major business areas are chemicals, oil and gas, automotive products, aerospace products and industrial and technology products (which include electronics and health care products). Fisher manufactures laboratory and biomedical instruments, appliances, and supplies, and reagent chemicals (used for experiments and analysis in laboratories). Sales for the Industrial and Technology Sector to which Fisher belongs were $1.7 billion in 1983; sales for the entire corporation were over $10 billion in 1983.

The Fisher plant is a packaging, formulating and manufacturing operation that breaks bulk quantities of a wide variety of chemicals into smaller units for sale primarily to clinical and industrial laboratories. Fisher told INFORM that this plant uses purification, mixing and formulating steps in its operations, but carries on no chemical synthesis. The plant's large distribution operations, serving primarily scientific and educational customers, are also used by outside companies. The Fair Lawn plant also receives damaged, expired or discontinued hazardous and non-hazardous products from other Fisher facilities to determine if they are reusable or if they are wastes. Chemicals received at Fair Lawn are either tested and resold as is, reworked through distillation or other purification steps before distribution, or disposed of. The laboratory grade reagent chemicals and solutions manufactured at the plant are produced in batch operations.

WASTE GENERATION AND WASTE REDUCTION

AIR EMISSIONS

Generation

According to the New Jersey Department of Environmental Protection's air permit files, **trichloroethylene** emissions from a tank which is part of a closed-loop (recirculating) process cooling system are 1.68 pounds per year. Air emissions from two tanks whose content varies over time are 2,200 pounds per year of one or more of the following substances: petroleum ether, **acetonitrile**, tetrahydrofuran, **methanol** and hexane.

WASTEWATER DISCHARGES

Generation

Fisher currently discharges an average of 72,250 gallons per day of wastewater which contains heavy metals and other process chemicals to the Passaic Valley Sewerage Commission (PVSC), according to information supplied by Fisher to INFORM. (PVSC wastewater discharge records for 1981 indicate that Fisher formerly discharged 222,000 gallons per day of wastewater to the PVSC.)

Wastewater contaminants originate primarily from equipment rinses in between batch operations. Fisher treats various wastestreams prior to discharge to the sewer and **cyanide** bearing wastes are oxidized and neutralized prior to discharge. Aqueous water-soluble dyes were formerly passed through an activated carbon column before being discharged, until PVSC allowed Fisher to discontinue this treatment step.

SOLID WASTES

Generation

Fisher stores and treats solid wastes on-site and sends wastes off-site for disposal. In its 1982 Hazardous

Waste Facility Annual Report, Fisher described its wastes as "essentially pure materials that must be discharged due to internal quality control procedures." Its wastes are comprised of lab reagent manufacturing process wastes, expired reagents, damaged or contaminated reagents and laboratory wastes, according to a 1981 RCRA inspection report.

Mixed flammable solvents are segregated and sent off-site to be burned by Marisol, Inc. in Middlesex, New Jersey, as a fuel supplement. These solvents, sometimes contaminated with heavy metals, were formerly burned as a fuel supplement at Chemical Pollution Control, Inc., Bay Shore, New York. Fisher reported to INFORM that heavy metal-contaminated solvent wastes are not currently generated. Federal and state environmental files gave no indication of what other off-site disposal methods are used to handle Fisher wastes. **Mercury**-bearing and **mercuric nitrate** wastes were once treated in a system installed at Fisher around 1981, according to the RCRA inspection report of that year. However, Fisher told INFORM that it does not currently treat its **mercury** wastes, which are typically sent off-site for disposal. Copper sulfate wastes, once treated at Fisher in an electroplating recovery system, are now sold for recovery. Among the largest hazardous solid waste streams generated at Fisher are **chloroform**-containing wastes, 177,400 pounds per year and halogenated solvents, 209,000 pounds per year. In 1984, Fisher also generated 544,000 pounds of non-halogenated flammable organic solvents, its largest wastestream.

OBSERVATIONS

Fisher told INFORM that there had not been any waste reduction efforts undertaken at the Fair Lawn plant because their operations are not suited to traditional waste minimization schemes. Non-traditional aspects of Fisher's operations include: the small quantities of chemicals being handled; the need for frequent equipment cleanouts between operations; and the limited shelf life of the materials handled at the plant. Wastes generated at the plant include off-specification products which cannot be resold, expired products, small quantities of products returned by customers, and equipment rinses, which are the largest source of wastewaters.

The absence of large, dedicated production systems has, to date, precluded the traditional sort of manufacturing approach which would enable waste minimization. Fisher reported that recent steps have been taken at the plant to modify its former focus, thus paving the way for waste minimization schemes. After taking over Fisher's operations in 1981, Allied Corporation has significantly altered the structure and orientation of plant management. For instance, a new materials manager has been hired to keep track of the flow of raw materials at the plant. The new manager's responsibilities include reducing inventory, improving yields, and minimizing wastage, for example, by making sure that materials leave the plant before they expire. Computerization of material flow records is now being carried out at branch and district corporate levels to better manage handling procedures and to minimize material expiration.

Fisher has also installed a process management program at the plant, using an outside consultant, which is intended to measure production costs and to better identify inefficiencies or problems (such as a spill), thereby facilitating quicker corrective actions. All of these changes will bear directly upon Fisher's near-term waste management objectives. The plant's goal for 1985 is to identify waste management alternatives, including waste reduction, and to complete the background research for subsequent implementation of these options. This study will serve to give Fisher a clearer picture of waste generation and disposal at the plant and related costs. Plans for 1986 include the actual implementation of any waste reduction measures identified during the study period.

SOURCES OF INFORMATION

SOURCE	LOCATION
1. Air permit files, 1984	New Jersey Department of Environmental Protection (NJDEP) Division of Environmental Quality Newark Field Office 1100 Raymond Boulevard Newark, NJ 07102 (201) 648-2075
2. Passaic Valley Sewerage Commission (PVSC) industrial discharge data, 1981 (in PVSC wastewater discharge files)	U.S. Environmental Protection Agency, (U.S. EPA), Region II Permits Administration Branch 26 Federal Plaza New York, NY 10278 (212) 264-9881
3. Revised RCRA Part A Application, 1983	NJDEP Division of Waste Management 32 East Hanover Street Trenton, NJ 08625 (609) 292-9879
4. Hazardous Waste Facility Annual Report, 1983	NJDEP Division of Waste Management (see #3)
5. Hazardous Waste Facility Annual Report, 1982	NJDEP Division of Waste Management (see #3)

6. Hazardous Waste Manifest Records, 1980*

    NJDEP
    Bureau of Hazardous Waste
    CN027
    Trenton, NJ 08625
    (609) 984-2302

7. Hazardous Waste Facility Annual Report, 1981

    NJDEP
    Division of Waste Management
    (see #3)

8. RCRA Inspection Report, June 1981

    U.S. EPA,
    Region II
    Permits Administration Branch
    (see #2)

9. Information supplied by Fisher to INFORM

* Fisher's internal records do not agree with NJDEP manifest records for 1980. Neither the company nor NJDEP could resolve the discrepancy, although NJDEP acknowledged considerable problems for that year in recording and managing manifest data.

FISHER SCIENTIFIC

WASTE GENERATION & DISPOSAL

Type of Waste ● Indicates Wastes Affected by Reduction	Generation or Disposal	Quantity of Waste	Sources of Information
Air Emissions			
HAZARDOUS Organic chemicals which may include petroleum ether, acetonitrile, tetrahydrofuran, methanol and hexane (content of tanks is variable)	Emissions from 2 tanks	2,200 lbs/yr (max.) (1984)	(1,9)
Trichloroethylene	Emissions from 1 tank	1.7 lbs/yr (1984)	(1)
Wastewater Discharges			
HAZARDOUS Wastewater containing processing chemicals including:	Discharged to the Passaic Valley Sewerage Commission	72,250 gals/day (1984)	(9)
	"	222,000 gals/day (1981)	(2)
Copper	"	.659 mg/l (1981)	(2)
Silver	"	.009 mg/l (1981)	(2)
Zinc	"	.689 mg/l (1981)	(2)

334

FISHER SCIENTIFIC

WASTE GENERATION & DISPOSAL

● Type of Waste Indicates Wastes Affected by Reduction	Generation or Disposal	Quantity of Waste	Sources of Information
Solid Wastes			
HAZARDOUS			
Acetonitrile	Stored on-site	2,200 lbs/yr (1983)	(3)
Benzene	"	440 lbs/yr (1983)	(3)
Carbon disulfide	Sent off-site for disposal at All County Environmental, Edgewater, NJ	250 lbs/yr (1891)	(7)
Chloroform waste	Stored on-site and sent off-site for disposal	177,400 lbs/yr (1983)	(4)
1,2-Dichloroethane	Stored on-site	330 lbs/yr (1983)	(3)
Dichloromethane	"	22,000 lbs/yr (1983)	(3)
Dichloromethane and dirt	Stored on-site and sent off-site for disposal	300 lbs/yr (1983)	(4)
1,4-Dioxane	Stored on-site	2,200 lbs/yr (1983)	(3)
Ethyl ether	"	4,400 lbs/yr (1983)	(3)
Formaldehyde	Stored and treated on-site	4,400 lbs/yr (1983)	(3)
Formic acid	"	330 lbs/yr (1983)	(3)
Isobutyl alcohol	Stored on-site	220 lbs/yr (1983)	(3)
Methanol	"	22,000 lbs/yr (1983)	(3)
Methyl ethyl ketone	"	330 lbs/yr (1983)	(3)
Methyl isobutyl ketone	"	275 lbs/yr (1983)	(3)
Phenol wastes	"	2,200 lbs/yr (1983)	(3)
	?	2,326 lbs/yr (1982)	(5)
	Stored on-site and sent off-site for disposal	353 lbs/yr (1983)	(4)

FISHER SCIENTIFIC

WASTE GENERATION & DISPOSAL

Type of Waste ● Indicates Wastes Affected by Reduction	Generation or Disposal	Quantity of Waste	Sources of Information
Solid Wastes			
HAZARDOUS (cont.)			
1,1,2,2-Tetrachloroethane	Stored on-site	330 lbs/yr (1983)	(3)
Tetrachloroethene	"	220 lbs/yr (1983)	(3)
Tetrachloromethane	"	330 lbs/yr (1983)	(3)
Toluene	"	2,200 lbs/yr (1983)	(3)
1,1,1-Trichloroethane	"	330 lbs/yr (1983)	(3)
1,1,2-Trichloroethane	"	352 lbs/yr (1983)	(3)
Trichloroethylene	Sent off-site for disposal	27,000 lbs/yr (1980)	(6)
	Stored on-site	550 lbs/yr (1983)	(3)
	Stored on-site and sent off-site for disposal	500 lbs/yr (1983)	(4)
2,4,6-Trichlorophenol	Stored on-site	4.4 lbs/yr (1983)	(3)
	Stored on-site and sent off-site for disposal	300 lbs/yr (1983)	(4)
Trinitrobenzene	Stored on-site	6,600 lbs/yr (1983)	(3)
Solvents, contaminated with cadmium, chromium, lead, mercury or silver*	Burned as fuel supplement by Chemical Pollution Control, Bay Shore, NY**	15,500 gals/yr (1981)	(7)

* Fisher told INFORM that solvents containing metals are rarely generated.
**Solvents are now sent to Marisol, Inc., Middlesex, NJ to be burned as fuel.

336

FISHER SCIENTIFIC

WASTE GENERATION & DISPOSAL

Type of Waste ● Indicates Wastes Affected by Reduction	Generation or Disposal	Quantity of Waste	Sources of Information
Solid Wastes			
HAZARDOUS (cont.)			
Solvents, halogenated	Stored on-site	44,000 lbs/yr (1983)	(3)
	Stored on-site and sent off-site for disposal	209,000 lbs/yr (1983)	(4)
	?	40,000 lbs/yr (1982)	(5)
	Sent off-site for disposal at Advanced Environmental Technology Corporation, Mount Olive, NJ	43,900 lbs/yr (1981)	(7)
	Sent off-site for disposal	16,400 lbs/yr (1980)	(6)
Solvents, non-halogenated	Stored on-site	220 lbs/yr (1983)	(3)
Toxic wastes, unspecified	Stored and treated on-site	220 lbs/yr (1983)	(3)
Arsenic waste	Stored on-site and sent off-site for disposal	121 lbs/yr (1983)	(4)
Barium waste	"	36,100 lbs/yr (1983)	(4)
Chromium sulfate	?	400 gals/yr (1982)	(5)
Cyanides	Sent off-site for disposal	400 lbs/yr (1980)	(6)
	Stored and treated on-site	220 lbs/yr (1983)	(3)
	Sent off-site for disposal at Advanced Environmental Technology Corporation, Mount Olive, NJ	400 lbs/yr (1981)	(7)
	Stored on-site and sent off-site for disposal	662 lbs/yr (1983)	(4)
Cyanogen bromide	Stored on-site	220 lbs/yr (1983)	(3)
Hydrofluoric acid	Stored and treated on-site	220 lbs/yr (1983)	(3)

FISHER SCIENTIFIC

WASTE GENERATION & DISPOSAL

●	Type of Waste Indicates Wastes Affected by Reduction	Generation or Disposal	Quantity of Waste	Sources of Information
	Solid Wastes			
	HAZARDOUS (cont.)			
	Lead waste	Stored on-site and sent off-site for disposal	1,000 lbs/yr (1983)	(4)
	Mercury waste	Stored on-site	110 lbs/yr (1983)	(3)
		Treated on-site	24,255 gals/yr* (1982)	(5)
		Stored on-site and sent off-site for disposal	384 lbs/yr (1983)	(4)
	Potassium cyanide	Stored and treated on-site	110 lbs/yr (1983)	(4)
		Stored on-site and sent off-site for disposal	200 lbs/yr (1983)	(3)
	Sodium cyanide	Stored and treated on-site	220 lbs/yr (1983)	(3)
	LIKELY TO CONTAIN HAZARDOUS CHEMICALS			
	Chlorinated organic resin	Sent off-site for disposal	900 lbs/yr (1980)	(6)
	Solvents, halogenated	"	94,500 lbs/yr (1980)	(6)
	MAY CONTAIN HAZARDOUS CHEMICALS			
	Packed lab chemicals	"	52,400 lbs/yr (1980)	(6)
		"	1,126 lbs/yr (1983)	(4)
	Poison/pesticide waste, unspecified	"	37,600 lbs/yr (1983)	(4)

* Fisher explains that its mercury treatment system was only used one time; mercury wastes are typically sent off-site for disposal.

FISHER
SCIENTIFIC

WASTE GENERATION & DISPOSAL

Type of Waste ● Indicates Wastes Affected by Reduction	Generation or Disposal	Quantity of Waste	Sources of Information
Solid Wastes			
MAY CONTAIN HAZARDOUS CHEMICALS (cont.)			
Solvents, non-halogenated flammable	Sent off-site for disposal	544,000 lbs/yr (1984)	(9)
		410,000 lbs/yr (1983)	(9)
		459,000 (1982)	(9)
		1,024,200 lbs/yr* (1980)	(6)
Waste chemicals, unspecified	Sent off-site for disposal	12,000 lbs/yr (1980)	(6)
	Stored on-site and sent off-site for disposal	24,800 lbs/yr (1983)	(4)
Waste poison solid, unspecified	?	153 lbs/yr (1982)	(5)
NON-HAZARDOUS			
Acetone	Stored on-site	1,100 lbs/yr (1983)	(3)
Cyclohexane	"	440 lbs/yr (1983)	(3)
Ethyl acetate	"	220 lbs/yr (1983)	(3)
Furan	"	330 lbs/yr (1983)	(3)
Fufural	"	154 lbs/yr (1983)	(3)
Menthonol in water	Sent off-site for disposal	8,200 lbs/yr (1980)	(6)
Tetrahydrofuran	Stored on-site	6,600 lbs/yr (1983)	(3)
Trichloroacetic acid	?	595 lbs/yr (1982)	(5)
Acid solution	Sent off-site for disposal	4,650 lbs/yr (1980)	(6)

* Fisher told INFORM that its records show this figure to be 368,000 lbs/yr.

FISHER SCIENTIFIC

WASTE GENERATION & DISPOSAL

Type of Waste ● Indicates Wastes Affected by Reduction	Generation or Disposal	Quantity of Waste	Sources of Information
Solid Wastes			
NON-HAZARDOUS (cont.)			
Alkaline solution	Sent off-site for disposal	3,400 lbs/yr (1980)	(6)
Corrosive wastes, unspecified	Stored and treated on-site	24,200 lbs/yr (1983)	(3)
	?	19,561 lbs/yr (1982)	(5)
	Stored on-site and sent off-site for disposal	30,600 lbs/yr (1983)	(4)
	Stored on-site	1,350 lbs/yr (1983)	(4)
Degreaser	Sent off-site for disposal	10,800 lbs/yr (1980)	(6)
Fatty acid, fester alcohol and glycol	"	1,350 lbs/yr (1980)	(6)
Flammable wastes, unspecified (including oxidizers)	"	2,250 lbs/yr (1980)	(6)
	Stored on-site	440,000 lbs/yr (1983)	(3)
	?	57,420 gals/yr (1982)	(5)
	Stored on-site and sent off-site for disposal	23,139 lbs/yr (1983)	(4)
	Stored on-site	23,500 lbs/yr (1983)	(4)
Lacramators, amines and mercaptan	Sent off-site for disposal	1,305 lbs/yr (1980)	(6)
Oil and oil sludges	"	1,350 lbs/yr (1980)	(6)
Reactive wastes, unspecified	Stored and treated on-site	220 lbs/yr (1983)	(3)
Solvents, flammable	Burned as fuel supplement at Chemical Pollution Control Bay Shore, NY*	38,500 gals/yr (1981)	(7)

* Solvents are now sent to Marisol, Inc., Middlesex, NJ to be burned as fuel.

FISHER
SCIENTIFIC

WASTE GENERATION & DISPOSAL

● Type of Waste Indicates Wastes Affected by Reduction	Generation or Disposal	Quantity of Waste	Sources of Information
Solid Wastes			
NON HAZARDOUS (cont.)			
Solvents, non-halogenated	Stored on-site	220 lbs/yr (1983)	(3)
Aluminum chloride	?	300 lbs/yr (1982)	(5)
Ammonium sulfide	?	150 lbs/yr (1982)	(5)
Potassium ferric cyanide	Sent off-site for disposal at Advanced Environmental Technology Corporation, Mount Olive, NJ	600 lbs/yr (1981)	(7)
Sodium hydrogen sulfite	Stored on-site and sent off-site for disposal	125 lbs/yr (1983)	(4)

341

Frank Enterprises, Inc.

700 Rose Lane
Columbus, Ohio 43219
(614) 253-5519

SUMMARY

Frank Enterprises (Columbus, Ohio) is a small custom manufacturer of specialty organic chemicals. Its products, made in bulk quantities, vary with each different contract. Raw materials have included **methanol, nitrobenzene, zinc** and **methyl isobutyl ketone**. Wastes, which vary with the raw materials used, are either discharged as wastewater to the Columbus Southerly municipal sewage treatment plant or to the atmosphere as air emissions. There is no indication from available information that Frank has adopted any waste reduction techniques.

PLANT OPERATIONS

Frank Enterprises, located in Columbus, Ohio, was established in 1970, employs five people and produces $500,000 of products annually. Frank custom manufactures specialty organic chemicals in quantities ranging from 100 to 100,000 pounds, on a contract-by-contract basis. Consequently, both raw materials and products vary with each new order. For example, environmental files indicate that Frank has manufactured cupferron (a polymerization inhibitor), butyl nitrite, 1-nonene, isoamyl nitrite and isoeugenol acetate over the past seven years.

NOTE: Chemicals in **boldface** type are hazardous, according to INFORM's criteria (see Appendix A).

Its batch operations utilize four chemical reactors and two tanks. Operating schedules vary with each new chemical being produced.

In 1978, Frank produced 75,000 pounds of cupferron, using **nitrobenzene, zinc**, ammonium formate, water, ammonia, butyl nitrite and **methyl isobutyl ketone (MIBK)** as raw materials, according to Ohio Environmental Agency (OEPA) files. Eighty thousand pounds of butyl nitrite were also produced that year, using butyl alcohol, sodium nitrite, sulfuric acid, water and sodium bicarbonate as raw materials. **Methyl isobutyl ketone**, a solvent used in cupferron manufacture, was recovered at a rate of 336,000 pounds per year.

In 1982, Frank's products included 20,000 pounds of isoeugenol acetate, 75 pounds of 1-nonene and 4,000 pounds of isoamyl nitrite, according to OEPA air permit files. Raw materials for isoeugenol acetate manufacture include isoeugenol, acetic anhydride and **methanol**. The chemical, 1-nonene, was produced through the separation and purification of mixed alkenes. Raw materials for isoamyl nitrite include isoamyl alcohol, sodium nitrite, sulfuric acid, water and sodium bicarbonate.

WASTE GENERATION AND WASTE REDUCTION

AIR EMISSIONS

Generation

Air emissions from Frank's four reactors and two tanks, primarily solvent vapors, have not been measured by the company, according to OEPA air permit files. All six pieces of equipment were granted "T-status" by OEPA in June 1982, meaning that they no longer require permits since the equipment emissions are considered minimal by the Agency.

WASTEWATER DISCHARGES

Generation

Frank discharges 3,400 gallons per day of process and other wastewaters to the Columbus Southerly municipal sewage treatment plant, according to OEPA records. Wastes are discharged directly to the sewer, via plant drains. In the past, wastes also migrated off-site through a storm conduit and into a ditch which eventually flowed into the sewer.

Following an October 1978 odor complaint investigation by the OEPA, Frank was ordered to cease discharge of volatile, hazardous, or flammable materials to the sewer (either directly or via the storm conduit). Based on observations made during this investigation, the City of Columbus Division of Sewerage and Drainage suspected that Frank was discharging drums of stored chemicals into the sewer. This practice was also prohibited, based on the possibility that these drums contained flammable or hazardous substances.

SOLID WASTES

Generation

Little information is available on solid waste generation or disposal at Frank. Some by-products are recycled or recovered, such as the in-process recovery of **MIBK** used in cupferron manufacture. The company notified the U.S. Environmental Protection Agency on its 1980 RCRA Notification Form that its solid wastes may contain the following chemicals: **acrylonitrile, methanol, quinones, toluene** and **cyanides.**

SOURCES OF INFORMATION

SOURCE	LOCATION
1. Air Permit files, 1984	Ohio Environmental Protection Agency (OEPA) Central District Office Air Division 361 East Broad Street Columbus, OH 43215 (614) 462-8266
2. Columbus Southerly Sewer Department "Industrial Waste Survey," 1982	OEPA Pretreatment Division 361 East Broad Street Columbus, OH 43215 (614) 462-6787
3. Ohio Department of Health water quality data, 1978	OEPA Emergency Response Group 361 East Broad Street Columbus, OH 43215 (615) 466-6542
4. Ohio Environmental Protection Agency Complaint Investigation Report #10-25-1615, October 1978	OEPA (see #2)
5. City of Columbus, Division of Sewerage and Drainage, letter to Frank Enterprises, October 1978	OEPA (see #2)
6. RCRA Notification form, 1980	U.S. Environmental Protection Agency, Region V Waste Management Branch 230 South Dearborn Street Chicago, IL 60604 (312) 886-6134

FRANK ENTERPRISES

WASTE GENERATION & DISPOSAL

Type of Waste ● Indicates Wastes Affected by Reduction	Generation or Disposal	Quantity of Waste	Sources of Information
Air Emissions			
HAZARDOUS			
Methyl isobutyl ketone	Emissions from 3 process reactors	Not determined by company (1984)	(1)
	Emissions from 1 process source	2.5 lbs/hour (1984)	(1)
Solvent vapors	Emissions from 4 process reactors and 2 tanks	?* (1984)	(1)
Wastewater Discharges			
Process wastewater, cooling water and floor washings, including:	Discharged to the Columbus Southerly sewage treatment plant, Columbus, OH	3,400 gals/day (1978, 1982)	(2,3,4,5)
HAZARDOUS			
Methyl isobutyl ketone	"	8,399 ppm (1978)	(3)
Nitrobenzene	"	140 ppm (1978)	(3)
NON-HAZARDOUS			
Acetone	"	63 ppm (1978)	(3)
1-Butanol	"	520 ppm (1978)	(3)
Butyl nitrite	"	? (1978)	(5)

* These are "T-Status" emissions sources. They are considered minor by the Ohio EPA, which does not require submission of their emission rates.

FRANK ENTERPRISES

WASTE GENERATION & DISPOSAL

Type of Waste ● Indicates Wastes Affected by Reduction	Generation or Disposal	Quantity of Waste	Sources of Information
Solid Wastes			
HAZARDOUS			
Acrylonitrile	?	? (1980)	(6)
Methanol	?	? (1980)	(6)
Quinones	?	? (1980)	(6)
Toluene	?	? (1980)	(6)
Residue from methyl isobutyl ketone recovery, including methyl isobutyl ketone, zinc sulfate and sodium sulfate	?	5 lbs/hour (1984)	(1)
Cyanides	?	? (1980)	(6)
MAY CONTAIN HAZARDOUS CHEMICALS			
Residue from 1-nonene manufacture	?	.20 lbs/hour (1984)	(1)
NON-HAZARDOUS			
Acetic acid from isoeugenol acetate manufacture	Recycled or disposed of	? (1984)	(1)
Alkenes from 1-nonene manufacture	?	? (1984)	(1)

347

J.E. Halma Company, Inc.

91 Dell Glenn Avenue
Lodi, New Jersey 07644
(201) 772-4464

J.E. Halma Company is located in Lodi, a city in northeastern New Jersey, about ten miles west of New York City. This seven-employee batch-manufacturing plant produces solvents, etchants, acids and cleaners, which are used by semiconductor and transistor manufacturers. A total absence of information on this plant from public records precluded analysis of its waste generation and handling methods. There is no indication from available information that Halma has adopted any waste reduction techniques.

Halma qualified its lack of waste reduction efforts by explaining, in a letter to INFORM, that its manufacturing operations simply generate no wastes. Since all of its products, once "completed in the master batch tank, are packaged up into one, five or 55 gallon containers and shipped to our customers," Halma explained that "the problem of waste generation and disposal therefore rests with our customers."

International Flavors and Fragrances Inc.

800 Rose Lane
Union Beach, New Jersey 07735
(201) 264-4500

SUMMARY

International Flavors and Fragrances' (IFF) plant (Union Beach, New Jersey) manufactures about 800 synthetic perfume and fragrance chemical ingredients. Raw materials used by IFF include **acetaldehyde, acrolein, benzene, benzyl chloride, carbon tetrachloride, chloroform, crotonaldehyde, diethyl phthalate, ethylene dichloride, formaldehyde, methyl bromide, methylene chloride, toluene** and **xylene.** Air emissions include **acrolein, benzene, carbon tetrachloride, chloroform, ethylene dichloride** and **toluene.** IFF annually discharges close to 80 million gallons of process and other wastewaters contaminated by more than 35 hazardous organic chemicals and metals to the Bayshore Regional Sewage Authority. New Jersey Department of Environmental Protection files also indicate that IFF discharges process wastewater to groundwater from three surface impoundments used in its treatment operations. Solid wastes are treated and stored on-site, sent off-site for disposal, incinerated on-site or burned on-site in boilers as supplemental fuel. Biological chemical sludge produced from wastewater treatment operations is one of the largest solid waste streams sent off-site for disposal. There is no indication that IFF has adopted any waste reduction techniques.

NOTE: Chemicals in **boldface** type are hazardous, according to INFORM's criteria (see Appendix A).

PLANT OPERATIONS

International Flavors and Fragrances (IFF) operates a plant in Union Beach, a city on the New Jersey coast, twenty miles south of Newark. Established in 1952, this plant employs 375 people. In addition to its manufacturing plant, IFF also operates a Research and Development Center in Union Beach, involved in flavor and fragrance chemical development, and a compounding plant in nearby Hazlet, New Jersey, which mixes and blends organic materials to form fragrance compounds.

IFF manufactures flavor and fragrance products used in a variety of consumer goods. Fragrance products, which account for two-thirds of its business, are sold to perfume, cosmetics, soap and detergent manufacturers. Flavor products, which comprise the remainder of its business, are sold to prepared foods, beverage, dairy foods, pharmaceutical, confectionery and tobacco product manufacturers. The Union Beach facility is one of the company's five U.S. manufacturing plants. Sales for IFF were $461 million in 1983.

This plant manufactures about 800 different organic chemical compounds, most of which are synthetic perfume and fragrance chemical ingredients. U.S. EPA air permit files indicate that IFF carries on mostly batch processes but that 200 of its chemical products are produced on a semi-continuous basis. As many as 500 different chemicals might be produced by the plant in any given year.

IFF reported to the New Jersey Department of Environmental Protection (NJDEP) in the 1978 Industrial Survey use of the following raw materials: **acetaldehyde,** a reactant; **acrolein,** a reactant; **benzene,** used as a solvent and a reactant; **benzyl chloride,** a reactant; **carbon tetrachloride,** a solvent; **chloroform,** a solvent; **crotonaldehyde,** a reactant; **diethyl phthalate,** a diluent; **ethylene dichloride,** a solvent; **formaldehyde,** a reactant; **methyl bromide,** and **methylene chloride,** preparants; propylene oxide, a reactant; **toluene,** used as a solvent (major use) and as a reactant (minor use);

and **xylene,** a diluent. Data on the quantity of these materials in use were treated as "Classified Business Information" by NJDEP (at the request of IFF) and was not made available to INFORM.

WASTE GENERATION AND WASTE REDUCTION

AIR EMISSIONS

Generation

IFF's air emissions in 1978 included 972 pounds of **toluene** and smaller quantities of **acrolein, carbon tetrachloride** and **ethylene dichloride.** NJDEP's Air Pollution Emissions Data System for 1982 indicates that the plant's emissions also include **benzene** and **chloroform.**

WASTEWATER DISCHARGES

Generation

IFF discharges 79,843,750 gallons per year of wastewater to the Bayshore Regional Sewage Authority (BRSA), according to its 1982 indirect wastewater discharge permit application to NJDEP. This total includes 55,845,000 gallons of process wastewater and 16,060,000 gallons of boiler and condenser water. A 1980 Bioassay Study of the IFF plant conducted by NJDEP indicated that its wastewater is treated prior to discharge to BRSA. Sludges produced from wastewater treatment are sent off-site for disposal (see "Solid Wastes" section). Wastewater discharged to BRSA is contaminated by more than 35 hazardous organic chemicals and metals, including, **benzene, ethylene dichloride, ethylhexyl phthalate, phenol, 1,1,2,2-tetrachloroethane, zinc** and **chromium.**

IFF's Groundwater Discharge Permit, issued by the NJDEP in 1983, indicates that IFF discharges 12,000 to 15,000 gallons per day of process wastewater to groundwater from three surface impoundments used in its treatment operations: an equalization basin, an aeration basin and a sludge lagoon. Groundwater which receives this discharge ultimately flows into the Raritan Bay.

SOLID WASTES

Generation

A variety of organic and inorganic solid wastes produced by IFF are treated and stored on-site, sent off-site for disposal, incinerated on-site or burned on-site in boilers as supplemental fuel. Among the largest wastestreams sent off-site for disposal is the hazardous biological chemical sludge produced by wastewater treatment operations, 32.4 million pounds in 1982 and 17.5 million pounds in 1983, according to IFF's Hazardous Waste Annual Reports. Government environmental files gave no indication of the off-site disposal methods used to handle IFF wastes. Solvent wastes are incinerated at IFF, including 80,000 pounds per year of **toluene,** according to the 1978 New Jersey Industrial Survey. Other waste solvents, including 1.3 million pounds per year of hazardous non-halogenated solvents, are burned by IFF as supplemental fuel in boilers.

SOURCES OF INFORMATION

SOURCE	LOCATION
1. New Jersey Industrial Survey, 1978 data	New Jersey Department of Environmental Protection (NJDEP) Office of Science and Research 190 West State Street Trenton, NJ 08625 (609) 292-6714
2. Computer printout of air emissions in New Jersey, from Air Pollution Emissions Data System (APEDS), 1982	NJDEP Bureau of Air Pollution Control CN027 Trenton, NJ 08625 (609) 633-7994
3. Indirect Wastewater Discharge (Significant Industrial User) Permit Application, 1982	NJDEP Division of Water Resources Pretreatment Section CN029 Trenton, NJ 08625 (609) 292-4860
4. IFF wastewater discharge data in "Preliminary Assessment of the Bayshore Regional Sewage Authority's Treatment Process by ATP Biomass Analysis," prepared by NJDEP, Division of Water Resources, Bureau of Systems Analysis and Wasteload Allocations, 1981	U.S. Environmental Protection Agency (U.S. EPA), Region II Office of Regional Counsel 26 Federal Plaza New York, NY 10278 (212) 264-4430

5. "Bioassay Study: On-Site 96 hour Flow-through, International Flavors and Fragrances, Inc., Union Beach, New Jersey," prepared by the NJDEP, Division of Water Resources, Bureau of Monitoring and Data Management, Biological Services Unit, 1980

    U.S. EPA, Region II
    (see #4)

6. IFF wastewater discharge data in "Biomonitoring Studies, Bayshore Regional Sewerage Authority, Union Beach, New Jersey," prepared by the U.S. EPA, 1981

    U.S. EPA, Region II
    (see #4)

7. Final groundwater discharge permit (filed with the NJDEP Division of Water Resources), 1983

    NJDEP
    Division of Waste Management
    32 East Hanover Street
    Trenton, NJ 08625
    (609) 292-9879

8. "Site Inspection of International Flavors and Fragrances, Union Beach, New Jersey," prepared by Ertec Atlantic, Inc., for the U.S. Environmental Protection Agency (Contract #68-01-6515), 1982

    U.S. EPA, Region II
    (see #4)

9. Hazardous Waste Annual Report, 1983

    NJDEP
    (see #7)

10. Hazardous Waste Annual Report, 1981

    NJDEP
    (see #7)

11. Hazardous Waste Annual   U.S. EPA, Region II
    Report, 1980   Permits Administration
                                  Branch
                                26 Federal Plaza
                                New York, NY 10278
                                (212) 264-9881

12. RCRA Part A Application,   U.S. EPA, Region II
    1980   (see #11)

13. Hazardous Waste Annual   NJDEP
    Report, 1982   (see #7)

WASTE GENERATION & DISPOSAL

Type of Waste ● Indicates Wastes Affected by Reduction	Generation or Disposal	Quantity of Waste	Sources of Information
Air Emissions			
HAZARDOUS			
Acrolein	Plant-wide emissions	18 lbs/yr (1978)	(1)
Benzene	Emissions from 12 sources	14.75 lbs/hour (1982)	(2)
Carbon tetrachloride	Plant-wide emissions	83.7 lbs/yr (1978)	(1)
	Emissions from 2 sources	.10 lbs/hour (1982)	(2)
Chloroform	Emissions from 10 sources	2.67 lbs/hour (1982)	(2)
Ethylene dichloride	Plant-wide emissions	83.7 lbs/yr (1978)	(1)
	Emissions from 1 source	.33 lbs/hour (1982)	(2)
Toluene	Plant-wide emissions	972 lbs/yr (1978)	(1)

WASTE GENERATION & DISPOSAL

Type of Waste ● Indicates Wastes Affected by Reduction	Generation or Disposal	Quantity of Waste	Sources of Information
Wastewater Discharges			
Treated process wastewater including:	Discharged to the Bayshore Regional Sewerage Authority (BRSA)	79.84 million gals/yr: ° 55.84 million gals/yr of process wastewater; ° 16.06 million gals/yr of boiler and condenser water; ° 7.94 million gals/yr of other waters (1982)	(3)
HAZARDOUS			
Benzene	"	284 ug/l (1982)	(3)
		1,200 ug/l (1981)	(6)
		4,940 ug/l (1980)	(5)
Chloroform	"	545 ug/l (1980)	(5)
Cumene	"	203 ppb (1981)	(4)
		44 ug/l (1980)	(5)
Dichlorobenzene, isomers m/p	"	16 ug/l (1981)	(6)
o-Dichlorobenzene	"	409 ppb (1981)	(4)
p-Dichlorobenzene	"	79 ppb (1981)	(4)
Diethyl phthalate	"	225 ug/l (1982)	(3)
		6.1 ug/l (1981)	(6)
Di-n-octyl phthalate	"	2.5 ug/l (1981)	(6)

WASTE GENERATION & DISPOSAL

Type of Waste ● Indicates Wastes Affected by Reduction	Generation or Disposal	Quantity of Waste	Sources of Information
\multicolumn{4}{c}{**Wastewater Discharges**}			
Treated wastewater (cont.) <u>HAZARDOUS</u> (cont.)	Discharged to the Bayshore Regional Sewerage Authority (BRSA)		
Ethyl benzene	"	12 ug/l (1981)	(6)
		30 ug/l (1980)	(5)
Ethylene dichloride	"	1,099 ppb (1981)	(4)
		24 ug/l (1981)	(6)
Ethylhexyl phthalate	"	229 ug/l (1982)	(3)
		2,400 ug/l (1981)	(6)
Isophorone	"	822 ug/l (1982)	(3)
Methylene chloride	"	12 ug/l (1981)	(6)
Phenols	"	0.88 mg/l (1981)	(4)
		13 ug/l (1981)	(6)
		1.03 mg/l (1980)	(5)
1,1,2,2-Tetrachloro-ethane	"	7,751 ug/l (1980)	(5)
Tetrachloroethylene	"	180 ug/l (1980)	(5)
Toluene	"	458 ug/l (1982)	(3)
		270 ug/l (1981)	(6)
		973 ug/l (1980)	(5)
Trichloroethylene	"	539 ug/l (1980)	(5)

WASTE GENERATION & DISPOSAL

Type of Waste ● Indicates Wastes Affected by Reduction	Generation or Disposal	Quantity of Waste	Sources of Information
\multicolumn{4}{Wastewater Discharges}		
Treated wastewater (cont.) HAZARDOUS (cont.)	Discharged to the Bayshore Regional Sewerage Authority (BRSA)		
Xylene	"	35 ug/l (1982)	(3)
m,p-Xylene	"	131 ppb (1981)	(4)
		739 ug/l (1980)	(5)
o-Xylene	"	161 ug/l (1980)	(5)
Antimony	"	64 ppb (1981)	(4)
Arsenic	"	163 ppb (1981)	(4)
		57 ug/l (1980)	(5)
Barium	"	0.31 mg/l (1982)	(3)
		47 ppb (1981)	(4)
Beryllium	"	2 ppb (1981)	(4)
Cadmium	"	1 ppb (1981)	(4)
		5 ug/l (1980)	(5)
Chromium, hexavalent	"	5 ug/l (1980)	(5)
Chromium, total	"	0.32 mg/l (1982)	(3)
		3,271 ppb (1981)	(4)
		476 ug/l (1980)	(5)

WASTE GENERATION & DISPOSAL

Type of Waste ● Indicates Wastes Affected by Reduction	Generation or Disposal	Quantity of Waste	Sources of Information
Wastewater Discharges			
Treated wastewater (cont.) HAZARDOUS (cont.) Copper	Discharged to the Bayshore Regional Sewerage Authority (BRSA) "	1,883 ppb (1981)	(4)
		1,636 ug/l (1980)	(5)
Cyanide	"	.005 mg/l (1981)	(4)
Lead	"	21 ppb (1981)	(4)
		18 ug/l (1980)	(5)
Mercury	"	0.5 ppb (1981)	(4)
		0.5 ug/l (1980)	(5)
Nickel	"	.295 mg/l (1982)	(3)
		2,480 ppb (1981)	(4)
		302 ug/l (1980)	(5)
Selenium	"	34 ppb (1981)	(4)
Silver	"	10 ppb (1981)	(4)
Zinc	"	.102 mg/l (1982)	(3)
		15,618 ppb (1981)	(4)
		758 ug/l (1980)	(5)

IFF

WASTE GENERATION & DISPOSAL

Type of Waste ● Indicates Wastes Affected by Reduction	Generation or Disposal	Quantity of Waste	Sources of Information
Wastewater Discharges			
Treated wastewater (cont.) NON-HAZARDOUS	Discharged to the Bayshore Regional Sewerage Authority (BRSA)		
Bromodichloromethane	"	198 ug/l (1980)	(5)
p-Bromofluorobenzene	"	64 ppb (1981)	(4)
Butyl benzene	"	24 ug/l (1980)	(5)
p-Cymene	"	26 ug/l (1980)	(5)
Heptane	"	46 ug/l (1980)	(5)
Mesitylene	"	61 ppb (1981)	(4)
		30 ug/l (1980)	(5)
Octane	"	817 ug/l (1980)	(5)
Propyl benzene	"	8 ug/l (1980)	(5)
Aluminum	"	14.4 mg/l (1982)	(3)
		238,500 ppb (1981)	(4)
Boron	"	2.4 mg/l (1982)	(3)
Iron	"	19,172 ppb (1981)	(4)
		4,000 ug/l (1980)	(5)
Magnesium	"	24.9 mg/l (1982)	(3)

WASTE GENERATION & DISPOSAL

Type of Waste ● Indicates Wastes Affected by Reduction	Generation or Disposal	Quantity of Waste	Sources of Information
Wastewater Discharges			
Treated wastewater (cont.) NON-HAZARDOUS (cont.)	Discharged to the Bayshore Regional Sewerage Authority (BRSA)		
Manganese	"	361 ppb (1981)	(4)
		117 ug/l (1980)	(5)
Tin	"	1,466 ppb (1981)	(4)
MAY CONTAIN HAZARDOUS CHEMICALS			
Process wastewater	Discharged to groundwater from aeration basin, equalization basin and sludge lagoon; ultimately discharged to Raritan Bay	12,000–15,000 gals/day (1982, 1983)	(7,8)

IFF

WASTE GENERATION & DISPOSAL

● Type of Waste Indicates Wastes Affected by Reduction	Generation or Disposal	Quantity of Waste	Sources of Information
Solid Wastes			
HAZARDOUS			
Benzene, recovered	Sent off-site for disposal	76,400 lbs/yr (1983)	(9)
Butyl phenol	"	38,000 lbs/yr (1981)	(10)
Chromium wastes:			
Biological chemical sludge	"	17,537,400 lbs/yr (1983)	(9)
	Stored on-site or sent off-site for disposal	32,408,000 lbs/yr (1982)	(13)
	Sent off-site for disposal	26,638,000 lbs/yr (1981)	(10)
	"	376,000 lbs/yr (1980)	(11)
Chromium/acetic acid waste	"	300,000 lbs/yr (1983)	(9)
		98,000 lbs/yr (1981)	(10)
Other	"	408,000 lbs/yr (1981)	(10)
		1,200 lbs/yr (1980)	(11)
	Stored on-site or sent off-site for disposal	8,120,000 lbs/yr (1980)	(12)
Lead wastes	Sent off-site for disposal	272,000 lbs/yr (1983)	(9)
		536,000 lbs/yr (1982)	(13)
		826,000 lbs/yr (1981)	(10)
Toluene	Incinerated on-site	80,000 lbs/yr (1978)	(1)

WASTE GENERATION & DISPOSAL

Type of Waste ● Indicates Wastes Affected by Reduction	Generation or Disposal	Quantity of Waste	Sources of Information
Solid Wastes			
HAZARDOUS (cont.)			
Solvents, halogenated	Sent off-site for disposal	45,800 lbs/yr (1983)	(9)
	"	7,600 gals/yr (1982)	(13)
	Stored on-site or sent off-site for disposal	40,000 lbs/yr (1980)	(12)
Solvents, non-halogenated	Stored on-site	841,400 lbs/yr (1983)	(9)
	Burned as supplemental fuel in boiler	1,321,000 lbs/yr (1983)	(9)
	Incinerated on-site	481,400 lbs/yr (1983)	(9)
	Stored on-site	160,000 lbs/yr (1981)	(10)
	Incinerated on-site	500,000 lbs/yr (1981)	(10)
	Stored on-site	16,000 lbs/yr (1980)	(11)
	Incinerated on-site	60,000 lbs/yr (1980)	(11)
	Stored on-site	800,000 lbs/yr (1980)	(12)
	"	400,000 lbs/yr (1980)	(12)
MAY CONTAIN HAZARDOUS CHEMICALS			
Waste liquids, unspecified	Sent off-site for disposal	12,800 lbs/yr (1983)	(9)
Waste solids and liquids, unspecified	"	15,600 lbs/yr (1983)	(9)
NON-HAZARDOUS			
Aluminum chloride solution	"	2,114,000 lbs/yr (1983)	(9)
Corrosive wastes, unspecified	"	164,800 lbs/yr (1982)	(13)

IFF

WASTE GENERATION & DISPOSAL

●	Type of Waste Indicates Wastes Affected by Reduction	Generation or Disposal	Quantity of Waste	Sources of Information
	Solid Wastes			
	NON-HAZARDOUS (cont.)			
	Ignitable wastes, unspecified	Sent off-site for disposal	105,600 lbs/yr (1982)	(13)
	Solvents, non-halogenated	Stored on-site	1,576,400 lbs/yr (1983)	(9)
		Burned as supplemental fuel in boiler	1,521,600 lbs/yr (1983)	(9)
		Incinerated on-site	642,200 lbs/yr (1983)	(9)
		Stored on-site	260,000 lbs/yr (1981)	(10)
		Incinerated on-site	1,390,000 lbs/yr (1981)	(10)
		Stored on-site	4,000 lbs/yr (1980)	(11)
		Incinerated on-site	104,000 lbs/yr (1980)	(11)
		Stored on-site	800,000 lbs/yr (1980)	(11)
		Stored on-site or incinerated on-site	7,200,000 lbs/yr (1980)	(12)
		Stored on-site	400,000 lbs/yr (1980)	(12)
		Stored on-site or sent off-site for disposal	100,000 lbs/yr (1980)	(12)
	Waste solvents, unspecified	Stored on-site	300,000 lbs/yr (1981)	(10)
		Incinerated on-site	6,822,000 lbs/yr (1981)	(10)
		"	64,000 lbs/yr (1980)	(11)

Max Marx Color and Chemical Company

192 Coit Street
Irvington, New Jersey 07111
(201) 373-7801

SUMMARY

Max Marx Color and Chemical Company (Irvington, New Jersey) is a small manufacturer of organic pigments. Raw materials in use include **aniline** and Rhodamine dyes. Little information about this plant's wastes is available from public records beyond the fact that they discharge wastewater to the Joint Meeting Maintenance, Union and Essex Counties, Elizabeth, New Jersey. There is no indication from available information that Max Marx has adopted any waste reduction techniques.

PLANT OPERATIONS

Max Marx Color and Chemical Company is a 25-employee organic pigment manufacturer in Irvington, New Jersey, two miles south of Newark. Its pigments are primarily used by the printing ink and graphic arts industries. The plant uses two types of pigment manufacturing processes: 1) the pyrazolone process, which yields water soluble dyes (acid dyes) used in silk and wool; and 2) the lake process, which yields non-soluble pigments.

According to the 1978 New Jersey Industrial Survey, Max Marx uses 92,000 pounds per year of **aniline** as a raw material but does not discard any as waste, since all of it is consumed in its production processes.

NOTE: Chemicals in **boldface** type are hazardous, according to INFORM's criteria (see Appendix A).

The plant also uses 1,500 pounds per year of Rhodamine 6G and 1,500 pounds per year of Rhodamine B as raw materials. These chemicals are also consumed entirely in production.

WASTE GENERATION AND WASTE REDUCTION

Max Marx discharges 24,000 gallons per day of process and other wastewaters to the Joint Meeting Maintenance, Union and Essex Counties, Elizabeth, New Jersey, according to the New Jersey Department of Environmental Protection's Pretreatment/Residual Waste Survey, November 1981. This discharge includes 12,000 gallons per day of pigment wastes.

SOURCES OF INFORMATION

SOURCE	LOCATION
1. New Jersey Industrial Survey, 1978 data	New Jersey Department of Environmental Protection (NJDEP) 190 West State Street Trenton, NJ 08625 (609) 292-6714
2. New Jersey Department of Environmental Protection Pretreatment/ Residual Waste Survey, November 1981	NJDEP Pretreatment Section 1474 Prospect Street Trenton, NJ 08625 (609) 292-4860

WASTE GENERATION & DISPOSAL

Type of Waste Indicates Wastes Affected by Reduction (●)	Generation or Disposal	Quantity of Waste	Sources of Information
Wastewater Discharges			
Process and other wastewaters, including: MAY CONTAIN HAZARDOUS CHEMICALS	Discharged to the Joint Meeting Maintenance, Union and Essex Counties, Elizabeth, NJ	24,000 gals/day (1978)	(1)
Pigment wastes	"	12,000 gals/day (1981)	(2)

Merck and Company, Inc.

126 East Lincoln Avenue
P.O. Box 2000
Rahway, New Jersey   07065
(201) 574-4000

SUMMARY

Merck and Company's operations (Rahway, New Jersey) are the site of a large chemical manufacturing plant along with research and development facilities and the corporation's headquarters. The plant manufactures bulk organic chemicals for use in animal and human health products and in pesticides. Thiabendazole (TBZ) is the plant's major product; many other products are produced in smaller quantities on an intermittent basis. Organic solvents are widely used during processing and include **benzene, chloroform, methylene chloride, tetrachloroethane, chlorobenzene, toluene** and **methyl ethyl ketone.** The solvents are major contributors to all of the plant's wastestreams, including solid waste generation of 1.8 million pounds per year of halogenated solvent wastes and 700,000 pounds per year of non-halogenated solvent wastes, most of which are sent off-site to be incinerated. Hydrogen chloride air emissions have been reduced through reuse in the TBZ manufacturing process after they are scrubbed. **Methylene chloride** and isoamyl alcohol air emissions and acetone in wastewaters were reduced through solvent recovery techniques added to the new Primaxin manufacturing process during the process development stage. Very little product material enters the wastestream at the plant, leaving little opportunity for waste reduction of these materials; waste reduction of non-product materials such as organic solvents has been limited by the economics and variable nature of the plant's operations.

NOTE: Chemicals in **boldface** type are hazardous, according to INFORM's criteria (see Appendix A).

PLANT OPERATIONS

Merck and Company's operations in Rahway, located in a heavily industrialized area of northern New Jersey, are the site of the corporation's international headquarters and major research facilities as well as a large-scale chemical manufacturing plant. Merck and Company, with sales in 1984 of $3.6 billion, is a major manufacturer of human and animal health products and agricultural chemicals. A segment of their overall sales (18.6 percent in 1984) also stems from the manufacture of specialty chemicals and equipment for use in environmental control products.

In addition to Merck's corporate headquarters and major research facilities, designed much like a college campus, the Rahway site also houses a manufacturing plant. Total employment at the site, which first began operations in 1903, currently stands at 4,000 people; 500 in manufacturing and 3,500 in corporate-wide research and management activities.

The manufacturing plant produces a wide variety of bulk organic chemicals that are shipped off-site for final formulation as animal health care products, pharmaceuticals for human use, and agricultural pesticides. The largest volume chemical manufactured at the Rahway plant, and one of the company's major products, is thiabendazole (TBZ). The Rahway plant produces 2.2 million pounds of TBZ a year for use as an antiparasitic drug for deworming livestock and as an agricultural pesticide. Other chemicals are manufactured on a campaign basis -- that is, production is highly flexible and varied depending on market demand for individual products. TBZ and butyramidine are manufactured in continuous process operations; all other production at Rahway is from batch operations.

The Rahway plant uses substantial quantities of organic chemicals as solvents during its pharmaceutical manufacturing operations, including **chloroform, methylene chloride, tetrachloroethane, chlorobenzene,** dichlorobenzene, **methyl ethyl ketone, acetonitrile, benzene** and **toluene.** Other chemicals used during manufacturing include **aniline, carbon disulfide, phosgene** and **cobalt.**

WASTE GENERATION AND WASTE REDUCTION

AIR EMISSIONS

Generation

Records from the New Jersey Department of Environmental Protection (NJDEP) allowed a comparison of 1978 and 1982 annual air emissions for three chemicals used at Rahway. Merck also supplied 1985 emissions projections for these three and other chemicals, facilitating a comparison with earlier NJDEP emissions figures:

	1978 (pounds)	1982 (pounds)	1985 (pounds)
Benzene	87,000	97,400	26,800
Chloroform	8,050	40,580	1.3
Dichlorobenzene	9,500	--	5,825
Methylene chloride	5,075	--	18,692
Phosgene	5	--	15.2
Tetrachloro-ethane	2,050	2,830	2.5
Toluene	17,200	--	2,833

Although emissions of all chemicals except **methylene chloride** and **phosgene** were lowered between 1978 and 1985, Merck did not report any waste reduction practices accounting for these changes. Instead, the lower emissions are the result of additional air pollution control equipment and variable production schedules, which call for different types and quantities of solvents each year. Emissions of **benzene** and **chloroform** were reduced in response to NJDEP regulation of toxic volatile organic substances (TVOS) and to improve employee safety in the workplace. Where possible, Merck substituted other solvents for **benzene** and **chloroform** as a means of lowering emissions. The substituted chemicals were also hazardous chemicals, but are not regulated in New Jersey as TVOS[*]. Merck attributed the increase in **methylene chloride** emissions to the start-up of a new process at the plant that uses this solvent.

Merck officials told INFORM that the 1982 NJDEP figures are of questionable accuracy for those chemicals, such as **chloroform** and **tetrachloroethane**, which are used in manufacturing processes run on a campaign basis. While NJDEP permit applications require the submission of annual emissions estimates, campaign-based processes are only run for part of a year. Consequently, emissions reported are those that would occur were a campaign-based process run for an entire year and may not reflect the actual annual emissions of a given chemical.

Reduction

Merck officials reported to INFORM that part of the scrubbed hydrogen chloride gas emissions from the TBZ manufacturing process are now recovered and reintroduced into another stage in the TBZ manufacturing process.

---

[*] Because the substitute chemicals are hazardous according to INFORM's criteria these practices were not counted as hazardous waste reduction in this report.

In the past, all of the hydrogen chloride solution formed by the scrubber was discharged to the sewage treatment plants.

WASTEWATER DISCHARGES

Generation

The Rahway plant discharges wastewater from manufacturing operations, cooling towers and surface runoff to three destinations:

-- Wastewaters from the Research and Development operations and the pilot plant are discharged to the Rahway Valley Sewerage Authority's (RVSA) sewage treatment plant without any prior treatment. According to NJDEP records, Merck discharges 1.57 million gallons per day of untreated wastewater containing 34.2 mg/l of "total toxic organics" (TTO -- a category used in New Jersey wastewater discharge regulations)[*] including **toluene** (23.5 mg/l); **benzene** (2.5 mg/l); **chloroform** (1.3 mg/l); **methylene chloride** (5.7 mg/l) and 17 other organic chemicals in smaller concentrations. The discharge also contains 25 mg/l of metals, mostly magnesium, aluminum and tin.

-- Process wastewaters from organic chemical manufacturing, along with cooling water and other non-process waters are treated on-site through equalization and neutralization and discharged to another sewage treatment plant operated by the Linden-Roselle Sewerage Authority (LRSA). The 1.34 million gallons per day discharge, according to NJDEP records, contains 152.0 mg/l of TTO[*] including **benzene** (71.3 mg/l); **toluene** (49.0 mg/l); **methylene chloride**

---

[*] In a 1.57 million gallons per day discharge, a TTO of 34.2 mg/l is roughly equal to a discharge of 500 pounds per day of total toxic organics. In a discharge of 1.34 million gallons per day, a TTO of 152 mg/l is approximately equal to 1,700 pounds.

15.5 mg/l); **chlorobenzene** (8.5 mg/l); and **ethylbenzene** (4.7 mg/l). Metal discharges amount to 41.9 mg/l, mostly aluminum and magnesium.

-- Cooling water and plant runoff are discharged into local surface waters without prior treatment. About 50,000 gallons per day are discharged to Kings Creek, and another 50,000 gallons per day to the Rahway River (via a storm drain) although both discharges are highly variable and can be considerably higher during periods of heavy rainfall.

Several of the wastewater streams contributing to surface water discharges have been diverted in the past 10 years and are now discharged to the sewage treatment plants, according to NJDEP wastewater discharge permit records. The diversions were made in order to eliminate two discharge points into Kings Creek, thus eliminating the need to apply for permits for these discharges, and to bring other surface water discharges into compliance with permit restrictions.

Reduction

INFORM has learned of one waste reduction practice affecting wastewater discharges at the Merck plant. Acetone is internally recovered during the Primaxin process from a process stream containing acetone and water. This recovery and reuse technique saves the need to discharge 229,600 pounds per year of acetone to the Linden-Roselle Sewerage Authority for disposal, saving Merck $47,750 a year in sewer discharge fees. This waste reduction step was implemented during the process development stage for Primaxin.

SOLID WASTES

Generation

A wide variety of organic solvents are used during pharmaceutical manufacturing at Merck, and these constitute a major source of solid waste generation.

About 1.8 million pounds of halogenated solvents and 700,000 pounds of non-halogenated solvents were disposed of in 1985, according to Merck estimates, including acetone, **chlorobenzene, chloroform, dichlorobenzene, methanol, methylene chloride** and **toluene.** These wastestreams are sent off-site to New Jersey waste management firms, such as Rollins Environmental Services and Inland Chemical, for disposal. According to Merck officials, waste solvents are either recovered and reused on-site, recovered off-site and sold by commercial solvent recovery services to other users, or disposed of off-site by burning them as supplemental fuel. In addition to spent solvents, solid wastes from the Rahway plant consist of solvent-contaminated materials such as sludges and filter cakes. Merck officials told INFORM that it considers disposal in a landfill as the waste management option of last resort, and prefers to incinerate wastes.

Prior to 1983, Merck had burned about 500 gallons per week of waste solvents as a supplemental fuel in an on-site incinerator. However, Merck officials told INFORM that this practice has since been discontinued for economic reasons: the cost of complying with requirements for test burns and emissions monitoring was greater than the cost of shipping these solvents off-site for disposal and buying fuel oil for the boiler.

Reduction

INFORM has learned of two waste reduction practices affecting solid waste generation at Merck. Both are internal solvent recovery steps that were added to the new Primaxin process in the process development stage as a means of minimizing solvent waste generation and have led to the recovery of 423,000 pounds per year of isoamyl alcohol and 2,610,00 pounds per year of **methylene chloride.**

OBSERVATIONS

Merck officials reported to INFORM that very little of the chemical products manufactured at Rahway enter the wastestreams; recovery rates for these chemicals are extremely high. Instead, waste generation stems chiefly from disposal of non-product chemicals, such as solvents, as well as from chemical disposal from the research and development operations. The company emphasized that the value of pharmaceutical products is so high relative to the cost of other materials in use at the plant that high rates of product recovery are essential. Much of the research and development activity at the company is geared towards improving the product recovery efficiencies of existing operations, as demanded by economic considerations.

Plant management told INFORM that their ability to reuse solvent materials is limited by their concern for quality control, the highly variable nature of Rahway's production and the relatively small volume of a solvent in use in any given process at any one time. Merck emphasized, however, that reuse of solvents has always been a standard part of some processes at Rahway.

The TBZ production process, for instance, employs a closed-loop system for reuse of **benzene** solvent. Similar steps in the production of Indomethacin allow for internal reuse of **toluene** and other process solvents. Unlike the solvent recovery steps reported earlier, which Merck reported were specifically selected over other options as a means of minimizing wastes, these reuse steps are necessary to the production processes and do not constitute waste reduction practices.

The variability in Rahway's operations also influences Merck's documentation of its wastestreams. The constant changeover of batch processes and chemicals at Rahway, and the frequent and thorough rinsing of all process equipment as processes change, renders it extremely difficult to know with precision the exact nature of the plant's waste material. Additionally, Merck officials explained that the identity of reaction by-products is often unknown and, even with state-of-the-art analytical procedures, difficult to identify.

Economics also plays a role in how the plant documents its wastestreams. Merck officials reported that the Rahway plant does not fully document the nature of the wastes from each batch run, nor does it detail the subsequent costs (on a per process basis) of treating these wastes, because there has not been an economic incentive to do so. The value of pharmaceutical products is so high relative to the costs of waste management that it has not been necessary to track environmental costs against individual processes.

Plant management commented to INFORM that in the manufacture of lower-priced commodity chemicals, where waste management costs can severely impact profitability, there have been greater economic incentives to initiate a process-by-process accounting of these costs. These economic pressures have not been as severe in the pharmaceutical industry, where the costs of research, raw materials, and quality control far exceed waste management costs.

However, Merck is pursuing a process-by-process cost accounting approach that documents waste generation and waste management costs for all new products, process changes or expansions introduced at Rahway to allow for a greater awareness and flexibility in future environmental management decisions. Such an assessment was performed for the new Primaxin process discussed earlier, where waste reduction measures were identified and implemented in the process development stage.

SOURCES OF INFORMATION

SOURCE	LOCATION
1. RCRA Part A Application (revised), February 1983	New Jersey Department of Environmental Protection (NJDEP) Division of Waste Management 32 East Hanover Street Trenton, NJ 08625 (609) 292-9879
2. RCRA Part A Application, November 1980	U.S. Environmental Protection Agency (U.S. EPA) Region II 26 Federal Plaza New York, NY 10278 (212) 264-9881
3. Hazardous Waste Manifest Records, 1980	NJDEP Bureau of Hazardous Waste CN027 Trenton, NJ 08625 (609) 984-2302
4. New Jersey Industrial Survey, 1978 data	NJDEP Office of Science and Research 190 West State Street Trenton, NJ 08625 (609) 292-6714
5. Air Pollution Enforcement Data System printout, July 1984	NJDEP Bureau of Air Pollution Control CN027 Trenton, NJ 08625 (609) 633-7994

6. Air Pollution Enforcement Data System Toxic Substances printout, July 1982
   NJDEP
   (see #5)

7. Air emissions file memos, January-March, 1979
   U.S. EPA, Region II
   Air Facilities Branch
   26 Federal Plaza
   New York, NY 10278
   (212) 264-9627

8. National Pollution Discharge Elimination System (NPDES) plant inspection and status report, February 1977
   U.S. EPA, Region II
   (see #2)

9. Letter from Merck to NJDEP, Division of Water Resources, December 9, 1977
   U.S. EPA, Region II
   (see #2)

10. NPDES Inspection Analysis Report, August 1979
    U.S. EPA, Region II
    (see #2)

11. NPDES Compliance Monitoring Report, 1980
    U.S. EPA, Region II
    (see #2)

12. Letter to Merck from NJDEP, Division of Water Resources, February 22, 1982
    U.S. EPA, Region II
    (see #2)

13. Discharge Surveillance Report, September 1982
    U.S. EPA, Region II
    (see #2)

14. Significant Industrial User Application, August 1981
    NJDEP
    Division of Water
      Resources
    Pretreatment Section
    CN029
    Trenton, NJ 08625
    (609) 292-4860

15. Computed by INFORM based on hourly emissions data (see #6) and hours per year of operation (see #16)

16. Air Permit files, 1982   NJDEP
    Division of Environmental
     Quality
    Newark Field Office
    1100 Raymond Boulevard
    Newark, NJ 07102
    (201) 648-2075

17. Summary information on VOS (volatile organic substances) emissions, February 1985   NJDEP (see #16)

18. Company provided information to INFORM

19. Hazardous Waste Generator Annual Report, 1981, filed to NJDEP, Division of Waste Management   Information supplied by Merck to INFORM

20. Hazardous Waste Generator Annual Report, 1983, filed to NJDEP, Division of Waste Management   Information supplied by Merck to INFORM

21. Hazardous Waste Generator Annual Report, 1984, filed to NJDEP, Division of Waste Management   Information supplied by Merck to INFORM

22. Hazardous Waste Generator Annual Report, 1982, filed to NJDEP, Division of Waste Management   Information supplied by Merck to INFORM

MERCK

WASTE REDUCTION

Type of Waste	G-Generation D-Disposal Q-Quantity (year)	Waste Reduction Practice	Results	Sources of Information
HAZARDOUS				
Methylene chloride	G - ? D - ? Q - ? (1985)	Internal solvent recovery step in the new Primaxin process PS (?)	Recovery of 2,610,000 pounds per year of methylene chloride	(18)
NON-HAZARDOUS				
Acetone	G - ? D - ? Q - ? (1985)	Internal solvent recovery and reuse step in the new Primaxin process PS (?)	Recovery and reuse of 229,600 pounds per year of acetone. Elimination of the need to discharge this process stream to the Linden-Roselle Sewerage Authority. Savings of $47,750 in sewer discharge fees.	(18)
Hydrogen chloride	G - ? D - ? Q - ? (1985)	Recovery and reuse of part of the scrubbed hydrogen chloride gas emissions from the TBZ process PS (?)	Elimination of need to send some of the hydrogen chloride solution produced by the scrubber to the sewage treatment plants	(18)
Isoamyl alcohol	G - ? D - ? Q - ? (1985)	Internal solvent recovery step in the new Primaxin process PS (?)	Recovery of 423,000 pounds per year of isoamyl alcohol	(18)

Key to Waste Reduction Changes
- PR- Product
- PS- Process
- EQ- Equipment
- CH- Chemical Substitution
- OP- Operational

WASTE GENERATION & DISPOSAL

MERCK

Type of Waste ● Indicates Wastes Affected by Reduction	Generation or Disposal	Quantity of Waste	Sources of Information
Air Emissions			
HAZARDOUS			
Acetonitrile	Emissions from finishing facilities	0-600 lbs/yr (1979)	(18)
Benzene	Plant-wide emissions	26,800 lbs/yr (1985)	(18)
	Emissions from process sources	97,400 lbs/yr (1982)	(6, 15, 16)
	Plant-wide emissions	87,000 lbs/yr (1978)	(4)
Chloroform	"	1.3 lbs/yr (1985)	(18)
	Emissions from process sources	40,580 lbs/yr (1982)	(6, 15, 16)
	Plant-wide emissions	8,050 lbs/yr (1978)	(4)
o-Chlorophenol	Emissions from 1 source	.03 lbs/hour (1984)	(5)
o-Dichlorobenzene	Emissions from 4 sources	.72 lbs/hour (1985)	(18)
1,4-Dichlorobenzene	Plant-wide emissions	5,825 lbs/yr (1985)	(18)
		4,500 lbs/yr (1978)	(4)
Ethylene dichloride	Emissions from 2 sources	.05 lbs/hour (1982)	(6)
		312 lbs/yr (1982)	(6, 15, 16)
Methylene chloride	Plant-wide emissions	18,692 lbs/yr (1985)	(18)
		5,075 lbs/yr (1978)	(4)

MERCK

WASTE GENERATION & DISPOSAL

Type of Waste ● Indicates Wastes Affected by Reduction	Generation or Disposal	Quantity of Waste	Sources of Information
Air Emissions			
HAZARDOUS (cont.)			
Phosgene	Plant-wide emissions	15.2 lbs/yr (1985)	(18)
		5 lbs/yr (1978)	(4)
1,1,2,2-Tetrachloroethane	"	2.5 lbs/yr (1985)	(18)
	Emissions from 10 sources	2,830 lbs/yr (1982)	(6, 15, 16)
	Plant-wide emissions	2,050 lbs/yr (1978)	(4)
Toluene	"	2,833 lbs/yr (1985)	(18)
		17,200 lbs/yr (1978)	(4)
Volatile organic chemicals*	Plant-wide emissions	1,944,000 lbs/yr (1985)	(17)
		2,484,000 lbs/yr (1980)	(17)
NON-HAZARDOUS			
p-Chlorophenol	Emissions from 2 sources	.002 lbs/hour (1984)	(5)
● Hydrogen chloride	Emissions from Thiabendazole manufacturing	? (1985)	(18)
Propylene oxide	Emissions from 3 sources	14.08 lbs/yr (1985)	(18)

* Includes many of the air emissions listed by specific chemical name.

WASTE GENERATION & DISPOSAL

Type of Waste ● Indicates Wastes Affected by Reduction	Generation or Disposal	Quantity of Waste	Sources of Information
\multicolumn{3}{c}{**Wastewater Discharges**}	
Process and other wastewaters	Treated and discharged to the Linden-Roselle Sewerage Authority (LRSA)	1,340,000 gals/day (1981)	(14)
Wastewater from research and pilot plant operations	Discharged to the Rahway Valley Sewerage Authority (RVSA)	1,570,000 gals/day (1981)	(14)
The above waste streams include:			
HAZARDOUS			
Benzene	LRSA	71.3 mg/l (1981)	(14)
	RVSA	2.5 mg/l (1981)	(14)
	Total discharge to both sewage treatment plants	22,000 lbs/yr (1978)	(4)
Carbon disulfide	?	? (1985)	(18)
Chlorobenzene	LRSA	8.5 mg/l (1981)	(14)
	RVSA	.10 mg/l (1981)	(14)
Chloroethane	LRSA	.90 mg/l (1981)	(14)
Chloroform	LRSA	.50 mg/l (1981)	(14)
	RVSA	1.3 mg/l (1981)	(14)
o-Dichlorobenzene	LRSA	.90 mg/l (1981)	(14)
	RVSA	.30 mg/l (1981)	(14)
1,4-Dichlorobenzene	Total discharge to both sewage treatment plants	3,000 lbs/yr (1978)	(4)

WASTE GENERATION & DISPOSAL

Type of Waste ● Indicates Wastes Affected by Reduction	Generation or Disposal	Quantity of Waste	Sources of Information
Wastewater Discharges			
Wastewaters (cont.)			
HAZARDOUS (cont.)			
Dichloroethylene	RVSA	.10 mg/l (1981)	(14)
Ethylbenzene	LRSA	4.7 mg/l (1981)	(14)
	RVSA	.04 mg/l (1981)	(14)
Methylene chloride	LRSA	15.5 mg/l (1981)	(14)
	RVSA	5.7 mg/l (1981)	(14)
	Total discharge to both sewage treatment plants	450 lbs/yr (1978)	(4)
1,1,2,2-Tetrachloro-ethane	RVSA	.20 mg/l (1981)	(14)
Toluene	LRSA	49 mg/l (1981)	(14)
	RVSA	23.5 mg/l (1981)	(14)
	Total discharge to both sewage treatment plants	4,000 lbs/yr (1978)	(4)
Trichloroethylene	LRSA	.43 mg/l (1981)	(14)
	RVSA	.24 mg/l (1981)	(14)
Xylene	LRSA	23.4 mg/l (1981)	(14)
	RVSA	.07 mg/l (1981)	(14)
Antimony	LRSA	.40 mg/l (1981)	(14)
	RVSA	.60 mg/l (1981)	(14)

WASTE GENERATION & DISPOSAL

Type of Waste ● Indicates Wastes Affected by Reduction	Generation or Disposal	Quantity of Waste	Sources of Information
Wastewater Discharges			
Wastewaters (cont.)			
HAZARDOUS (cont.)			
Barium	LRSA	.20 mg/l (1981)	(14)
	RVSA	.40 mg/l (1981)	(14)
Beryllium	LRSA	.10 mg/l (1981)	(14)
	RVSA	.20 mg/l (1981)	(14)
Chromium	LRSA	.10 mg/l (1981)	(14)
	RVSA	.17 mg/l (1981)	(14)
Copper	LRSA	.10 mg/l (1981)	(14)
	RVSA	.12 mg/l (1981)	(14)
Cyanide	LRSA	3.6 mg/l (1981)	(14)
	RVSA	1.2 mg/l (1981)	(14)
Lead	LRSA	.10 mg/l (1981)	(14)
	RVSA	.20 mg/l (1981)	(14)
Nickel	LRSA	.20 mg/l (1981)	(14)
	RVSA	.09 mg/l (1981)	(14)
Thallium	LRSA	.10 mg/l (1981)	(14)
	RVSA	.20 mg/l (1981)	(14)

WASTE GENERATION & DISPOSAL

Type of Waste ● Indicates Wastes Affected by Reduction	Generation or Disposal	Quantity of Waste	Sources of Information
Wastewater Discharges			
Wastewaters (cont.)			
HAZARDOUS (cont.)			
Zinc	LRSA	.80 mg/l (1981)	(14)
	RVSA	.70 mg/l (1981)	(14)
	Total discharge to both sewage treatment plants	1,600 lbs/yr (1978)	(4)
NON-HAZARDOUS			
● Acetone	LRSA	? (1985)	(18)
Aluminum	LRSA	19.4 mg/l (1981)	(14)
	RVSA	2.1 mg/l (1981)	(14)
Magnesium	LRSA	18.6 mg/l (1981)	(14)
	RVSA	16.1 mg/l (1981)	(14)
Other metals	LRSA	1.75 mg/l (1981)	(14)
	RVSA	3.6 mg/l (1981)	(14)

MERCK
WASTE GENERATION & DISPOSAL

Type of Waste ● Indicates Wastes Affected by Reduction	Generation or Disposal	Quantity of Waste	Sources of Information
Wastewater Discharges			
Cooling and storm waters, including:	Discharged to Kings Creek	50,000-75,000 gals/day (up to 800,000 gals/day during storms) (1977, 1979, 1982)	(8,9, 10, 12)
HAZARDOUS			
Chromium	"	.01 mg/l (1982)	(13)
		.006 mg/l (1980)	(11)
Zinc	"	.278 mg/l (1982)	(13)
		.12 mg/l (1980)	(11)
Cooling and storm waters, including:	Discharged to Rahway River	20,000-65,000 gals/day (up to 125,000 gals/day during storms) (1977, 1982)	(8,9, 13)
HAZARDOUS			
Chromium	"	.027 mg/l (1982)	(13)
		.006 mg/l (1980)	(11)
Zinc	"	.354 mg/l (1982)	(13)
		.094 mg/l (1980)	(11)

MERCK

WASTE GENERATION & DISPOSAL

●	Type of Waste Indicates Wastes Affected by Reduction	Generation or Disposal	Quantity of Waste	Sources of Information
	Solid Wastes			
	HAZARDOUS			
	Acetaldehyde	Stored on-site	1,300 lbs/yr (max.) (1983)	(1,2)
	Acetonitrile	"	37,000 lbs/yr (max.) (1983)	(1,2)
	Benzene	Sent off-site for disposal	226,197 lbs/yr (1980)	(3)
		Burned as fuel supplement at Earthline Inc., Newark, NJ	760,000 lbs/yr (1978)	(4)
		Incinerated at Scientific Chemical Processing, Newark, NJ	597,000 lbs/yr (1978)	(4)
	Benzene waste	Sent off-site for disposal	1,160,000 lbs/yr (1981)	(19)
	Benzene, dichloromethane and xylene waste	"	810,660 lbs/yr (1983)	(20)
			75 cubic yds/yr (1983)	(20)
	Bromomethane, chloroform, 1,2-transdichloroethylene and dichloromethane waste	"	99 cubic yds/yr (1984)	(21)
	Carbon disulfide	Stored on-site	10,000 lbs/yr (max.) (1983)	(1,2)

NOTE: Merck told INFORM that all of the chemical quantities contained in the RCRA Application (INFORM Information Sources 1 and 2) represent worst-case maximum discharge quantities (denoted by max. on this and following pages) and that actual discharge quantities might be much smaller.

NOTE: Wastes sent off-site for disposal may be sent to any of the following off-site disposal facilities: A.B.C.O. Industries, Inc., Roebuck, SC; Advanced Environmental Technology Corp., Morris Plains and Flanders, NJ; Bethlehem Apparatus Co., Hellertown, PA; Browning Ferris Industries, Glen Burnie, MD; CECOS International Inc., Cincinnati and Williamsburg, OH and Niagara Falls, NY; Chemical Waste Management, Inc., Emelle, AL; General Electric Company, Tonawanda, NY and Philadelphia, PA; Industrial Solvents, Inc., Emigsville, PA; Leonetti Oil Recovery, Old Bridge, NJ; McKesson Envirosystems, Newcastle, KY and Newark, NJ; Radiac Research Corp., Brooklyn, NY; Recycling Industries, Braintree, MA; Rollins Environmental Services, West Bridgeport, NJ and Houston, TX; Solvents Recovery Services of New Jersey, Linden, NJ; and Spectron Inc., Elkton, MD.

WASTE GENERATION & DISPOSAL

Type of Waste ● Indicates Wastes Affected by Reduction	Generation or Disposal	Quantity of Waste	Sources of Information
Solid Wastes			
HAZARDOUS (cont.)			
Carbon disulfide waste	Sent off-site for disposal	39,590 lbs/yr (1982)	(22)
		382,000 lbs/yr (1981)	(19)
Chlorobenzene	Stored on-site	10,034 lbs/yr (1985)	(18)
Chlorobenzene and other solvents	Sent off-site for disposal	10,000 gals/yr (1983)	(20)
		7,702 gals/yr (1982)	(22)
Chloroform	Incinerated at Rollins Environmental Services	20,187 lbs/yr (1985)	(18)
		86,950 lbs/yr (1978)	(4)
Chloroform, chloromethane and dichloromethane	Sent off-site for disposal	75 cubic yds/yr (1983)	(20)
Chloroform, dichloromethane, diethyl phthalate and trichloroethene waste	"	25 cubic yds/yr (1983)	(20)
2-Chlorophenol	Stored on-site	1,100 lbs/yr (max.) (1983)	(1,2)
Cyclohexane and other solvents	Sent off-site for disposal	2,180 gals/yr (1982)	(22)
o-Dichlorobenzene	Stored on-site	28,059 lbs/yr (1985)	(18)
1,4-Dichlorobenzene	Disposed of in an off-site landfill at Glen Burnie, MD	297,500 lbs/yr (1978)	(4)
Ethyl ether	Stored on-site	1,300 lbs/yr (max.) (1983)	(1,2)

WASTE GENERATION & DISPOSAL

MERCK

● Type of Waste Indicates Wastes Affected by Reduction	Generation or Disposal	Quantity of Waste	Sources of Information
Solid Wastes			
HAZARDOUS (cont.)			
Formaldehyde, phthalic anhydride, and 1,2-dichlorobenzene waste	Sent off-site for disposal	80,000 lbs/yr (1981)	(19)
Malononitrile	Stored on-site	1,800 lbs/yr (max.) (1983)	(1,2)
Methanol	"	46,000 lbs/yr (max.) (1983)	(1,2)
Methanol and other solvents	Sent off-site for disposal	5,000 gals/yr (1984)	(21)
Methyl ethyl ketone	"	25 cubic yds/yr (1983)	(20)
● Methylene chloride	Sent off-site for disposal or recovery	1,716,595 lbs/yr (1985)	(18)
	Incinerated at Scientific Chemical Processing, Newark, NJ	130,475 lbs/yr (1978)	(4)
Methylene chloride and other solvents	Sent off-site for disposal	4,500 gals/yr (1984)	(21)
Polychlorinated biphenyls	"	4,926 lbs/yr (1984)	(21)
		13,375 lbs/yr (1983)	(20)
		1,297 lbs/yr (1982)	(22)
		90 gals/yr (1982)	(22)
Pyridine	Stored on-site	1,600 lbs/yr (max.) (1983)	(1,2)
1,1,1,2-Tetrachloroethane	Incinerated off-site	2,800 lbs/yr (max.) (1983)	(1,2)

WASTE GENERATION & DISPOSAL

MERCK

Type of Waste • Indicates Wastes Affected by Reduction	Generation or Disposal	Quantity of Waste	Sources of Information
Solid Wastes			
HAZARDOUS (cont.)			
1,1,2,2-Tetrachloroethane	Incinerated off-site	22,970 lbs/yr (1985)	(18)
	Incinerated at Scientific Chemical Processing, Newark, NJ	59,950 lbs/yr (1978)	(4)
Toluene	Incinerated off-site	376,677 lbs/yr (1985)	(18)
	Incinerated at Scientific Chemical Processing, Newark, NJ	375,000 lbs/yr (1978)	(4)
Toluene and other solvents	Sent off-site for disposal	13,000 gals/yr (1984)	(21)
Toluene waste	"	5,000 lbs/yr (1983)	(20)
		4,000 gals/yr (1983)	(20)
		39,193 gals/yr (1982)	(22)
Xylene	Stored on-site	29,000 lbs/yr (max.) (1983)	(1,2)
Corrosive wastes	Treated and stored on-site	4,578,660,000 lbs/yr (max.)* (1983)	(1,2)
Solvents:			
Compressed gas	Sent off-site for disposal	10 lbs/yr (1982)	(22)
Halogenated	"	1,800,000 lbs/yr (1985)	(18)
Non-halogenated	"	700,000 lbs/yr (1985)	(18)
Unspecified	"	101,100 lbs/yr (1984)	(21)

* The origin and fate of this waste stream is not clearly identified, but it appears to refer to Merck's wastewater discharges contaminated with a wide variety of hazardous chemicals.

393

MERCK
WASTE GENERATION & DISPOSAL

Type of Waste ● Indicates Wastes Affected by Reduction	Generation or Disposal	Quantity of Waste	Sources of Information
Solid Wastes			
HAZARDOUS (cont.)			
Solvents (cont.)			
Unspecified (cont.)	Sent off-site for disposal	106,408 gals/yr (1984)	(21)
		76,263 lbs/yr (1983)	(20)
		114,630 gals/yr (1983)	(20)
		300 cubic yds/yr (1983)	(20)
		500 lbs/yr (1982)	(22)
		164,775 gals/yr (1982)	(22)
		3,461,400 lbs/yr (1981)	(19)
Asbestos	"	37,700 lbs/yr (1980)	(3)
Thallium (I) sulfate	"	25 cubic yds/yr (1983)	(20)
		10 cubic yds/yr (1982)	(22)
LIKELY TO CONTAIN HAZARDOUS CHEMICALS			
Solvents, halogenated	"	984,699 lbs/yr (1980)	(3)
MAY CONTAIN HAZARDOUS CHEMICALS			
Acid solution	"	258,867 lbs/yr (1980)	(3)
Alkaline solution	"	56,790 lbs/yr (1980)	(3)
Lacramators, amines and mercaptan	"	43,920 lbs/yr (1980)	(3)

WASTE GENERATION & DISPOSAL

●	Type of Waste Indicates Wastes Affected by Reduction	Generation or Disposal	Quantity of Waste	Sources of Information
	Solid Wastes			
	MAY CONTAIN HAZARDOUS CHEMICALS (cont.)			
	Oil and oil sludges	Sent off-site for disposal	75,195 lbs/yr (1980)	(3)
	Organic wastes, miscellaneous	"	189,060 lbs/yr (1980)	(3)
	Packed laboratory chemicals	"	1,126 lbs/yr (1983)	(20)
			12,190 lbs/yr (1982)	(22)
			176,322 lbs/yr (1980)	(3)
	Pesticides	"	43,200 lbs/yr (1980)	(3)
	Pharmaceutical wastes	"	151,912,515 lbs/yr (1980)	(3)
	Solvents, non-halogenated	"	4,498,241 lbs/yr (1980)	(3)
	Waste, unspecified	"	77,901 lbs/yr (1984)	(21)
			83,516 lbs/yr (1983)	(20)
			1,554 gals/yr (1983)	(20)
			35,000 lbs/yr (1982)	(22)
			5,150 gals/yr (1982)	(22)
			31 cubic yds/yr (1982)	(22)
	Waste oils and tank bottoms	"	42,200 lbs/yr (1984)	(21)
			1,280 gals/yr (1984)	(21)
			14,100 lbs/yr (1983)	(20)
			40 cubic yds/yr (1983)	(20)

MERCK
WASTE GENERATION & DISPOSAL

●	Type of Waste Indicates Wastes Affected by Reduction	Generation or Disposal	Quantity of Waste	Sources of Information
	\multicolumn{4}{c}{**Solid Wastes**}	

	Type of Waste	Generation or Disposal	Quantity of Waste	Sources
	NON-HAZARDOUS			
	Acetone	Sent off-site for disposal	87,480 lbs/yr (1980)	(3)
	n-Butyl alcohol	Stored on-site	1,500 lbs/yr (max.) (1983)	(1,2)
	Cyano thiazole waste	Sent off-site for disposal	36,920 lbs/yr (1983)	(20)
			8,505 gals/yr (1982)	(22)
			1 cubic yd/yr (1982)	(22)
	Cyclohexane	Stored on-site	1,300 lbs/yr (1980)	(3)
	Dichlorophenamide	Sent off-site for disposal	7,200 lbs/yr (1983)	(20)
	Ethyl acetate	Stored on-site	1,500 lbs/yr (max.) (1983)	(1,2)
	Hexyl resorcinol waste	Sent off-site for disposal	5,560 lbs/yr (1983)	(20)
●	Isoamyl alcohol	?	? (1985)	(18)
	Sulfur waste	Sent off-site for disposal	2,000 lbs/yr (1983)	(20)
	Tetrahydrofuran	Stored on-site	2,000 lbs/yr (max.) (1983)	(1,2)
	Trichlorofon waste	Sent off-site for disposal	5,810 lbs/yr (1982)	(22)
	Corrosive and reactive wastes	"	9,821 lbs/yr (1984)	(21)
			12,931 lbs/yr (1983)	(20)
			3,000 gals/yr (1983)	(20)

WASTE GENERATION & DISPOSAL

Type of Waste ● Indicates Wastes Affected by Reduction	Generation or Disposal	Quantity of Waste	Sources of Information
Solid Wastes			
NON-HAZARDOUS (cont.)			
Corrosive and reactive wastes (cont.)	Sent off-site for disposal	4,926 lbs/yr (1982)	(22)
		95 gals/yr (1982)	(22)
		700 cubic yds/yr (1982)	(22)
		24,840 lbs/yr (1981)	(19)
Flammable wastes	"	15,779 lbs/yr (1984)	(21)
		28,965 gals/yr (1984)	(21)
		38,658 lbs/yr (1983)	(20)
		22,090 gals/yr (1983)	(20)
		9,377 lbs/yr (1982)	(22)
		4,715 gals/yr (1982)	(22)
		116,372 lbs/yr (1981)	(19)
Solvents, non-halogenated	?	1,400,000 lbs/yr (max.) (1983)	(1)
Solvents, unspecified	Sent off-site for disposal	131,600 lbs/yr (1981)	(19)
Waste treatment sewerage	"	3,980,000,000 lbs/yr* (1982)	(22)
		4,060,000,000 lbs/yr* (1981)	(19)

* These figures probably correspond to the corrosive wastes (over 4 billion pounds) listed above.

Additional Wastes Reported by Merck Not Included in the Chart

SOLID WASTES*

HAZARDOUS

Acrolein
Bis(chloro methyl)ether
Cadmium waste
Cadmium and lead waste
Copper cyanide
Cyanides
Dimethyl sulfate
Fluoroacetic acid, sodium salt
Hydrogen sulfide
Lead waste
Maleic anhydride
Mercury waste
Nickel catalyst
Phthalic anhydride
Phosgene
Potassium cyanide

MAY CONTAIN HAZARDOUS CHEMICALS

Antibiotics waste
Catalyst residues
Drug and medicine waste

NON-HAZARDOUS

Aprinocid
Compressed gas
Furan
Sodium Cefoxiden

---

* All wastes listed here are generated in quantities of less than 1,000 pounds per year. Wastes above 1,000 pounds per year are listed on the waste generation table.

Monsanto Company

Port Plastics Plant
River Road
Addyston, Ohio 45001
(501) 467-2400

SUMMARY

The Monsanto Company's Port Plastics Plant (Addyston, Ohio) manufactures a wide variety of plastics, adhesives, and resins from such raw materials as **acrylonitrile, phenol, maleic anhydride,** styrene, and butadiene. Monsanto reports that, since beginning operations in 1952, its change from batch production to continuous production of polystyrene has reduced organic air emissions by over 99 percent (even though production of the chemical increased 13-fold during that time) and that a program of fugitive emissions control has reduced **acrylonitrile** emissions. Monsanto has reduced its solid wastes by eliminating a filtering step in the formulation of an adhesive and by training specialists to segregate hazardous from non-hazardous trash. A plant-wide focus on quality control has in some instances reduced the amount of wastes generated from accidental losses of materials, but in other instances it has hampered waste reduction efforts.

PLANT OPERATIONS

The Port Plastics Plant in Addyston, Ohio, is a large (800-employee) facility located along the banks of the Ohio River in an industrialized area east of Cincinnati. Built in 1952, it is owned by the Monsanto Company, a major international corporation that manufactures

NOTE: Chemicals in **boldface** type are hazardous, according to INFORM's criteria (see Appendix A).

pesticides, synthetic fibers, plastics, resins and other industrial chemicals, and such products as "AstroTurf" artificial grass surfaces and "Wear-Dated" synthetic carpets. Total company sales in 1983 were $6.3 billion. The plant is operated by Monsanto's Polymer Products Division -- with 29 percent of total sales, the company's largest division.

The Port Plastics Plant manufactures plastics, resins, latexes, and adhesives from raw materials and specialty additives mixed in various combinations. The plant formulates -- by grinding, pelletizing, and extruding -- its plastic products before marketing them. Major products are:

> --Acrylonitrile-butadiene-styrene (ABS) plastic and styrene-acrylonitrile (SAN) plastic (annual production, 280 and 48 million pounds, respectively) --both widely used to make pipes and other construction materials, appliance housings, and automotive parts;

> --Styrene-maleic anhydride (SMA) resin (annual production, 22 million pounds) -- a hardening and thickening agent used in adhesives, paints, floor waxes, and detergents;

> --Phenol-formaldehyde resins, used in adhesives, for appliance housings and in the manufacture of plywood and other housing materials. Monsanto told INFORM that it discontinued these operations in March 1984;

> --Polystyrene plastic (annual production, 100 to 400 million pounds) used to make styrofoam (for cups and packaging materials), toys, and housewares.

Raw materials used at the plant include styrene (a major component of many of the plastics manufactured), **acrylonitrile, phenol, maleic anhydride,** and butadiene.

WASTE GENERATION AND WASTE REDUCTION

AIR EMISSIONS

Generation

The U.S. EPA estimates that in 1978, annual plant-wide air emissions for **formaldehyde** were 85,300 pounds and for **acrylonitrile** were 359,300 pounds. Monsanto contested the latter estimate and submitted its own figure of 185,200 pounds.

A U.S. EPA plant-wide study of fugitive emissions of **formaldehyde** from leaking pumps, valves, seals, and other equipment found that no such emissions existed. Monsanto believes that **formaldehyde** may have formed a self-sealing solid within the process equipment, thus preventing leaks.

Reduction

Monsanto officials told INFORM that organic air emissions have fallen sharply since the plant began operating in 1952, the result of what they call "evolutionary" or long-term chemical process changes affecting waste generation and reduction. The original process for manufacturing polystyrene, which involved a large number of batch reactors with limited reaction-control capabilities, produced 90,000 pounds per day of polystyrene.

Air emissions of styrene and other process chemicals from the original batch process amounted to nearly 4,500 pounds per day, roughly five percent of total production. However, polystyrene production is now continuous. Although the new closed-system process manufactures 1.2 million pounds per day of polystyrene (a 13-fold increase over the earlier process), polystyrene emissions have dropped to 200 pounds per day, or 0.017 percent of total production. Put another way, the original process resulted in 50 pounds of emissions for every 1,000 pounds of product, whereas the current process emits 0.17 pounds for every 1,000 pounds of product -- an overall reduction of 99.7 percent.

In 1978, plant management studied the fugitive air emissions of **acrylonitrile** and other chemicals from the ABS plastic manufacturing units that were escaping from leaking equipment and from the open trenches used to convey wastewater discharges containing the chemical to plant sewers. The study -- which was prompted by Monsanto's concern over the possible carcinogenicity of **acrylonitrile,** as well as by its anticipation of federal standards restricting workers' exposure to this chemical -- identified sources of **acrylonitrile** emissions. However, technological difficulties in accurately measuring emissions at the plant -- in particular, interference by other types of vapors -- prevented Monsanto from determining either the quantity of these emissions or the extent of their reduction. Monsanto did report to the U.S. EPA that the study's findings led to emissions control and reductions in the levels of workers' exposure to **acrylonitrile.** The company did not specify the steps it took to minimize fugitive emissions, however.

Monsanto reported to the U.S. EPA in 1978 that a series of projects have resulted in what plant officials call a "major reduction of whatever hydrocarbons have been present" in the plant's air emissions. The principal means of lowering emissions was to vent them to on-site boilers, where they are burned as supplemental fuel, rather than to reduce them at the source.

WASTEWATER DISCHARGES

Generation

The Port Plastics Plant discharges an average of one million gallons per day of process water and 10 million gallons per day of cooling water to the Ohio River after treatment at an on-site plant that was constructed in 1977. Although such organic chemicals as **acrylonitrile, toluene,** and **ethylbenzene** are present in the water before treatment, recent independent sampling by both Monsanto and the U.S. EPA found that **phenol** and **methylene chloride** are the only hazardous organic chemicals remaining in the wastewater in detectable quantities after treatment. According to the Ohio EPA, the average daily amount of **phenol** in the plant's final discharge to the Ohio River fell from 400 pounds in the mid-1970s to 0.4 pounds in 1982 (a 1,000-fold decrease), as a result of the wastewater treatment plant. Inorganic pollutants are also present in trace quantities.

SOLID WASTES

Generation

The Port Plastics Plant generated more than 400,000 pounds per year of hazardous solid wastes in 1982 and 1983, and over a million pounds in 1981, as a result of disposing of obsolete phenol-formaldehyde resins. These wastes are sent to commercial waste handling firms in Ohio and Alabama.

Reduction

Monsanto reported two waste reduction practices which affect its solid waste generation. The first practice is a product change which took about two years to implement. Monsanto reduced wastes by modifying a process that manufactured specialized industrial adhesives. The modification allowed hazardous particulate matter to remain in the product (rather than being filtered out) without affecting the adhesive properties of the product, thus improving product yield and avoiding solid waste generation. Before the process change could be implemented, however, Monsanto's marketing division had to overcome the initial reluctance of their customer who was concerned that the change might adversely affect the adhesive's quality.

The second waste reduction practice is the segregation of waste types to avoid contaminating large amounts of non-hazardous wastes with small amounts of hazardous materials. Non-process trash (including discarded clothing, rags, used chemical containers, and cleanup from small spills) is routinely generated at the plant and discarded. In the past, small quantities of hazardous materials, such as contaminated absorbents, were sometimes included in the trash, requiring that entire loads be handled as hazardous wastes. Now, however, three recently trained trash specialists keep hazardous and non-hazardous trash separate, thus preventing the unnecessary contamination of entire waste loads.

OBSERVATIONS

Plant management told INFORM that two factors have influenced the Port Plastics Plant's waste reduction efforts. One factor, concern over quality control, is both an incentive and a deterrent to waste reduction. A second factor, the existence of alternative waste management options, is, however, a deterrent only.

According to Monsanto, the company's increasingly strong focus on quality control has sharpened plant officials' efforts to reduce air emissions, wastewater discharges, and solid waste generation. By "quality control" the company means not only the specifications of final products but also the procedures used to handle -- and insure the quality of -- chemicals throughout their life cycles. In one plant official's words, "doing things right the first time" allows more materials to become part of the plant's final products and allows fewer materials to enter its wastestream. In particular, Monsanto has attempted to reduce the number of accidents that generate unnecessary waste.

Chemical spills reported by the company to the Ohio EPA illustrate the relationship between quality control and waste reduction. In 1983, a **phenol** spill produced hazardous waste when the failure of an automatic cut-off device and the plugging of an overflow pipe caused a tank being filled with **phenol** to spill over and release 1,300 pounds of the chemical onto the plant grounds during a heavy rainstorm. About 1,000 pounds washed directly into the Ohio River (the equivalent of nearly seven years of **phenol** discharges at Monsanto's normal level of 0.4 pounds per day); the remaining **phenol** was absorbed into the soil, subsequent excavation of which generated 60,000 pounds of contaminated solid waste. Similar spills had occurred earlier.

As a result of such accidents, the plant instituted the following operational changes to make recurrences less likely: 1) revision of inspection procedures to insure proper functioning of the automatic cut-off device and overflow pipe, 2) installation of an automatic alarm that alerts personnel to potential spills, and 3) observation of tanks while they are being filled with hazardous materials. These changes are designed to reduce the number of spills and other accidental losses of materials. The amount of waste released declines as a result.

In contrast, concern over the quality of final products has inhibited, rather than encouraged, waste reduction and recycling efforts at the plant, Monsanto told INFORM. On one occasion, Monsanto intended to reuse as a raw material some of the solid wastes collected during the plant's wastewater treatment. However, the company's uneasiness about the effect of recycling on the quality of its products led management to shelve the plan.

Also affecting the plant's waste reduction efforts are alternative waste management options, Monsanto told INFORM.

> --Many of the plant's process air emissions containing concentrated (and potentially reusable) levels of organic chemicals such as **formaldehyde** and **acrylonitrile** are vented to on-site boilers, where they are burned as supplemental fuels. The availability of this fuel-use option minimizes the incentive to consider alternatives -- including waste reduction.

> --**Ethylbenzene,** which is routinely captured as a by-product of polystyrene manufacture, can be reintroduced directly into the manufacturing

process, or it can be burned on-site or off-site as a supplemental fuel. The fluctuating costs of **ethylbenzene** and of fuels determine whether Monsanto disposes of the chemical by the former practice (waste reduction) or by the latter (energy recovery).

SOURCES OF INFORMATION

SOURCE	LOCATION
1. Ohio Environmental Protection Agency Air Permit files, 1983	Southwestern Ohio Air Pollution Control Agency 2400 Beekman Street Cincinnati OH 45214 (513) 251-8777
2. Correspondence from Monsanto to U.S. Environmental Protection Agency (U.S. EPA), Office of Air Quality Planning and Standards, February 13, 1978	Southwestern Ohio Air Pollution Control Agency (see #1)
3. "Human Exposure to Atmospheric Concentrations of Selected Chemicals," prepared by Systems Applications, Inc. (Contract #68-02-3066), March 1983[*]	U.S. Environmental Protection Agency (U.S. EPA) Office of Air Quality Planning and Standards Research Triangle Park, NC 27111 (919) 541-5315
4. Information supplied by Monsanto to the U.S. EPA for a draft version of source #3	U.S. EPA, Office of Air Quality Planning and Standards (see #3)
5. Wastewater Discharge Permit Renewal Application, January 1981	U.S. EPA, Region V Water Division 230 South Dearborn Street Chicago, IL 60604 (312) 886-6112

[*] This report gives 1978 estimates of air emissions from individual plants. The U.S. EPA considers these figures unreliable because of the estimation techniques used.

6. Priority Pollutant U.S. EPA, Region V
   Effluent Guidelines Eastern District Office
   Survey, July 1978 25089 Center Ridge Road
   Westlake, OH 44145
   (216) 835-5200

7. Compliance Inspection Ohio Environmental Pro-
   Report, February 1983 tection Agency (OEPA)
   Southwest District Office
   Industrial Wastewater
   Section
   7 East Fourth Street
   Dayton, OH 45402
   (513) 461-4670

8. Briefing document, 1973 OEPA
   Industrial Wastewater
   Section
   361 East Broad Street
   Columbus, OH 43215
   (614) 466-2390

9. Compliance Evaluation U.S. EPA, Region V,
   Report, April 1980 Eastern District Office
   (see #6)

10. Generator and Facility OEPA, Division of
    Annual Hazardous Waste Hazardous Waste
    Report, 1981 Management
    361 East Broad Street
    Columbus, OH 43215
    (614) 466-1598

11. RCRA "Part A" Applica- OEPA, Division of
    tion, November 1980, Hazardous Waste
    and revisions, Management
    December 1982 (see #10)

12. Information provided by
    Monsanto to INFORM

WASTE REDUCTION

MONSANTO

Type of Waste	G-Generation D-Disposal Q-Quantity (year)	Waste Reduction Practice	Results	Sources of Information
HAZARDOUS				
Acrylonitrile	Plant-wide air emissions 185,200 lbs/yr (1978)	Plant-wide program to identify and minimize all fugitive emission sources OP (1978)	?	(2)
Hazardous trash (discarded clothing, rags, containers, etc.)	Off-site disposal Q - ? (1983)	Employment of trained specialists to segregate hazardous and non-hazardous wastes OP (year unknown)	Prevention of contamination of large quantities of solid wastes from small volumes of hazardous materials	(12)
Solid wastes from a filtration step during industrial adhesives manufacture	G - ? D - ? Q - 0 (1983)	Hazardous particulate matter retained in adhesives product instead of being filtered out PR (year unknown)	Elimination of solid waste generation from filtration step and improved product yield	(12)
NON-HAZARDOUS				
Styrene	Air emissions from polystyrene manufacturing process 170 lbs/day (1983)	Transition from batch to continuous polystyrene manufacturing process PS (1952 to present)	99.7 percent reduction of emissions from 50 to 0.17 lbs per 1,000 lbs of product	(12)
UNKNOWN				
Spills and leaks	G - ? D - ? Q - ? (1983)	Company-wide program to reduce accidents and spills OP (year unknown)	?	(12)

Key to Waste Reduction Changes
PR - Product
PS - Process
EQ - Equipment
CH - Chemical Substitution
OP - Operational

409

MONSANTO

WASTE GENERATION & DISPOSAL

Type of Waste ● Indicates Wastes Affected by Reduction	Generation or Disposal	Quantity of Waste	Sources of Information
Air Emissions			
HAZARDOUS			
● Acrylonitrile	Plant-wide emissions	359,300 lbs/yr* (1978)	(3)
		185,200 lbs/yr (1978)	(4)
Formaldehyde	"	85,300 lbs/yr* (1978)	(3)
Formaldehyde, methanol, and other organic vapors from formaldehyde manufacture	Burned in an on-site boiler	?** (1983)	(1)
Phenol	Emissions from "T-Status" tanks***	? (1983)	(1)
LIKELY TO CONTAIN HAZARDOUS CHEMICALS			
Organic emissions from phenol-formaldehyde resin manufacture****	Process emissions	trace amounts (1983)	(1)
MAY CONTAIN HAZARDOUS CHEMICALS			
Particulates from plastic-pellet processing	"	1,071,000 lbs/yr (1983)	(1)
NON-HAZARDOUS			
Butadiene	"	? (1983)	(1)
● Styrene	Emissions from the polystyrene manufacturing process	170 lbs/day (1983)	(12)
	Process emissions	118,114 lbs/yr (1983)	(1)
	Emissions from "T-Status" tanks***	? (1983)	(1)

 * The U.S. EPA considers these figures unreliable because of the estimation techniques used.

 ** Trace quantities of formaldehyde are present in a waste stream comprised of nitrogen and hydrogen.

 *** "T-Status" emissions sources are considered minor by the Ohio EPA, which does not require submission of their emission rates.

 **** Monsanto discontinued phenol-formaldehyde resin manufacture in March 1984.

MONSANTO

WASTE GENERATION & DISPOSAL

Type of Waste ● Indicates Wastes Affected by Reduction	Generation or Disposal	Quantity of Waste	Sources of Information
Wastewater Discharges			
Process water, contaminated stormwater and cooling water, including:	Treated and discharged to the Ohio River (cooling water and uncontaminated stormwater are directly discharged to the Ohio River)	11,000,000 gals/day (1 million gals/day of process water and 10 million gals/day of cooling water) (1980, 1981)	(5,9)
HAZARDOUS			
Methylene chloride	"	1.76 lbs/day (max.) (1981)	(5)
Phenol	"	400 lbs/day (1973)	(8)
		.60 lbs/day (1981)	(5)
		.40 lbs/day (1983)	(7)
Antimony	"	.20 lbs/day (max.) (1981)	(5)
Arsenic	"	.17 lbs/day (max.) (1981)	(5)
Chromium	"	.45 lbs/day (max.) (1981)	(5)
Copper	"	.10 lbs/day (max.) (1981)	(5)
Cyanide	"	.002 lbs/day (max.) (1981)	(5)
Mercury	"	trace amounts (1978)	(6)
Zinc	"	? (1978)	(6)

MONSANTO

WASTE GENERATION & DISPOSAL

Type of Waste ● Indicates Wastes Affected by Reduction	Generation or Disposal	Quantity of Waste	Sources of Information
Solid Wastes			
HAZARDOUS			
Acrylonitrile wastes	?	80,000 lbs/yr (1982)	(11)
Diethyl phthalate wastes	?	4,000 lbs/yr (1982)	(11)
Formaldehyde wastes	Off-site disposal at Chemical Waste Management, Inc., Emelle, AL	52,250 lbs/yr (1981)	(10)
Isobutyl alcohol wastes	?	40,000 lbs/yr (1982)	(11)
Maleic anhydride wastes	Off-site disposal at CECOS/CER Co., Williams- burg, OH	3,000 lbs/yr (1981)	(10)
	Off-site disposal at Chemical Waste Management, Inc., Emelle, AL	7,650 lbs/yr (1981)	(10)
Methanol wastes	?	20,000 lbs/yr (1982)	(11)
Methyl ethyl ketone wastes	?	80,000 lbs/yr (1982)	(11)
Organic solvent wastes	Off-site disposal at Chemical Waste Management, Inc., Emelle, AL	36,000 lbs/yr (1981)	(10)
Phenol wastes	Off-site disposal at CECOS/CER Co., Williams- burg, OH	390,800 lbs/yr (1981)	(10)
Quinone wastes	?	16,000 lbs/yr (1982)	(11)
Toluene wastes	?	80,000 lbs/yr (1982)	(11)
Xylene wastes	?	80,000 lbs/yr (1982)	(11)
● Hazardous trash (discarded clothing, rags, con- tainers, etc.)	Off-site disposal	? (1983)	(12)

MONSANTO
WASTE GENERATION & DISPOSAL

Type of Waste ● Indicates Wastes Affected by Reduction	Generation or Disposal	Quantity of Waste	Sources of Information
Solid Wastes			
HAZARDOUS (cont.)			
● Solid wastes from a filtration step during industrial adhesives manufacture	?	0* (1983)	(12)
Total hazardous wastes	Sent off-site for disposal to sites in Ohio and Alabama	469,000 lbs/yr (1983)	(12)
		427,000 lbs/yr (1982)	(12)
		1,100,000 lbs/yr (1981)	(12)
MAY CONTAIN HAZARDOUS CHEMICALS			
Flammable wastes, unspecified	Off-site disposal at CECOS/ CER Co., Williamsburg, OH	136,400 lbs/yr (1981)	(10)
	Off-site disposal at General Portland Systech, Paulding, OH	89,080 lbs/yr (1981)	(10)
	Off-site disposal at Chemical Waste Management, Inc., Emelle, AL	171,750 lbs/yr (1981)	(10)
Organic solvent wastes, unspecified	?	80,000 lbs/yr (1982)	(11)
NON-HAZARDOUS			
Acetone wastes	?	80,000 lbs/yr (1982)	(11)
Styrene wastes	Incinerated at General Portland Systech, Paulding, OH	57,400 lbs/yr (1981)	(10)
Varnish wastes	Off-site disposal at Chemical Waste Management Inc., Emelle, AL	40,000 lbs/yr	(10)

* No waste generated since reduction technique was implemented.

Perstorp Polyols, Inc.

600 Matzinger Road
Toledo, Ohio 43612
(419) 729-5448

SUMMARY

Perstorp Polyols, Inc. (Toledo, Ohio) manufactures pentaerythritol and sodium formate from the raw materials **formaldehyde** and **acetaldehyde**. Both the addition of a filter to one source of pentaerythritol dust and the elimination of a mixing step that had produced pentaerythritol dust have reduced air emissions of that chemical. Equipment and process changes have reduced by 50 to 75 percent the level of organic chemicals discharged in the plant's wastewater. The confidential nature of information on products and discharges preclude a detailed assessment of waste reduction at the plant.

PLANT OPERATIONS

Perstorp Polyols, Inc. is a U.S. subsidiary of Perstorp AB, a large Swedish chemical company, which manufactures **formaldehyde**, formaldehyde-based chemicals, polyalcohols, resins, plastics, laminates and a variety of other products. Total 1983 sales for Perstorp AB -- which has plants in Europe, Brazil, and four in the United States -- were $346 million. Its Toledo plant is a moderate-sized facility, with less than 50 employees, located in an industrialized district on the outskirts of Toledo. Perstorp purchased the plant in 1977 from the Pan American Chemical Corporation (the owner since 1971), which manufactured the same

> NOTE: Chemicals in **boldface** type are hazardous, according to INFORM's criteria (see Appendix A).

products as Perstorp but was unable to operate the plant profitably in the face of rising energy and raw material prices.

Perstorp's Toledo plant operates 24 hours a day, seven days a week, 49 weeks a year. In its continuous manufacturing process, the raw materials **formaldehyde** (70 million pounds a year, obtained from a Du Pont plant adjacent to Perstorp) and **acetaldehyde** (15 million pounds a year) react in the presence of sodium hydroxide to yield two products, pentaerythritol (PE) and sodium formate, which are then separated from solution. Annual production of PE (used in paints, printing inks and synthetic lubricants) is 25 to 35 million pounds; annual production of sodium formate (used in the leather, textile, paper, and chemical industries) is 10 to 15 million pounds.

WASTE GENERATION AND WASTE REDUCTION

AIR EMISSIONS

Generation

Perstorp's raw materials -- **formaldehyde** and **acetaldehyde** -- are organic liquids that can readily vaporize to become air emissions. PE and sodium formate are solid powders that can generate dust during processing.

According to a U.S. EPA report, annual **formaldehyde** and **acetaldehyde** emissions in 1978 were estimated to be 602,000 pounds and 129,000 pounds, respectively. Officials at the Toledo Environmental Services Agency (TESA) believe that, because **formaldehyde** odors are detectable at low concentrations, the absence of odors at the plant suggests that the U.S. EPA figure for that raw material is an extreme overstatement. Although exact emissions figures are not available, Perstorp estimates that emissions are about 1,000 pounds per year of **formaldehyde** and 100 pounds per year of **acetaldehyde.**

Reduction

TESA files report that various improvements at Perstorp have brought about a reduction in PE emissions. During the initial stages of upgrading the plant, Perstorp produced a low grade of PE and blended it with higher-grade imported PE -- a process that resulted in PE-dust emissions. Unspecified process and equipment changes improved the quality of PE manufactured at the plant and eliminated the need to mix both grades of PE -- and, as a result, the emissions produced by that step. A second waste reduction practice -- equipping one emissions source with a baghouse filter, which captures PE dust for reuse (thereby preventing its discharge as waste) -- recovered an estimated 99 percent of PE dust from that source.

WASTEWATER DISCHARGES

Generation

Perstorp discharges between 30,000 and 75,000 gallons a day of wastewater to Toledo's sewage treatment plant. According to TESA, which operates that plant, each month's discharge contains an average of about 400,000 pounds of unspecified organic materials. The agency routinely monitors the wastewater's overall level of these materials, but collects no information on specific chemicals. There are no specific restrictions on the type or quantity of contaminants that can be discharged in Perstorp's wastewater. TESA does, however, classify Perstorp as a "high strength discharger," meaning that the plant releases levels of chemicals high enough for the agency's High Strength Surcharge Program to assess fees in addition to those already imposed on routine discharges.

Reduction

Perstorp reported to INFORM that a variety of changes introduced at the plant since its purchase from Pan American Chemical in 1977 have increased the efficiency of the plant's operations, thereby reducing by 50 to 70 percent the level of organic chemicals discharged in the plant's wastewater. The company would not specify the nature of these changes. Perstorp explained to INFORM that, because the goal of its improvements was greater efficiency in running the plant, it has not documented the impact of individual changes on waste reduction. However, as an overall result of all changes, large quantities of organic materials now remain in the product stream instead of leaving the plant as wastes.

TESA files did include mention of changes that had been made at the Perstorp plant to improve production efficiency. The changes included improvement of an existing crystallization unit and installation of a recrystallization unit and three distillation towers. However, TESA files do not indicate whether these changes led to reduced waste generation.

Information from TESA files on the levels of organic contamination in Perstorp's wastewater discharges are suggestive of the impact of waste reduction practices. Records of chemical oxygen demand (COD) -- a commonly used index of organic chemical contaminants -- in Perstorp's wastewater, show a decrease since 1981:

	Average Monthly Discharge (pounds of COD)
1981	485,611 pounds
1982	449,066 pounds
1983	390,323 pounds
1984*	396,794 pounds

The overall decrease of 18 percent from 1981 to 1984 may have resulted from waste reduction at the plant, but the numbers are difficult to interpret with precision. Decreased quantities of COD could be related

---
* Based on January-May data.

to decreased production at the plant rather than to actual waste reduction practices. Alternatively, larger reductions in the quantity of COD generated for every pound of product produced could be masked by increases in production. An absence of more detailed information on the actual levels of production and longer-term trends in waste generation precludes a full evaluation of the relation of the decline in COD to actual waste reduction practices.

Perstorp told INFORM that it is looking for process improvements which would further reduce waste loads discharged to the sewer.

SOLID WASTES

Generation

Public files do not report any routine generation of solid hazardous waste. TESA files indicate that Perstorp occasionally discards off-specification products as solid wastes.

OBSERVATIONS

Perstorp maintains that the changes in the plant operations introduced since 1977 have resulted in lower energy costs, a higher-quality product, and greater efficiency and overall yields than those of its predecessor, Pan American Chemical Corporation. As a result, Perstorp says, it can -- unlike Pan American Chemical -- show a profit. Perstorp's plant manager noted* that Pan American Chemical was not able to manufacture polyalcohols efficiently or with consistent quality in order to sell them profitably.

According to Perstorp officials, their improvements also brought about waste reduction, because many

---
* Reported in a Toledo newspaper, The Blade, February 6, 1979.

materials that previously entered the plant's wastestreams now remain in its product streams. But precisely because these improvements are essential to Perstorp's success in the marketplace, detailed publicly available information about them -- including their effects on hazardous waste generation and reduction -- is scanty. Perstorp's General Manager noted in a 1981 letter to TESA,* "Process know-how in this somewhat mature technology is the keystone to profitability and to our future in Toledo." The letter goes on to request confidentiality for two kinds of information supplied by Perstorp to TESA in response to an emissions-inventory questionnaire:

> "Question No. 10 is a direct reference to the <u>amounts of material produced</u> in our Toledo plant. This information is not so isolated in any other report. It can be used in calculations to determine market share; raw material usage; cost of raw materials; net cost of produced product; product margin estimates; etc. In short, reverse marketing and accounting procedures could provide a 'numbers' basis upon which to base a strategy designed to blunt, or eliminate, Perstorp from the marketplace in this technology."

> "Question No. 11...requests the supplied actual emission rates per ton of produced product. Any knowledgeable Production Chemical Engineer can use these rates to determine closely parallel processes to achieve these excellent results...."

TESA has honored these requests to protect company secrets. The absence of data on amounts of waste and emissions rates makes it impossible to assess in detail the full effect of Perstorp's waste reduction practices.

---

* The letter, dated July 29, 1981, is part of TESA's air permit files.

SOURCES OF INFORMATION

SOURCE	LOCATION
1. "Human Exposure to Atmospheric Concentrations of Selected Chemicals," prepared by Systems Applications, Inc. (Contract #68-02-3066), March 1983* (based on 1978 data)	U.S. Environmental Protection Agency (U.S. EPA) Office of Air Quality Planning and Standards Research Triangle Park, NC 27111 (919) 541-5315
2. Ohio Environmental Protection Agency Air Permit files, 1983	Toledo Environmental Services Agency (TESA) 26 Main Street Toledo, OH 43605 (415) 693-0350
3. City of Toledo Sewage Treatment Plant files, 1984	TESA (see #2)
4. Information provided by Perstorp to INFORM	

---

* This report gives 1978 estimates of air emissions from individual plants. The U.S. EPA considers these figures unreliable because of the estimation techniques used.

WASTE REDUCTION

PERSTORP

Type of Waste	G-Generation D-Disposal Q-Quantity (year)	Waste Reduction Practice	Results	Sources of Information
LIKELY TO CONTAIN HAZARDOUS CHEMICALS				
Organic chemicals in wastewater	Discharged to the City of Toledo sewage treatment plant 396,794 lbs/month (avg.) (1984)	Unspecified general improvements to reduce organics in wastewater ? (year unknown)	50 to 70 percent reduction in level of organic chemicals in wastewater	(2)
NON-HAZARDOUS				
Pentaerythritol (PE)	Process air emissions .64 lbs/yr (1983)	Improvements in product quality eliminating need to mix high- and low-grade PE PS, EQ (year unknown)	Elimination of PE emissions produced by the mixing step	(2)
		Addition of baghouse filter to process source EQ (1978)	99 percent recovery of particulates	(2)

Key to Waste Reduction Changes
 PR- Product
 PS- Process
 EQ- Equipment
 CH- Chemical Substitution
 OP- Operational

PERSTORP

WASTE GENERATION & DISPOSAL

Type of Waste ● Indicates Wastes Affected by Reduction	Generation or Disposal	Quantity of Waste	Sources of Information
Air Emissions			
HAZARDOUS			
Acetaldehyde	Plant-wide emissions	129,000 lbs/yr* (1978)	(1)
		100 lbs/yr (1984)	(4)
Formaldehyde	"	602,000 lbs/yr* (1978)	(1)
		1,000 lbs/yr (1984)	(4)
NON-HAZARDOUS			
● Pentaerythritol (PE)	Process emissions	.64 lbs/hour (1983)	(2)
Sodium formate	"	1.5 lbs/hour (1983)	(2)
Wastewater Discharges			
Wastewater, including: LIKELY TO CONTAIN HAZARDOUS CHEMICALS	Discharged to the City of Toledo sewage treatment plant	30,000-75,000 gals/day (1984)	(3)
● Organic chemicals**	"	396,794 lbs/month (1984 average)	(3)
		390,323 lbs/month (1983 average)	(3)
		449,066 lbs/month (1982 average)	(3)
		485,611 lbs/month (1981 average)	(3)
Solid Wastes			
NON-HAZARDOUS			
Off-specification products and other solid wastes	Disposed of in an off-site landfill	? (1983)	(2)

* The U.S. EPA considers these figures unreliable because of the estimation techniques used.
** As indicated by chemical oxygen demand (COD), a commonly used index of organic chemical contaminants.

Polyvinyl Chemical Industries

Beatrice Chemical Company
501 Green Island Road
Vallejo, California 94590
(707) 552-3500

SUMMARY

Polyvinyl Chemical Industries (Vallejo, California) manufactures varnishes, lacquers, resins and coatings. Raw materials in use include **toluene, xylene, formaldehyde** and **maleic anhydride.** The plant emits 20,000 pounds per year of organic chemicals to the atmosphere and generates hazardous solid wastes. Little detailed information was available on the types and quantities of wastes generated. There is no indication from available information that Polyvinyl has adopted any waste reduction techniques.

PLANT OPERATIONS

Polyvinyl Chemical Industries is an 18-employee plant located in Vallejo, California, a city on San Pablo Bay about 10 miles north of San Francisco. Established in 1970, this plant was called the California Resin and Chemical Company until Beatrice Foods Company (Chicago, Illinois) purchased it in 1982. Polyvinyl is a division of the Beatrice Chemical Company, a subsidiary of Beatrice Foods. In December 1984, Beatrice Company agreed to sell Polyvinyl Chemical along with the rest of Beatrice's chemical operations to Imperial Chemical Industries.

NOTE: Chemicals in **boldface** type are hazardous, according to INFORM's criteria (see Appendix A).

Polyvinyl manufactures polymers used in coatings, inks, adhesives, concrete sealers, furniture finishes and floor polishes. The Polyvinyl plant in Vallejo produces varnishes, lacquers, resins (alkyd, polyvinyl acetate and asphalt), defoamers, driers and waterproofing compounds for the paint industry. Raw materials used on-site include **toluene, xylene, formaldehyde** and **maleic anhydride.**

WASTE GENERATION AND WASTE REDUCTION

AIR EMISSIONS

Generation

The 1983 Bay Area Air Quality Management District records indicate that air emissions from Polyvinyl include 20,720 pounds per year of organics from the plant's 81 emissions sources. Chemicals stored in the plant's 61 mix and storage tanks include **toluene** and **xylene.**

SOLID WASTES

Generation

Hazardous paint manufacturing wastes are handled on-site, according to Polyvinyl's 1980 RCRA Notification Form, although no information is available as to quantity or disposal methods. In addition to hazardous wastes, the company also reported a list of hazardous chemicals it handles on-site which may be wastes, including **di-n-butyl phthalate, formic acid, isobutyl alcohol, maleic anhydride, methanol, methyl ethyl ketone, phthalic anhydride, toluene** and **xylene.**

SOURCES OF INFORMATION

SOURCE	LOCATION
1. Bay Area Air Quality Management District printout, June 1983	Bay Area Air Quality Management District 939 Ellis Street San Francisco, CA 94109 (415) 771-6000
2. RCRA Notification of Hazardous Waste Activity form, August 1980	U.S. Environmental Protection Agency, Region IX Toxics and Waste Management Division 215 Fremont Street San Francisco, CA 94105 (415) 974-8119

POLYVINYL

WASTE GENERATION & DISPOSAL

●	Type of Waste Indicates Wastes Affected by Reduction	Generation or Disposal	Quantity of Waste	Sources of Information
	**Air Emissions**			
	LIKELY TO CONTAIN HAZARDOUS CHEMICALS			
	Organic chemicals	Plant-wide emissions	20,720 lbs/yr (1983)	(1)
	**Solid Wastes**			
	HAZARDOUS			
	Air pollution control dust or sludge	?	? (1980)	(2)
	Caustic wastes from equipment cleaning	?	? (1980)	(2)
	Solvent wastes from equipment cleaning	?	? (1980)	(2)
	Wastewater treatment sludges from paint manufacture	?	? (1980)	(2)
	Di-n-butyl phthalate	These chemicals are handled on-site and may be wastes	? (1980)	(2)
	Formaldehyde	"	? (1980)	(2)
	Formic acid	"	? (1980)	(2)
	Isobutyl alcohol	"	? (1980)	(2)
	Maleic anhydride	"	? (1980)	(2)
	Methanol	"	? (1980)	(2)
	Methyl ethyl ketone	"	? (1980)	(2)
	Phthalic anhydride	"	? (1980)	(2)

WASTE GENERATION & DISPOSAL

●	Type of Waste Indicates Wastes Affected by Reduction	Generation or Disposal	Quantity of Waste	Sources of Information
	Solid Wastes			
	HAZARDOUS (cont.)			
	Toluene	These chemicals are handled on-site and may be wastes	? (1980)	(2)
	Xylene	"	? (1980)	(2)
	NON-HAZARDOUS			
	Acetone	"	? (1980)	(2)
	Acrylic acid	"	? (1980)	(2)
	n-Butyl alcohol	"	? (1980)	(2)
	Ethyl acetate	"	? (1980)	(2)

Rhone-Poulenc Inc.

297 Jersey Avenue
New Brunswick, New Jersey 08903
(201) 846-7700

SUMMARY

Rhone-Poulenc Inc. (New Brunswick, New Jersey) manufactures aroma chemicals, organic intermediates and bulk pharmaceutical chemicals. Air emissions from this plant include more than 270,000 pounds per year of **toluene**, and smaller quantities of **diethyl phthalate, formaldehyde, phenol** and **zinc chloride**. Process wastewater containing **toluene, zinc chloride** and almost 600,000 pounds per year of **phenol** is discharged to the Middlesex County Sewerage Authority, Sayreville, New Jersey. Contaminated stormwater and rinsewater, formerly discharged to Mile Run Brook, have been sent to the New Brunswick Municipal Sewer System since 1982. Solid wastes, including **phenol** and **zinc chloride**, are disposed of through off-site land burial. The disposal methods for **coumarin** wastes and other hazardous wastes are not known. **Methanol** is recycled on-site. There is no indication from available information that Rhone-Poulenc has adopted any waste reduction techniques.

PLANT OPERATIONS

Rhone-Poulenc Inc. operates a plant in New Brunswick, New Jersey, 25 miles southwest of Newark, which is one of the 11 U.S. plants of this American subsidiary of Rhone-Poulenc S.A., France's largest chemical company. Rhone-Poulenc produces chemical specialties,

NOTE: Chemicals in **boldface** type are hazardous, according to INFORM's criteria (see Appendix A).

pharmaceuticals, agricultural chemicals, and fertilizers and manufactures textiles and communication systems.

Rhone-Poulenc Inc. had sales of $300 million in 1981; sales for the entire corporation, the 12th largest chemical company in the world, were $5.5 billion in 1982. Since nationalization in early 1982, Rhone-Poulenc S.A. has been owned and operated by the French government.

The New Brunswick plant, established in 1949 and now employing 140 people, manufactures aroma chemicals, organic intermediates and bulk pharmaceuticals. It manufactures four chemical specialties for the food and fragrance markets: **coumarin,** ethyl vanillin, rhonaldehyde and cyclamenaldehyde. Rhone-Poulenc reported in a March 1984 letter to the New Jersey Department of Environmental Protection that it is the only U.S. producer of these four chemicals and that only two or three other companies in the world manufacture them.

According to the 1978 New Jersey Industrial Survey, hazardous chemicals used as raw materials at this plant include: 3.5 million pounds per year of **phenol,** used in the manufacture of salicylaldehyde; 304,940 pounds per year of **toluene,** used as a solvent in salicylaldehyde manufacture; 372,390 pounds per year of **formaldehyde,** used in methacrolein manufacture; 7,500 pounds per year of **diethyl phthalate,** used in the preparation of fragrance products; and 5,996 pounds per year of **zinc chloride,** used as a catalyst in **coumarin** and methacrolein diacetate manufacture.

WASTE GENERATION AND WASTE REDUCTION

AIR EMISSIONS

Generation

Rhone-Poulenc's plant-wide air emissions include 270,951 pounds per year of **toluene**, according to the Industrial Survey. Thus, 89 percent of the **toluene** used as a raw material is discharged as air emissions. Smaller amounts of **diethyl phthalate, formaldehyde, phenol** and **zinc chloride** are also emitted to the atmosphere.

WASTEWATER DISCHARGES

Generation

Process, non-contact cooling water and other waters are discharged after neutralization of acid wastes to the Middlesex County Sewerage Authority, Sayreville, New Jersey. The Industrial Survey reports that the 161,000 gallons-per-day discharge includes 598,230 pounds per year of **phenol** (almost 17 percent of the amount of **phenol** used as a raw material at the plant). Annual discharges also contain 28,980 pounds of **toluene** (10 percent of raw material use) and 5,958 pounds of **zinc** (99 percent of raw material use). Contaminated stormwater and rinsewater were discharged directly to Mile Run Brook, which runs through the plant, until January 1982, when the company began sending this wastewater to the New Brunswick Municipal Sewer System.

SOLID WASTES

Generation

Solid wastes are generated from distillation processes, tank bottom cleanouts and occasional spill cleanups, according to a 1981 U.S. EPA inspection report. **Phenol** and **zinc chloride** wastes are sent off-site for land burial. **Methanol** and ethyl vanillin intermediates are recycled and reused on-site. Guaethol FR409 is sent off-site to be recycled at Rhone-Poulenc's Texas plant. The quantities and disposal methods of **coumarin** wastes, unspecified waste solvents and various other wastes are not known.

SOURCES OF INFORMATION

SOURCE	LOCATION
1. New Jersey Industrial Survey, 1978	New Jersey Department of Environmental Protection (NJDEP) Office of Science and Research 190 West State Street Trenton, NJ 08625 (609) 292-6714
2. Correspondence from Rhone-Poulenc to U.S. Environmental Protection Agency, February 1982	NJDEP Division of Water Resources 1474 Prospect Street Trenton, NJ 08625 (609) 292-5602
3. Hazardous Waste Inspection Form, June 1981	U.S. Environmental Protection Agency (U.S. EPA), Region II Permits Administration Branch 26 Federal Plaza New York, NY 10278 (212) 264-9881
4. Correspondence from New Jersey Department of Environmental Protection (NJDEP) to Rhone-Poulenc, March 1984	NJDEP Division of Waste Management 32 East Hanover Street Trenton, NJ 08625 (609) 292-9879
5. RCRA Part A Application, November 1980	U.S. EPA, Region II (see #3)
6. Correspondence from Rhone-Poulenc to NJDEP, Division of Waste Management, March 1984	NJDEP (see #4)

RHONE-POULENC

WASTE GENERATION & DISPOSAL

●	Type of Waste Indicates Wastes Affected by Reduction	Generation or Disposal	Quantity of Waste	Sources of Information
	Air Emissions			
	HAZARDOUS			
	Diethyl phthalate	Plant-wide emissions	7 lbs/yr (1978)	(1)
	Formaldehyde	"	3.1 lbs/yr (1978)	(1)
	Phenol	"	81 lbs/yr (1978)	(1)
	Toluene	"	270,951 lbs/yr (1978)	(1)
	Zinc chloride	"	10 lbs/yr (1978)	(1)
	Wastewater Discharges			
	Process and other waters, including:	Discharged to the Middlesex County Sewerage Authority, Sayreville, NJ	161,000 gals/day (1978)	(1)
	HAZARDOUS			
	Phenol	"	598,230 lbs/day (1978)	(1)
	Toluene	"	28,980 lbs/yr (1978)	(1)
	Zinc chloride	"	5,958 lbs/yr (1978)	(1)
	MAY CONTAIN HAZARDOUS CHEMICALS			
	Contaminated stormwater and rinsewater	Discharged to the New Brunswick Municipal Sewer System (prior to 1982 was discharged to Mile Run Brook)	2,500 gals/day (1978, 1982)	(1,2)

WASTE GENERATION & DISPOSAL

RHONE-POULENC

●	Type of Waste Indicates Wastes Affected by Reduction	Generation or Disposal	Quantity of Waste	Sources of Information
	Solid Wastes			
	HAZARDOUS			
	Coumarin wastes	?	? (1981)	(3)
	Methanol	Recycled on-site	? (1984)	(4)
	Phenol	Stored on-site	2,000 lbs/yr (1980)	(5)
		Land burial at Waste Management Inc. of Alabama	4,290 lbs/yr (1978)	(1)
	Solvents, non-halogenated	?	50,000 lbs/yr (1980)	(5)
	Zinc chloride	Land burial at Waste Management Inc. of Alabama	1,295 lbs/yr (1978)	(1)
	NON-HAZARDOUS			
	Corrosive wastes, unspecified	Stored on-site	85,860 lbs/yr (1980)	(5)
	Ethyl vanillin intermediates	Recycled on-site	? (1981)	(3)
	Ethyl vanillin residue	?	? (1981)	(3)
	Guaethol FR409	Recycled at Rhone-Poulenc's Texas plant	? (1981)	(3)
	Ignitable wastes, unspecified	Stored on-site	70,000 lbs/yr (1980)	(5)
	Lauryl acetate	?	? (1981)	(3)
	Methacrolein residue	?	? (1981)	(3)
	Methyl dichloroacetate	?	? (1981)	(3)
	Orthotrifluoromethyl aniline	?	? (1981)	(3)

RHONE-POULENC

WASTE GENERATION & DISPOSAL

Type of Waste Indicates Wastes Affected by Reduction	Generation or Disposal	Quantity of Waste	Sources of Information
Solid Wastes			
NON-HAZARDOUS (cont.)			
Salicylaldehyde residue	?	? (1981)	(3)
Succinic anhydride	?	? (1981)	(3)
Trimethyl tetrahydrobenzyl acetate residue	?	? (1981)	(3)

Scher Chemicals, Inc.

Industrial and Styertowne Roads
Allwood P.O. Box 1236
Clifton, New Jersey 07012
(201) 471-1300

SUMMARY

Scher Chemicals, Inc. (Clifton, New Jersey) manufactures a variety of specialty chemicals chiefly for use in the cosmetics and textile industries. Raw materials at the plant include **acrylonitrile, epichlorohydrin, maleic anhydride** and **tetrachloroethylene.** No waste reduction practices were reported at the plant; existing waste generation is already considered to be at a minimum due to the extremely high efficiencies characteristic of the plant's operations.

PLANT OPERATIONS

The Scher Chemicals plant in Clifton, New Jersey, six miles north of Newark, was built in 1956 and currently employs 31 people. The company began its operations in the 1930s as a chemical distributor and did not begin manufacturing chemicals until the Clifton plant was built. Annual sales for the company are between one million and five million dollars.

NOTE: Chemicals in **boldface** type are hazardous, according to INFORM's criteria (see Appendix A).

Scher Chemicals is a manufacturer of a wide variety of additives and other specialty chemicals for industrial use. Among its more than 200 products are: surfactants, emollients, water softeners and sunscreens for the cosmetics industry; conditioners, detergents, foaming and wetting agents for shampoos and other cleaning products; lubricants, solvents and scouring agents for the textile industry; spot removers and water repellents for dry cleaning operations; and specialty coatings for use in manufacturing glass bottles.

Scher's products are entirely batch-manufactured. Raw materials in use at the plant include **acrylonitrile, epichlorohydrin, tetrachloroethylene**, and **maleic anhydride**, as well as a wide variety of fatty acids, oils and alcohols. In addition to its own manufactured products, a small part of Scher's overall business stems from acting as a distributor for cosmetics chemicals manufactured by a Swiss firm.

WASTE GENERATION AND WASTE REDUCTION

AIR EMISSIONS

Generation

Scher reported to the New Jersey Department of Environmental Protection (NJDEP) miniscule annual emissions of organic chemicals: less than 20 pounds of **acrylonitrile,** 10 pounds of **tetrachloroethylene**[*], and five pounds of **maleic anhydride.**

---

[*] Total losses of **tetrachloroethylene** were reported by Scher as 10 pounds per year of combined air emissions and wastewater discharges.

WASTEWATER DISCHARGES

Generation

Similarly small quantities of organic chemicals were included as part of Scher's 49,000 gallon-per-day discharge of wastewater to the Passaic Valley Sewerage Commission's treatment plant, according to NJDEP records. Twenty pounds per year of **acrylonitrile** and a smaller amount of **tetrachloroethylene*** were lost in the plant's wastewater, which includes rinsewater used to clean equipment in between batches and non-contact cooling water.

SOLID WASTES

Generation

The plant does not routinely generate any hazardous solid wastes although Scher reported to INFORM that occasional processes in use at the plant can generate small quantities of potentially hazardous solid wastes which are shipped off-site to a commercial waste management firm.

OBSERVATIONS

The available information on waste generation at Scher indicates only very small discharges of potentially hazardous chemicals. The company reported to INFORM that there is little or no opportunity to pursue waste reduction for such small wastestreams. Management at Scher gave several reasons for their minimal waste generation:

---

* Total losses of **tetrachloroethylene** were reported by Scher as 10 pounds per year of combined air emissions and wastewater discharges.

1) Extremely high yields are characteristic of the type of chemistry that they undertake, with minimal waste generation as a consequence. Company management reported to INFORM that yields well above 95 percent are generally obtained in the plant's operations.

2) Although the products being manufactured change frequently, the basic raw materials and processes used stay the same. The plant personnel's familiarity with these materials and processes minimizes the likelihood of process upsets that could generate wastes. In addition, there are no "environmental surprises" that might arise from handling a new and unfamiliar chemical.

3) Scher will not ordinarily accept orders for new products that involve the use of toxic or noxious chemicals, or of materials that might endanger or discomfort workers at the plant.

Scher also painted an interesting portrait of environmental management at a small chemical company. The company reported that their small size was a distinct advantage in responding to environmental regulations due to overall simplicity and flexibility of their operations relative to a larger corporation. Even in the absence of a staff person with the sole responsibility for environmental management (the Director of Technical Services at the company handles the environmental paperwork such as permits; the company President makes decisions regarding waste management[*]), the company feels it has the in-house technical expertise to properly address all environmental concerns. In addition, the plant also has adequate economic resources to finance any necessary waste mangagement changes.

---

[*] Scher told INFORM that it plans to hire a Safety Director and Plant Manager who will together handle the plant's environmental responsibilities.

Plant management at Scher challenged the common assumption that large companies, with greater capital and technical resources, can better respond to environmental concerns than smaller companies. Management reported that Scher's small size and familiarity with its operations allow the company to respond as effectively, if not more so, than a larger plant.

SOURCES OF INFORMATION

SOURCE	LOCATION
1. New Jersey Industrial Survey, 1978 data	New Jersey Department of Environmental Protection Office of Science and Research CN402 Trenton, NJ 08625 (609) 292-6714
2. User Survey Data from the Passaic Valley Sewerage Commission (PVSC) Semiannual Report (in PVSC's wastewater discharge permit file, 1981)	U.S. Environmental Protection Agency, Region II Permits Administration Branch 26 Federal Plaza New York, NY 10278 (212) 264-9881

SCHER CHEMICALS

WASTE GENERATION & DISPOSAL

● Type of Waste Indicates Wastes Affected by Reduction	Generation or Disposal	Quantity of Waste	Sources of Information
Air Emissions			
HAZARDOUS			
Acrylonitrile	Plant-wide emissions	20 lbs/yr (1978)	(1)
Maleic anhydride	"	5.2 lbs/yr (1978)	(1)
Tetrachloroethylene	"	10.2 lbs/yr or less* (1978)	(1)
Wastewater Discharges			
HAZARDOUS			
Process and cooling water including: HAZARDOUS	Discharged to the Passaic Valley Sewerage Commission	49,000 gals/day (1978, 1981)	(1,2)
Acrylonitrile	"	20 lbs/yr (1978)	(1)
Tetrachloroethylene	"	10 lbs/yr or less* (1978)	(1)

* Combined loss of tetrachloroethylene to both air and wastewater totalled 10 lbs/yr.

Shell Chemical Company

Martinez Manufacturing Complex
Chemical Operations-East
2850 Willow Pass Road
West Pittsburg, California 94596
(415) 458-0400

SUMMARY

The Shell Chemical Company (West Pittsburg, California) operates a small chemical manufacturing facility on the site of the Shell Oil Company's Martinez oil refinery. It manufactures metallic catalysts for dehydrogenation and hydrotreating processes. These catalysts contain **chromium, copper, nickel**, iron and various other metals. Shell's non-contact cooling water is discharged to an on-site ditch and then to Suisun Bay. There is no indication from available information that Shell has adopted any waste reduction techniques.

PLANT OPERATIONS

The Shell Chemical operations in West Pittsburg, California are part of Shell Oil Company's Martinez manufacturing complex, considered by the company to be "the first modern, continuous refinery in America." The Martinez complex was built in 1916 and is located in a highly industrialized area about 30 miles northeast of San Francisco. The refinery processes crude oil and produces fuels and lubricating oils. The West Pittsburg plant is one of Shell Oil Company's nine U.S. chemical manufacturing facilities.

NOTE: Chemicals in **boldface** type are hazardous, according to INFORM's criteria (see Appendix A).

Shell Oil Company is owned by the Royal Dutch/Shell Group, a holding company owned by the Royal Dutch Petroleum Company (60 percent), a Netherlands company, and the Shell Transport and Trading Company, Ltd. (40 percent), a British company. Shell Oil's Chemical Division is the seventh largest U.S. chemical company, with 1983 sales of $3.2 billion.

The Martinez chemical plant, established in 1931, employs 59 people. It manufactures inorganic metallic catalysts for dehydrogenation and hydrotreating processes which contain **chromium, copper, nickel**, bismuth, cerium, cobalt, iron, molybdenum, potassium, tungsten and vanadium. Major raw materials include iron oxide for dehydrogenation catalyst manufacture and alumina for hydrotreating catalyst manufacture. Although the plant is categorized in U.S. EPA files as an organic chemical manufacturer, there is no indication that organic chemicals are currently being produced at the Shell Chemical facility.

WASTE GENERATION AND WASTE REDUCTION

Shell discharges 30,000 gallons per day of non-contact cooling water to an on-site ditch which feeds to Suisun Bay. During wet weather, this ditch also receives stormwater runoff from Shell and other nearby plants. Shell's discharge permit for this flow indicates that the California Regional Water Quality Control Board (RWQCB) has set limitations on the plant's **chromium, nickel** and **zinc** discharges. Shell reported to INFORM that these contaminants have not been detected in the plant's current discharge.

Until February 1980, 67,000 gallons per day of non-contact cooling water and 4,000 gallons per day of process wastewater were discharged to a 72-acre evaporation pond just north of the plant, formerly owned by Shell, but sold to the Pacific Gas and Electric (PG&E) Corporation in 1973. PG&E ordered Shell to cease discharge to the pond by 1980, according to Shell's November 1979 wastewater discharge permit application. The PG&E pond was coming under closer regulatory scrutiny as a potential threat to surface and ground waters because of the contaminated sediment which had built up in the pond over the years as a result of past discharges from Shell and from other plants. Cleanup of this pond has since been placed on the state's Superfund priority list, according to a June 1982 PG&E letter to the California Department of Health Services. Shell's non-contact cooling water is now discharged to an on-site ditch and then to Suisun Bay. Process wastewater has been treated on-site since 1980, and is no longer discharged by the plant.

Information sources reviewed by INFORM do not distinguish between discharges from Shell's chemical operations and its refinery located at the same site, and thus give no detailed information about other wastewater discharges, air emissions or solid wastes from the chemical plant alone.

SOURCES OF INFORMATION

1. Regional Water Quality Control Board Order #80-29 (also in Citizens for a Better Environment files), June 1980

   Regional Water Quality Control Board
   1111 Jackson Street, Room 6040
   Oakland, CA 94607
   (415) 464-1255

   Citizens for a Better Environment
   88 First Street
   San Francisco, CA 94105
   (415) 777-1984

2. Shell Chemical, National Pollution Discharge Elimination System Permit Application (also in Citizens for a Better Environment files), November 1979

   Regional Water Quality Control Board
   (see #1)

3. Pacific Gas and Electric Company letter to the California Department of Health Services, June 1982

   Regional Water Quality Control Board
   (see #1)

4. Regional Water Quality Control Board files, "Documentation or Modification of Recommendation" report, March 1980

   Regional Water Quality Control Board
   (see #1)

5. Information supplied by Shell to INFORM

SHELL

WASTE GENERATION & DISPOSAL

Type of Waste ● Indicates Wastes Affected by Reduction	Generation or Disposal	Quantity of Waste	Sources of Information
Wastewater Discharges			
Non-contact cooling water, including:	Discharged to an on-site ditch and then to Suisun Bay	30,000 gals/day (1979, 1980)	(1,2)
HAZARDOUS			
Chromium*	"	.02–.04 mg/l (1979)	(2)
Copper	"	.02–.03 mg/l (1979)	(2)
Nickel*	"	.18–1.7 mg/l (1979)	(2)
NON-HAZARDOUS			
Aluminum	"	.43–.91 mg/l (1979)	(2)
Iron	"	.51–4.7 mg/l (1979)	(2)

* These metals were not detected in Shell's current discharge. The other metals, although not measured, are also not expected by Shell to be present in the current discharge, since process wastewater is no longer included in this discharge.

Sherwin-Williams Company

Chemical Division
501 Murray Road
Cincinnati, Ohio 45217
(513) 242-3300

SUMMARY

The Sherwin-Williams Company's plant (Cincinnati, Ohio) produces a wide variety of organic chemicals, including saccharin, corrosion inhibitors, intermediates and additives and miscellaneous specialty chemicals. Raw materials in use include **methanol, phthalic anhydride,** and more than a dozen other hazardous chemicals. Wastes are either discharged as air emissions, disposed of off-site or discharged to the Metropolitan Sewer District (MSD) as wastewater. Sherwin-Williams has reduced **1,2,4-trichlorobenzene** discharges to the sewer since 1981. These reduction efforts, including process improvements, chemistry changes and internal recycling were prompted by MSD pressure to reduce the plant's effluent odor and associated health risks to MSD employees. Sodium sulfide discharges have been eliminated through process changes. The impact of these practices cannot be fully assessed, however, since the public file references to waste reduction are vague and fragmented.

PLANT OPERATIONS

The Sherwin-Williams Company's Cincinnati, Ohio, plant is one of five U.S. manufacturing facilities of the company's Chemical Division. Established in 1966, this plant now employs over 200 people.

NOTE: Chemicals in **boldface** type are hazardous, according to INFORM's criteria (see Appendix A).

Sherwin-Williams manufactures paints, coatings and related supplies in the U.S. and in eight foreign countries. Its architectural coatings, industrial finishes, wall and floor coverings and paint-related equipment are sold in 1,417 company-owned stores throughout the U.S., under well-known labels such as "Dutch Boy," "Sherwin-Williams" and "Martin-Senour." Sherwin-Williams' 402 "Gray Drug" and "Drug Fair" stores retail prescription drugs, health and beauty aids, cosmetics and various sundries in 10 states. Sales for the entire company were $2.0 billion in 1983; Chemical Division sales were $94 million.

The Cincinnati plant manufactures a wide variety of organic chemicals. Saccharin and sodium saccharin, produced exclusively by Sherwin-Williams in the U.S., are synthetic sweeteners used in dietetic foods and beverages, snacks, toothpastes, cosmetics, pharmaceuticals and tobacco. Tarnish inhibitors produced at this plant are used in lubricating and cutting oils, antifreeze, cleaners and metal polishes; specific inhibitors include anthranilic acid, benzotriazole, tolytriazole and various solutions of these chemicals. Intermediates and additives made by Sherwin-Williams are used by the pigment, food colorant, azo dye, flavor and fragrance chemical, textile processing, electroplating, oil drilling and polymer processing industries.

The plant carries on both batch (70 percent) and continuous (30 percent) production of organic chemicals. There are usually 15 batch processes per day. Raw materials include: **methanol,** used in the methyl anthranilate and photograde benzotriazole production processes; **ortho nitroaniline,** a reactant in the ortho phenylene diamine production process; **phthalic anhydride,** a reactant in the phthalimide, monosodium 4-chlorophthalate and isatoic anhydride production processes; **1,2,4-trichlorobenzene (TCB),** used in the saccharin production process. Other chemicals in use at the plant include **benzyl chloride, dimethyl sulfate, ethylene oxide, formaldehyde,** hydrazine, **methylene chloride, monochlorobenzene, ortho toluene diamine, p-cresol, perchloroethylene** and **xylene**.

WASTE GENERATION AND WASTE REDUCTION

AIR EMISSIONS

Generation

The Southwest Ohio Air Pollution Control Agency files identify process air emissions from 13 sources at the Sherwin-Williams plant. Chemicals emitted include 24 pounds per hour of **methanol** and 0.02 pounds per hour of **TCB**.

WASTEWATER DISCHARGES

Generation

Sherwin-Williams discharges 1,330,000 gallons per day of process and other wastewaters to the Metropolitan Sewer District (MSD), according to the 1984 MSD Wastewater Permit Application. Highly acidic or alkaline wastestreams are neutralized prior to sewer discharge. Chemicals discharged to the sewage treatment plant include **TCB** and the heavy metals **cadmium, chromium, copper, lead, nickel** and **zinc**.

Reduction

MSD files cite numerous efforts at the Sherwin-Williams plant to lower its wastewater discharges of **TCB** -- a chemical used in large quantity in the plant's saccharin manufacturing processes. Sherwin-Williams reported to MSD in 1982 that they had instituted an "improved practices program," which had resulted in 25 to 30 percent lowering of **TCB** wastes present in the wastewater discharges to the sewage treatment plant. The program included operational changes and process alterations which constituted waste reduction, as well as alternative waste disposal practices for **TCB**. The detailed nature of these changes was not reported to MSD, so that the extent to which waste reduction contributed to the drop in **TCB** discharges cannot be determined.

The company reported to MSD, in 1982, that it was considering isolating and incinerating wastestreams highly contaminated with **TCB** as a further means of lowering wastewater discharges, although there is no indication as to whether this practice was actually adopted. The plant also reported, in 1983, plans to install a "coalescence system" (equipment to separate small quantities of organic liquids from other liquids) to further lower **TCB** discharges and predicted that "current evaluations/studies should result in recovery and recycling of some organics now present in our effluent."

These efforts at lowering **TCB** wastewater discharges were developed to supplement earlier **TCB** recycling efforts. Although the plant reported to MSD, in its 1981 Wastewater Permit Application, that 97 percent of the **TCB** used in manufacturing was separated, distilled and recycled directly back to the process, the company admitted to a "poor understanding" of **TCB** "lost elsewhere" during processing.

Sherwin-Williams' concerted effort to reduce the concentration of **TCB** in its wastewater is a clear result of pressure from sewage treatment plant officials to do so. MSD officials told INFORM that, in the early 1980s, management at the sewage treatment plant became increasingly concerned about worker exposure to organic fumes. MSD undertook a detailed investigation of the source of these fumes after several workers became ill from exposure to vapors with the distinctly characteristic smell of **TCB**. As a result of this investigation, the Sherwin-Williams plant was cited by MSD officials early in 1982 for discharges which caused adverse health effects on several MSD personnel.

Their findings were followed up, in April 1982, with a legal order from MSD which noted "a substantial probability that the fumes, vapors and odors emanating from (Sherwin-Williams') discharge...continue to pose an immediate threat to the health and safety of the Metropolitan Sewer District employees and to the public at large." (See "Sources of Information," #10). Sherwin-Williams was ordered to cease and desist the discharge of any materials which would continue to put workers at risk, or which would cause nuisance or pollution problems at the sewage treatment plant. In addition, MSD altered Sherwin-Williams' discharge permit, in their April 14, 1982 letter to the company, to include a restriction on the total concentration of organic vapors that could be emitted from the company's wastewater discharges.

Although **TCB** was not specifically cited in the order or the revised permit, MSD officials reported to INFORM that this was the chemical of key concern, and it therefore became the major focus of Sherwin-Williams' efforts to lower the quantity of organics in its wastewater discharges.

Sherwin-Williams reported another waste reduction measure to MSD in 1982. Discharges of sodium sulfide, which the company referred to as a "potentially hazardous, toxic, malodorous, or noxious material," were totally eliminated through an improvement in process chemistry. The details of the change, and the manufacturing process involved, were not reported. Sherwin-Williams also reported, in 1983, efforts to find a means of reducing a concentrated bleach discharge to the sewage treatment plant.

SOLID WASTES

Generation

According to the plant's Generator Hazardous Waste Annual Reports for 1981 and 1982, solid wastes contaminated with **ortho amino para cresol, ortho nitroaniline, phthalic anhydride** and **toluene diamine** are shipped off-site for disposal. Non-hazardous waste solvents are incinerated at MSD.

SOURCES OF INFORMATION

SOURCE	LOCATION
1. Air Permit files, 1984	Southwestern Ohio Air Pollution Control Agency 2400 Beekman Street Cincinnati, OH 45214 (513) 251-8777
2. Metropolitan Sewer District (MSD) Wastewater Discharge Permit Application, April 1981	City of Cincinnati-Metropolitan Sewer District (MSD) 1600 Gest Street Cincinnati, OH 45204 (513) 352-4829
3. MSD metals analyses, June 1981 to May 1983	MSD (see #2)
4. MSD Questionnaire, February 1977	MSD (see #2)
5. Correspondence from Sherwin-Williams to MSD, September 1982	MSD (see #2)
6. Correspondence from Sherwin-Williams to MSD, January 1983	MSD (see #2)
7. Generator Hazardous Waste Annual Report, 1982	Ohio Environmental Protection Agency (OEPA) Division of Hazardous Waste Management Permits and Manifest Records Section 361 E. Broad Street Columbus, OH 43215 (614) 466-1598

8. Generator Hazardous   OEPA
   Waste Annual Report,   (see #7)
   1981

9. Correspondence from MSD   MSD
   to Sherwin-Williams,   (see #2)
   April 14, 1982

10. Correspondence from MSD   MSD
    to Sherwin-Williams,   (see #2)
    April 7, 1982

SHERWIN-WILLIAMS
WASTE REDUCTION

Type of Waste	G–Generation D–Disposal Q–Quantity (year)	Waste Reduction Practice	Results	Sources of Information
HAZARDOUS Trichlorobenzene (TCB)	Discharged in wastewater to the Metropolitan Sewer District of Greater Cincinnati 100-140 ppm (1981)	Unspecified changes in process and handling of TCB to capture and recycle TCB PS/OP (1982)	?	(5)
NON-HAZARDOUS Sodium sulfide	Discharged in wastewater to the Metropolitan Sewer District of Greater Cincinnati Q - 0 (1982)	Unspecified process change PS (year unknown)	Elimination of discharge of sodium sulfide solution to sewer	(5)

Key to Waste Reduction Changes
- **PR**– Product
- **PS**– Process
- **EQ**– Equipment
- **CH**– Chemical Substitution
- **OP**– Operational

SHERWIN-WILLIAMS

WASTE GENERATION & DISPOSAL

Type of Waste ● Indicates Wastes Affected by Reduction	Generation or Disposal	Quantity of Waste	Sources of Information
Air Emissions			
HAZARDOUS			
Methanol	Emissions from 2 process sources	24.05 lbs/hour (1984)	(1)
Trichlorobenzene vapors	Emissions from 1 process source	.02 lbs/hour (1984)	(1)
NON-HAZARDOUS			
Anthranilic acid	"	4,158 lbs/yr (1984)	(1)
Isatoic anhydride	"	0.30 lbs/hour (1984)	(1)
Methyl anthranilate	"	1 lb/hour (1984)	(1)
Ortho-phenylene diamine	Emissions from 2 process sources	225.6 lbs/yr (1984)	(1)
Phthalimide and ammonia	Emissions from 1 process source	0.10 lbs/hour (1984)	(1)
Phthalimide dust	"	2,491 lbs/yr (1984)	(1)
Wastewater Discharges			
Process and other wastewater from pollution control, cooling, boiler use and equipment washdown, including:	Discharged to the Metropolitan Sewer District of Greater Cincinnati	1,330,000 gals/day (1981)	(2)
HAZARDOUS			
● Trichlorobenzene	"	100-140 ppm (1981)	(2)
Cadmium	"	.001-.005 ppm (1977, 1981-1983)	(3,4)
Chromium	"	.047-.194 ppm (1977, 1981-1983)	(3,4)

WASTE GENERATION & DISPOSAL

● Type of Waste Indicates Wastes Affected by Reduction	Generation or Disposal	Quantity of Waste	Sources of Information
Wastewater Discharges			
Process wastewater...including (cont.) HAZARDOUS (cont.)	Discharged to the Metropolitan Sewer District of Greater Cincinnati		
Copper	"	15.85-35 ppm (1981-1983)	(2,3)
Lead	"	.01-.036 ppm (1981-1983)	(3)
Nickel	"	.071-.396 ppm (1981-1983)	(3)
Zinc	"	.13-1.84 ppm (1977, 1981-1983)	(3,4)
NON-HAZARDOUS			
Bleach, concentrated	"	? (1983)	(6)
● Sodium sulfide	"	0 (1982)	(5)
Solid Wastes			
HAZARDOUS			
Ortho amino para cresol	Off-site disposal at CECOS/CER Co., Williamsburg, OH	500 lbs/yr (1982)	(7)
Ortho nitroaniline	"	2,200 lbs/yr (1981)	(8)
Phthalic anhydride	"	15,300 lbs/yr (1981)	(8)
Toluene diamine	"	500 lbs/yr (1981)	(8)
Toluene diamine waste	"	3,400 lbs/yr (1981)	(8)
Clothes, wood and waste chemicals contaminated with toluene diamine and phthalic anhydride	"	27,920 lbs/yr (1982)	(7)

SHERWIN-WILLIAMS

WASTE GENERATION & DISPOSAL

Type of Waste ● Indicates Wastes Affected by Reduction	Generation or Disposal	Quantity of Waste	Sources of Information
Solid Wastes			
NON-HAZARDOUS			
Acetone and tetrahydro-furan	Incinerated at Metropolitan Sewer District of Greater Cincinnati	22,300 lbs/yr (1981)	(8)
Acetone/silica flour mixture	Off-site disposal at CECOS/CER Co., Williamsburg, OH	4,900 lbs/yr (1981)	(8)
Wood, filters and other materials contaminated with ortho phenylene diamine	"	24,800 lbs/yr (1981)	(8)

Smith and Wesson Chemical Company, Inc.

Lear Siegler, Inc.
2399 Foreman Road
Rock Creek, Ohio 44084
(216) 563-3660

SUMMARY

The Smith and Wesson Chemical Company's plant (Rock Creek, Ohio) manufactures chemical weapons, crowd control products, line-throwing rockets and marine safety products. The only wastes identified were hydrocarbons and particulate air emissions. There is no indication from available information that Smith and Wesson has adopted any waste reduction techniques.

PLANT OPERATIONS

The Smith and Wesson Chemical Company has operated a 100-employee plant since 1968 in Rock Creek, a city in Ashtabula, the northeasternmost county in Ohio. It manufactures chemical weapons, such as tear gas and mace, gun stock finishes, crowd control products, marine safety products and line-throwing rockets. Smith and Wesson has been owned by Lear Siegler, Inc. (San Francisco, California) since 1984.

WASTE GENERATION AND WASTE REDUCTION

Smith and Wesson annually emits to the air more than 40,000 pounds of organic chemicals from its gun stock finishing operations, according to Ohio Environmental Protection Agency air permit files, along with smaller quantities of hydrocarbons and particulates from other operations. There is no information in public records pertaining to wastewater discharges or solid waste generation.

SOURCES OF INFORMATION

SOURCE	LOCATION
1. Air permit files, 1984	Ohio Environmental Protection Agency North East District Office 2110 E. Aurora Road Twinsburg, OH 44087-1969 (216) 425-9171

WASTE GENERATION & DISPOSAL

●	Type of Waste Indicates Wastes Affected by Reduction	Generation or Disposal	Quantity of Waste	Sources of Information
	Air Emissions			
	MAY CONTAIN HAZARDOUS CHEMICALS Hydrocarbons	Emissions from 4 process sources	41,182 lbs/yr (1984)	(1)
	Hydrocarbons and particulates	Emissions from 1 process source	3,120 lbs/yr (1984)	(1)

Stauffer Chemical Company

1415 South 47th Street
Richmond, California 94804
(415) 231-1328

SUMMARY

Stauffer Chemical Company's plant (Richmond, California) manufactures and formulates pesticides; separate operations at the same site carry out corporate-wide research activities. Materials in use at the plant include **carbon tetrachloride, chloroform** and **toluene**. Waste generation from using kerosene to rinse equipment has been reduced by 800 gallons per year by reusing the solvent several times before discarding it. By replacing packed seals on plant transfer pumps with mechanical seals, leaks were minimized, which prevented the loss of 2,600 gallons of product and raw materials per year. A process change reduced the amount of disposable filters needed to purify VAPAM, a weed killer, by 535 drums per year. Through a process modification, Stauffer was able to improve raw material yields and reduce aqueous waste volumes in the manufacture of DEVRINOL, a soil fumigant, resulting in savings of $200,000 per year. Confidentiality concerns, which are of particular importance to pesticide manufacturers, prevented Stauffer from releasing certain details about these waste reduction practices.

NOTE: Chemicals in **boldface** type are hazardous, according to INFORM's criteria (see Appendix A).

PLANT OPERATIONS

Stauffer Chemical Company's 500-employee facility, in operation since the early 1900s, is located in Richmond, California, across the bay from San Francisco. A number of operations are conducted at this site in addition to chemical manufacturing, including research and development, toxicology, and patent operations. The chemical plant portion of the facility, which employs 95 people, manufactures, formulates and packages agricultural pesticides. It is operated by Stauffer's Agricultural Chemical Division and is one of 10 Stauffer plants in the country manufacturing pesticides. Total U.S. sales for Stauffer in 1982 were $1.4 billion, about a third of which stemmed from pesticide sales.

The Richmond plant manufactures DEVRINOL and VAPAM, Stauffer's trade names for, respectively, a selective weed killer used in a variety of crops and a soil fumigant. Aluminum sulfate, a chemical used for water treatment and in paper manufacture, is also produced at Richmond. In addition to its manufacturing operations, the Richmond plant formulates agricultural chemicals; that is, prepares them for commercial use by a variety of operations that includes grinding chemicals into fine powders, forming pellets, or dissolving and blending pesticides in a variety of solvents. Five herbicides belonging to the chemical category of pesticides known as thiocarbamates are formulated at Richmond. Stauffer's trade names for these are EPTAM, ORDRAM, RO-NEET, SUTAN and TILLAM. A non-thiocarbamate fungicide is also formulated.

Chemical manufacturing and formulating operations at Richmond are carried out primarily in batch operations; DEVRINOL purification is the only continuous process at the plant.

The Richmond operations include a corporate-wide research facility which develops and tests products and processes for the Richmond plant as well as for other agricultural and non-agricultural operations within Stauffer.

Chemicals handled at the plant in manufacturing, formulating and research operations include **carbon disulfide, carbon tetrachloride, chloroform, monochlorobenzene** (50,000 gallons per year), **toluene, phenol,** and alpha naphthol. The research facility uses hundreds of other chemicals in quantities ranging from only a few ounces to several thousands of pounds per year.

WASTE GENERATION AND WASTE REDUCTION

AIR EMISSIONS

Generation

Stauffer told INFORM that almost all of its organic emissions are vented to an on-site incinerator or to carbon absorbers. Total emissions of organic chemicals to the atmosphere amount to 5,580 pounds per year, according to records of the Bay Area Air Quality Management District, but specific chemical contaminants were not identified in these records.

WASTEWATER DISCHARGES

Generation

The plant generates 2,400 gallons per day of wastewater from its manufacturing and formulating processes, according to Stauffer. All contaminated process waters are either reused or hauled off-site for disposal. Other wastewater streams include cooling water, boiler water, pollution control device wastewater, equipment and floor washings, groundwater, rinsewater and other waters from the research laboratory, and surface runoff from production and handling areas. These non-process sources contribute 100,000 gallons per day to the wastewater flow, but can range as high as 1.5 million gallons during heavy rainfalls. Non-process wastewater is neutralized and discharged to on-site evaporation ponds prior to final discharge to San Francisco Bay. Contaminated non-process wastewaters are pumped through carbon absorption columns before they are neutralized.

Wastewater contaminated with herbicides and other organic chemicals has also inadvertently been discharged from the Richmond plant to local groundwater, according to records of the San Francisco Regional Water Quality Control Board (SFRWQCB). Leaks at the plant have resulted in detectable levels of **toluene,** DEVRINOL and EPTAM in the groundwater. The source of the leaks is uncertain, but on-site evaporation ponds, underground storage tanks, and chemical spills at the plant are considered the most likely sources by SFRWQCB. Stauffer has reported to SFRWQCB that it has taken steps to prevent future leaks to groundwater. The plant has discontinued use of underground storage tanks for raw material or waste storage and has installed a groundwater intercept system to recover contaminated groundwater for treatment and eventual discharge to San Francisco Bay.

Stauffer also discharges domestic and non-domestic wastewater to the City of Richmond sewage treatment facility, according to a user survey conducted by the treatment plant.

Stauffer told INFORM about three wastewater reuse and recovery practices that it implemented at the Richmond plant. These practices do not affect wastes at their source (hence, are not waste reduction practices in the context of this study) but have had a pronounced impact on the total quantity of wastewater discharges from the plant.

Stauffer installed a pond which is used to recirculate aluminum sulfate (alum) process wastewater back to the plant for recovery of the wastewater's alum content. Stauffer estimates that since the pond's installation in 1973, 1,500,000 gallons of alum wastewater have been recirculated back to the process. A second measure, the installation of a sump in 1984 to collect alum drips and water used to wash down the process area, has also lowered the amount of alum wastewater being discharged by the plant. Stauffer estimates that these two practices, used to recover alum wastewater, save the plant $150,000 to $200,000 each year.

The third measure was put into effect in 1984. It involves the reuse of drips and surface runoff from the VAPAM manufacturing area and of water used to wash down this area, which are collected in a catch basin. Before 1984, the waters collected in this basin were hauled off-site for disposal. By reusing this wastewater, Stauffer estimates that it saves $46,000 each year in disposal costs.

Reduction

INFORM has learned of one waste reduction practice affecting wastewater discharges at the Richmond plant. Stauffer implemented a modification in its DEVRINOL manufacturing process in 1976 which improved raw material yields and reduced aqueous waste volumes. Stauffer estimates that this process change saves the company $200,000 each year. However, further analysis of the impact of this waste reduction practice is not possible, since Stauffer did not provide any details about the nature of the process change.

SOLID WASTES

Generation

Hazardous solid wastes are disposed of off-site at commercial waste management facilities. Wastes include **carbon tetrachloride, chloroform, monochlorobenzene** and **toluene** from manufacturing, formulating and research operations. The research operations produce a total of 11,000 to 15,000 gallons per year of highly variable chemical wastes which are disposed of off-site.

Stauffer told INFORM of a procedure it implemented to lower the number of drums of wastes it sends off-site to landfills each year. Waste solids from dust collectors, formerly disposed of separately, are now used to solidify liquid wastes prior to disposal. Fewer drums are needed to transport the combined wastes than are required for separate wastestreams. By mixing these wastes, Stauffer lowered the number of drums requiring disposal in a landfill by 120 drums each year, saving the company $6,600 annually.

Reduction

Stauffer reported to INFORM three waste reduction practices affecting solid waste generation. A change in operating procedures reduced the quantity of kerosene wastes generated at the plant. Kerosene, used as a solvent to rinse tanks and equipment, was formerly used once and then discarded. In 1980, the plant began to save and reuse kerosene several times before discharging it as waste as a means of reducing material costs and the expense of waste disposal. The company estimates that this practice reduced kerosene wastes by 800 gallons each year and saves $1,000 annually.

The second waste reduction practice reduced the loss of product and raw materials from transfer pumps by minimizing leaks with an equipment change. The pump seals, which keep the liquid in the pumps isolated from the pump mechanism itself, were switched from "packed seals" (which contain densely packed material such as graphite and synthetic fibers, and are prone to leaking) to "mechanical seals" (rotating elements which virtually eliminate leaks). Stauffer estimates that this change reduced product and raw material losses by 2,600 gallons each year, corresponding to annual savings of $37,000.

The third waste reduction practice reduced the number of disposable filters needed to purify VAPAM. Before VAPAM is shipped off-site for sale, it first passes through disposable filters that take out unwanted particles from the product stream. Through an unspecified process change, Stauffer was able to reduce the amount of residual particles left in the VAPAM product stream prior to filtration. This, in turn, reduced the number of filters needed to purify the VAPAM, reducing wastes by 535 drums each year and saving Stauffer $28,085 in annual off-site disposal costs.

OBSERVATIONS

Officials at Stauffer's Richmond plant told INFORM waste reduction has been implemented in numerous instances at the plant and included a broad range of practices, as in the examples given above. However, most of the details regarding these practices were considered confidential business information and were not made available to INFORM.

Stauffer explained that confidentiality is particularly important in pesticide manufacturing and that virtually all information related to the Richmond plant is considered proprietary. For instance, the plant is the only one in the world manufacturing DEVRINOL and confidentiality is considered a vital means of maintaining this herbicide's competitive position in the marketplace. The management at the Richmond plant also speculated that the highly proprietary nature of pesticide manufacture makes the industry more prone to waste reduction than others. Once patents expire, and competitors are free to manufacture identical products, the overall efficiency and cost-effectiveness of the operations at Stauffer are a key element to maintaining a competitive edge. This pressure makes it second nature within the industry to continually look for means to improve process efficiency and, consequently, minimize waste.

SOURCES OF INFORMATION

SOURCE	LOCATION
1. Computer printout of air emissions, June 1983	Bay Area Air Quality Management District 939 Ellis Street San Francisco, CA 94109 (415) 771-6000
2. Notice of Wastewater Discharge Requirements, February 1978*	San Francisco Regional Water Quality Control Board (SFRWQCB) 1111 Jackson Street Oakland, CA 94607 (415) 464-1255
3. Correspondence from Stauffer to San Francisco Regional Water Quality Control Board, January 20, 1978*	SFRWQCB (see #2)
4. Wastewater Discharge Permit Application, August, 1977	U.S. Environmental Protection Agency, Region IX Water Management Division 215 Fremont Street San Francisco, CA 94105 (415) 974-8119
5. Field sampling sheets for 1980-1983	City of Richmond Water Pollution Control Plant (RWPCP) 601 Canal Boulevard Richmond, CA 94804 (415) 231-2145

* Also on file with Citizens for a Better Environment, 88 First Street, San Francisco, CA 94105, (415) 777-1984

6. Industrial Waste Application and User Survey, February 1983

   RWPCP (see #5)

7. Permit for the Disposal of Extremely Hazardous Waste, September 1979*

   Department of Health Services
   2151 Berkeley Way
   Berkeley, CA 94704
   (415) 231-2145

8. Internal memo: Summary of Groundwater Pollution Problem and Status of correction, April 15, 1980

   SFRWQCB (see #2)

9. Information provided by Stauffer to INFORM

---

* Also on file with Citizens for a Better Environment, 88 First Street, San Francisco, CA 94105, (415) 777-1984

WASTE REDUCTION

STAUFFER

Type of Waste	G-Generation D-Disposal Q-Quantity (year)	Waste Reduction Practice	Results	Sources of Information
MAY CONTAIN HAZARDOUS CHEMICALS				
DEVRINOL process wastewater	Off-site disposal Q - ? (1985)	Unspecified change in the DEVRINOL manufacturing process PS (1976)	Improved raw material yields and reduced wastewater volume; $200,000 annual savings	(9)
Disposable filters used to purify VAPAM	Off-site disposal 800 drums (1985)	Unspecified change in the VAPAM manufacturing process PS/PR (year unknown)	Fewer disposable filters needed to purify product, reducing solid wastes by 535 drums; $28,085 annual savings in disposal costs	(9)
Variable chemicals, unspecified	Off-site disposal Q - ? (1985)	Packed seals on transfer pumps replaced with mechanical seals, reducing leaks EQ (year unknown)	Product and raw material losses reduced by 2,600 gallons per year; $37,000 annual savings	(9)
NON-HAZARDOUS				
Kerosene	Off-site disposal Q - ? (1985)	Reused several times before being discarded OP (1980)	Kerosene losses reduced by 800 gallons per year for material savings of $1,000	(9)

Key to Waste Reduction Changes
PR – Product
PS – Process
EQ – Equipment
CH – Chemical Substitution
OP – Operational

STAUFFER

WASTE GENERATION & DISPOSAL

Type of Waste ● Indicates Wastes Affected by Reduction	Generation or Disposal	Quantity of Waste	Sources of Information
Air Emissions			
LIKELY TO CONTAIN HAZARDOUS CHEMICALS Organic chemicals	Plant-wide emissions	5,580 lbs/yr (1983)	(1)
Wastewater Discharges			
Organic contaminants, including:	Discharged to groundwater through leaks in on-site evaporation ponds and other sources*	? (1980)	(8)
HAZARDOUS Toluene	"	? (1980)	(8)
NON-HAZARDOUS DEVRINOL	"	? (1980)	(8)
EPTAM	"	? (1980)	(8)
LIKELY TO CONTAIN HAZARDOUS CHEMICALS Process wastewater	Off-site disposal	2,400 gals/day (1985)	(9)
	Reused on-site	? (1985)	(9)
MAY CONTAIN HAZARDOUS CHEMICALS Cooling, pollution control and other wastewaters	Discharged to on-site treatment and evaporation ponds prior to discharge to San Francisco Bay	100,000 gals/day (can increase to 1.5 million gallons during storms) (1977, 1978, 1985)	(2,3,4,9)
MAY CONTAIN HAZARDOUS CHEMICALS ● DEVRINOL process wastewater	Off-site disposal	? (1985)	(9)
Other non-domestic wastewater	Discharged to the City of Richmond Water Pollution Control Plant	972,000 gals/yr (1983)	(6)

* A groundwater intercept system has been installed.

STAUFFER

WASTE GENERATION & DISPOSAL

● Type of Waste Indicates Wastes Affected by Reduction	Generation or Disposal	Quantity of Waste	Sources of Information
Solid Wastes			
HAZARDOUS			
Carbon tetrachloride	Off-site disposal	12,000 lbs/yr* (1983)	(6)
Chloroform	Off-site disposal or recycled	1,300 gals/yr* (1983)	(6)
Monochlorobenzene	Off-site disposal	500 gals/yr (1985)	(9)
Phenol	"	1,000 lbs/yr* (1983)	(6)
Toluene laboratory wastes	"	15,000 lbs/yr* (1983)	(6)
Toluene processing wastes	"	15,000 gals/yr* (1983)	(6)
Research laboratory wastes	"	11,000-15,000 gals/yr (1979)	(7)
Mercury	"	54 lbs/yr* (1983)	(6)
	Recycled off-site	16 lbs/yr* (1983)	(6)
MAY CONTAIN HAZARDOUS CHEMICALS			
● Disposable filters used to purify VAPAM	Off-site disposal	? (1985)	(9)
● Variable chemicals, unspecified	?	? (1985)	(9)
NON-HAZARDOUS			
● Kerosene	?	? (1985)	(9)

* The Richmond Industrial Waste Survey (6) lists these figures as "quantity used or discharged." Stauffer told INFORM that not all of these quantities necessarily end up as waste.

Union Chemicals Division

Union Oil Company of California
(a subsidiary of Unocal Corporation)
14445 Alondra Boulevard
La Mirada, California 90638

SUMMARY

The Union Chemicals plant (La Mirada, California) carries out solvent blending operations and manufactures numerous latex polymers. Raw materials used include **benzene, dichloromethane, methyl ethyl ketone** and a wide variety of other organic chemicals. Solvent operations result in small quantities of organic air emissions that were vented to a burner prior to 1979, but are currently discharged directly to the atmosphere. Polymer operations produce a large wastewater stream and solid waste sludges. The plant discontinued use of a mercury-containing biocide in 1971, thus eliminating **mercury** wastes. Union reported little recent waste reduction activity at the plant. Solvent operations are already highly efficient and there is little opportunity for waste reduction. The plant is considering installing dedicated piping in the latex operations as a means of reducing wastes and increasing profitability.

PLANT OPERATIONS

Union Chemicals' La Mirada plant, east of Los Angeles, was built in 1949 and currently employs 100 people. Since 1969, the plant has been fully owned by Union Oil Company of California, a company well-known for

NOTE: Chemicals in **boldface** type are hazardous, according to INFORM's criteria (see Appendix A).

its petroleum products under the "Union 76" label. Union's Chemicals Division, which operates the plant, had sales of $1.1. billion in 1983, 50 percent of which came from the manufacture of polymers and solvents, the company's two major petrochemical product areas.

The La Mirada plant is one of six Union plants manufacturing petrochemical products. It carries out two separate operations.

Solvent Blending -- The plant markets a wide variety of solvents including petroleum solvents, naphthas, alcohols and ketones, primarily to customers in the paint industry. Raw materials are stored on-site and either blended in batches or sold without any additional processing. Current production levels range from 20 to 25 million gallons per year. Blended solvents account for 30 percent of production. The solvent facility operates 10 hours per day, five days per week. The organic chemicals handled include: **benzene, methanol, methyl ethyl ketone, methyl isobutyl ketone, perchloroethylene, toluene, dichloromethane, 1,1,1 trichloroethane,** acetone, cyclohexane, hexane, textile spirits, and rubber solvent.

Latex Polymer Operations -- The plant began polymer production in 1967 and produces polyvinyl acetate and styrene-butadiene latexes (used for carpet backing materials, concrete adhesives and in paint formulations) and acrylic resins (used in paint formulations). Fifty million pounds of polymers are produced annually entirely by batch processes. The polymer facility operates 24 hours per day, seven days per week. Raw materials at the plant include vinyl acetate, butadiene, styrene, and butyl acrylate. Small quantities of biocides are added to many of the polymer products to inhibit microbial growth.

WASTE GENERATION AND WASTE REDUCTION

AIR EMISSIONS

Generation

The key sources of organic chemical emissions at the La Mirada plant, according to air permit file information, are the vapors released during loading and unloading of the 91 underground tanks used in both the solvent and polymer operations for storage of raw materials and finished products. Annual emissions from these tanks include 1,191 pounds of **methyl ethyl ketone;** 1,530 pounds of **dichloromethane;** 1,570 pounds of lactol spirits; 1,583 pounds of textile spirits; 2,657 pounds of acetone and 3,783 pounds of vinyl acetate. Fugitive emissions of the organic chemicals handled at the plant come from pumps and other process equipment and amounted to another 4,045 pounds per year.

The quantity of emissions is a small fraction of the overall volume of organic chemicals handled at the plant. For example, the plant annually uses 12,600 barrels of acetone (approximately 3.5 million pounds) and loses 2,657 pounds to the atmosphere for an overall loss rate of less than 0.08 percent. Total organic emissions (underground tanks plus fugitive emissions) are 20,671 pounds per year according to air permit records.

From 1976 to 1979, twenty-five underground tanks which were the largest emissions sources were vented to an afterburner, a pollution control device which converted 90 percent of the organics to simple gases such as carbon dioxide and water vapor. In 1979, the plant submitted an application to the Southern California Air Quality Management District (SCAQMD) to discontinue use of the afterburner. The basis for Union's request was that their volatile organic chemical (VOC)* use had fallen well below the 20,000 gallon per day level established by SCAQMD as the minimum quantity of VOC use requiring a pollution control device. Their application was accepted and use of the afterburner was discontinued. All vapors are now vented directly to the atmosphere.

WASTEWATER DISCHARGES

Generation

The latex operations produce large quantities of wastewater from washdown of equipment such as sieves, reactor vessels and tank trucks, and from a styrene concentrator-stripper unit which produces wastewater during the course of its normal operation. The solvent operations do not generate any wastewater during routine operations.

Wastewater generation from the latex operations is 20,000 to 30,000 gallons per day and contains about one percent (200 to 300 gallons) of polymer materials. Low levels of biocides in use at the latex plant might also be present in the wastewaters. Currently, this water is treated at an on-site treatment system constructed in 1974 and then discharged to the Los Coyotes Water Reclamation Plant where it receives additional treatment before being reused for landscape irrigation.

Prior to 1974, wastewaters were stored in underground tanks and then hauled off-site to be land spread. Due to odor problems caused by styrene at the land-

---

* VOCs, as defined by both federal and California state air pollution laws, are those organics with a vapor pressure (a measure of the ease with which a liquid evaporates) greater than or equal to 1.5 pounds per square inch. Approximately one-fifth of the volume of organic solvents at Union are VOCs.

spreading site, the wastewater from the styrene concentrator-stripper (10,000 gallons per day) was segregated from the equipment washdown water and hauled to a more distant site. This segregation was discontinued once the on-site wastewater treatment system was put in place.

Reduction

In 1971, the plant discontinued the use of biocides containing **mercury** due to potential environmental and toxicity problems related to continuing use of this metal.

The plant is considering installing dedicated piping -- that is, pipes which are used exclusively for a single type of material -- to avoid the need to rinse out piping each time a different chemical is introduced into the process. If adopted, this would constitute a waste reduction practice since materials currently lost to the wastewater stream would be retained in the production process.

SOLID WASTES

Generation

Non-hazardous solid waste sludges are generated from the operation of the wastewater treatment plant and are hauled off-site for disposal. There is no solid waste generation associated with the solvent operations.

OBSERVATIONS

In interviews with plant personnel, the picture that emerged of Union's solvent blending operations was that of processes that generated few wastes to begin with and that had long ago been "fine tuned" to the point of maximizing yield and minimizing waste. Hence, there is little current opportunity for waste reduction.

No bulk wastestreams are generated during Union's solvent blending operations. It is not necessary to water-rinse equipment, hence no wastewater is generated. The operations are, in a sense, "self-rinsing" -- the use of a new set of solvents for a particular batch will rinse out any residual solvents from a prior batch. By careful tracking of equipment use, the many solvents handled can be mixed in a manner that avoids cross-contamination problems. Products that do not meet specification (too viscous, for example) can be blended back into another batch and do not end up as waste material.

The plant's reliance on underground tanks for the storage of all solvents acts to minimize air emissions. Underground tanks are much less prone to temperature fluctuations than above ground tanks. Such fluctuations cause gases in the tanks to repeatedly expand and contract, and the tank "breathes" with the consequent venting and loss of some of its contents. The plant's exclusive use of underground storage tanks, although adopted due to space considerations, minimizes these breathing losses.

For the vapors that do escape from underground tanks or as fugitive emissions, there is little regulatory incentive to pursue additional means of minimizing these emissions because control of VOC emissions is not required when less than 20,000 gallons per day are used. (Above 20,000 gallons, SCAQMD regulations require control of 95 percent of VOC emissions). More than 75 percent of this plant's 20,671 pounds per year of organic emissions stems from losses of the non-VOC materials in use at Union, according to SCAQMD air permit records. These are not subject to requirements to reduce emissions further.

SOURCES OF INFORMATION

SOURCE	LOCATION
1. Union Chemicals' Air Emissions Summary, 1982 information	South Coast Air Quality Management District (SCQAMD) 9150 Flair Drive El Monte, CA 91731 (213) 572-6321
2. Air Permit files, 1983	SCQAMD (see #1)
3. Wastewater Discharge files, 1983	Los Angeles County Sanitation District 1955 Workman Mill Road Whittier, CA 90607 (213) 685-5217
4. Information supplied by Union Chemicals to INFORM	

UNION
WASTE REDUCTION

Type of Waste	G–Generation D–Disposal Q–Quantity (year)	Waste Reduction Practice	Results	Sources of Information
HAZARDOUS Mercury	Discharged in wastewater to Los Coyotes Water Reclamation Plant after on-site treatment Q – 0 (1983)	Use of biocides containing mercury discontinued CH (1971)	All wastes containing mercury eliminated	(3)

Key to Waste Reduction Changes
  PR– Product
  PS– Process
  EQ– Equipment
  CH– Chemical Substitution
  OP– Operational

UNION

WASTE GENERATION & DISPOSAL

● Type of Waste Indicates Wastes Affected by Reduction	Generation or Disposal	Quantity of Waste	Sources of Information
Air Emissions			
HAZARDOUS			
Benzene	Emissions from underground storage tanks	16.3 lbs/hour (1983)	(2)
Dichloromethane	"	1,530 lbs/yr (1982)	(1)
Methanol	"	386 lbs/yr (1982)	(1)
Methyl ethyl ketone	"	1,191 lbs/yr (1982)	(1)
Methyl isobutyl ketone	"	26 lbs/yr (1982)	(1)
Perchloroethylene	"	3 lbs/yr (1982)	(1)
Toluene	"	459 lbs/yr (1982)	(1)
1,1,1 Trichloroethane	"	500 lbs/yr (1982)	(1)
Fugitive emissions*	Emissions from compressors, pumps and other equipment	4,045 lbs/yr (1982)	(1)
MAY CONTAIN HAZARDOUS CHEMICALS			
Lactol spirits**	Emissions from underground storage tanks	1,570 lbs/yr (1982)	(1)
Naphtholite	"	220 lbs/yr (1982)	(1)
Rubber solvent**	"	330 lbs/yr (1982)	(1)
Textile spirits	"	1,583 lbs/yr (1982)	(1)

* Contains organic chemicals in use at Union.

** Union Chemicals' term for solvent containing a broad mixture of organic chemical components.

WASTE GENERATION & DISPOSAL

UNION

● Type of Waste Indicates Wastes Affected by Reduction	Generation or Disposal	Quantity of Waste	Sources of Information
Air Emissions			
NON-HAZARDOUS			
Acetone	Emissions from underground storage tanks	2,657 lbs/yr (1982)	(1)
N-Butyl acetate	"	53 lbs/yr (1982)	(1)
N-Heptane	"	337 lbs/yr (1982)	(1)
Hexane	"	1,484 lbs/yr (1982)	(1)
Isobutyl acetate	"	95 lbs/yr (1982)	(1)
Isopropanol	"	307 lbs/yr (1982)	(1)
N-Propanol	"	54 lbs/yr (1982)	(1)
N-Propyl acetate	"	58 lbs/yr (1982)	(1)
Vinyl acetate	"	3,783 lbs/yr (1982)	(1)
Wastewater Discharges			
Process wastewater and rinsewater, including:	Discharged to Los Coyotes Water Reclamation Plant after on-site treatment	20,000-30,000 gals/day (1983)	(3,4)
● HAZARDOUS Mercury	"	None* (1983)	(3)
NON-HAZARDOUS Polymers	"	200-300 gals/day (1983)	(3,4)
Solid Wastes			
NON-HAZARDOUS Sludges from the wastewater treatment plant	Off-site disposal	? (1983)	(3)

* Mercury-containing biocides used prior to 1971 may have been present in wastewater discharges.

484

USS Chemicals

Haverhill Plant
P.O. Box 127
Ironton, Ohio 45638
(614) 532-3420

SUMMARY

USS Chemicals' Haverhill plant (Ironton, Ohio) manufactures the major products **phenol, aniline,** acetone and bisphenol A and the secondary products alpha methyl styrene and diphenylamine. Since 1981, USS Chemicals has reduced **cumene** emissions by 1,115,000 pounds per year by means of an adsorber and a condenser; use of the condenser resulted from a corporate program that identifies cost-reduction measures. Other equipment additions have reduced air emissions of acetone. Management has decided to continue the injection of several tons a day of **phenol** waste into deep wells, having found this technique preferable to either recovery or reduction at the source. USS Chemicals cites the use of a novel process to manufacture **aniline**, the use of new facilities for **phenol** production and improved diphenylamine production as practices that have reduced the generation of hazardous solid wastes, but the absence of detailed information precludes an assessment of the impact of these changes.

PLANT OPERATIONS

USS Chemicals' Haverhill, Ohio, plant is located along the Ohio River in an industrial and agricultural area. (The plant is located in Haverhill but uses Ironton, a larger town, as its mailing address.) The plant,

NOTE: Chemicals in **boldface** type are hazardous, according to INFORM's criteria (see Appendix A).

which employs 224 people, was built in 1962 and purchased in 1965 by USS Chemicals, the Industrial Chemicals Division of the United States Steel Corporation. The Division's sales in 1983 were $715 million and total corporate sales were over $17.5 billion in 1983.

The plant's large-scale, integrated operations manufacture four major products (see flowchart):

-- **Phenol** is widely used to manufacture synthetic resins and other industrial chemicals. Annual **phenol** production is 500 to 600 million pounds, making Haverhill the largest phenol-producing plant in the country. About half of this amount is used to manufacture other products at the plant. A facility known as Phenol I unit (operating since 1969) yields industrial-grade **phenol**, while a newer facility, the Phenol II unit (operating since 1979), produces a higher-grade **phenol**.

-- Acetone, co-produced with **phenol**, is used as a solvent in paints, pharmaceuticals and nail polish remover. Annual acetone production is 356 million pounds, part of which is also used to manufacture other products at the plant. Acetone operations yield the secondary product alpha methyl styrene (AMS), which is used in the manufacture of industrial paints and resins. Annual AMS production is 40 million pounds.

-- Bisphenol A (BPA) is also used in the manufacture of industrial paints and resins. BPA is produced from acetone and **phenol** at Haverhill; annual BPA production is 120 million pounds.

-- **Aniline** (annual production, 200 million pounds) is used by manufacturers of polyurethane foams, dyes, photographic chemicals, and other chemical products. **Aniline** operations yield the secondary product diphenylamine (DPA), which is used to process synthetic rubber and foam. Annual DPA production is two million pounds.

Annual Flow of Chemicals at USS Chemicals' Haverhill Plant

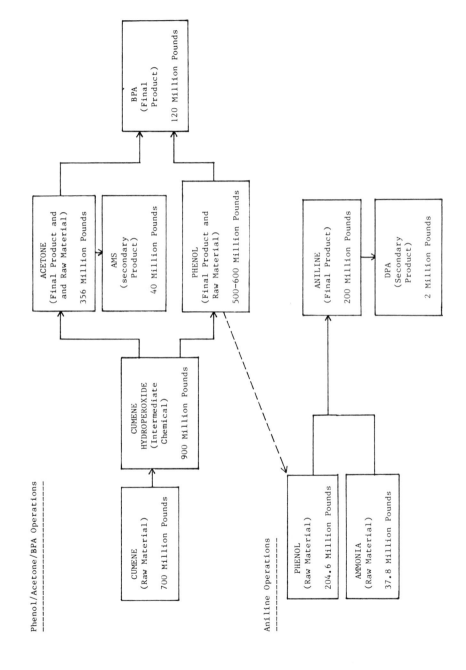

All manufacturing operations at Haverhill are relatively new. The **aniline** facility started operating in 1982, the Phenol II unit in 1979; BPA production started in 1979. The oldest facility still in operation, the Phenol I unit, was built in 1969. In the early 1980s USS Chemicals permanently discontinued its older, large-scale operations that manufactured polystyrene and industrial alcohols.

Production at the plant is continuous; all processes operate 24 hours a day, 365 days a year.

WASTE GENERATION AND WASTE REDUCTION

AIR EMISSIONS

Generation

Emissions of organic vapors from the plant's **cumene** oxidation process (an intermediate stage in the manufacture of **phenol**) are 290,000 pounds per year, according to USS Chemicals' estimates given to the U.S. EPA in 1978. A U.S. EPA report estimated that **phenol** emissions were 581,400 pounds a year in 1978*. Little current emissions information is available from government environmental files, and most air permit applications list **phenol** and **aniline** emissions from individual sources as "undetermined."

Reduction

USS Chemicals reports that two equipment changes have reduced **cumene** emissions from the Phenol I unit by 1,115,000 pounds per year. The plant has also adopted measures designed to reduce acetone vapors.

---

* USS Chemicals told INFORM that the U.S. EPA **phenol** emissions estimate is inaccurate and considerably overstated, but it did not have an alternative figure.

According to information supplied by USS Chemicals officials to the U.S. EPA, **cumene** emissions from the plant's Phenol I unit have been reduced 715,000 pounds per year (from 900,000 to 185,000 pounds) since 1979. The reduction resulted from the addition of resin adsorption systems (which soak up organic vapors for subsequent reuse) installed as part of the Phenol II unit in 1979, but which also service the older unit.

Plant management told INFORM that the following technological, financial, and regulatory factors influenced USS Chemicals' decision to use the resin adsorption system in order to reduce waste:

-- The commercial availability of new types of resins that can effectively and economically adsorb **cumene** vapors -- thereby allowing their reuse as a process material. A 1978 letter to the U.S. EPA from U.S. Steel's Associate Director for Environmental Control explained the value of the resin adsorber:

> "The new adsorber is an integral part of the process and is being installed to recover valuable raw material which will pay off the cost of the adsorber, and to clean the spent air.... This new adsorber is the equivalent of BACT* technology as only the most modern phenol plants will have this type of process system because of very recent technological improvements with new resins."

-- Dramatic changes in the economics of **phenol** production since construction of the Phenol I unit. Environmental staff cited the rise in the cost of **cumene** from "five cents a pound ten years ago to thirty cents a pound" in recent years as a major factor motivating the company to consider the costs of its wastestreams. The 715,000 pounds of **cumene** lost to the atmosphere had a value of $178,750.

---

* BACT = Best Available Control Technology, a requirement of the federal Clean Air Act (1977) for controlling emissions.

-- Environmental regulations. The adsorbers were seen as a way to insure that the plant would not be subject to more stringent air emission regulation as a result of expanding their operations, since the overall level of cumene emissions would be decreasing rather than increasing.

USS Chemicals told INFORM that use of the resin adsorbers was suggested by U.S. Steel's Technical Center in Monroeville, Pennsylvania. This research facility reviews and suggests modifications for all new processes proposed by various U.S. Steel plants, and looks for solutions to problems which plants are unable to solve themselves. After reviewing the process and engineering drawings for the Haverhill plant's new Phenol II unit, the Technical Center suggested the use of resin adsorbers based on earlier tests it had run to measure the ability of resin materials to adsorb organics from various waste streams. The plant subsequently evaluated and adopted the use of these adsorbers in order to prevent the loss of "economically valuable raw material."

The resin adsorbers were installed with the Phenol II unit, but have also been used to reduce emissions from the original **phenol** unit. However, the Technical Center's study of the use of these adsorbers was independent of plans to add the new **phenol** unit at Haverhill. USS Chemicals told INFORM that, owing to the motivating factors discussed above, resin adsorbers would have been added to the Phenol I unit even if the Phenol II unit had not been built.

USS Chemicals implemented a second equipment change that reduced **cumene** emissions after a plant operator reported **cumene** odors in the Phenol I process area. These vapors were discovered to be the result of uncontrolled vapors escaping from a pressure control vent. At the plant operator's suggestion, USS Chemicals added a surplus condenser in 1983 to return these emissions directly to the Phenol I process. The resulting recovery of 400,000 pounds of **cumene** in the first year after the change produced a savings of

$100,000 a year worth of **cumene**, compared with the $5,000 cost of installing the condenser.

The employee's idea arose through Suggestions for Cost Reduction (SCORE), a corporate program developed for U.S. Steel's chemical-manufacturing, steel-manufacturing and other operations. SCORE provides a direct financial incentive for non-management employees to identify cost reduction opportunities by rewarding them with a percentage (up to a maximum of $15,000 apiece) of the money that implementation of these measures saves the company.* Adoption of the program is at the discretion of U.S. Steel's managers. Haverhill, which evaluates almost 400 SCORE suggestions each year, has saved over half a million dollars from successful suggestions from its employees, management has estimated.

Acetone vapors have also been minimized through the use of floating roofs on storage tanks in both the Phenol I and Phenol II units to minimize evaporative losses. According to USS Chemicals, the floating roofs were part of the original design for acetone storage; their use is considered a standard industrial practice and was not prompted by any particular environmental or regulatory factor. Acetone is the only material handled at the plant which management considers volatile enough to warrant floating roofs. USS Chemicals did not report the efficiency rate of these roofs, which is typically about 90 percent.

WASTEWATER DISCHARGES

Generation

The Haverhill plant disposes of wastewater in two ways. It discharges into the Ohio River around one million

---

* SCORE suggestions are not restricted to waste reduction. Savings can result from energy conservation, reduced labor costs, process improvements and other measures.

gallons per day of wastewater that contains small quantities of organic chemicals and metals. It also injects several hundred thousand gallons of process wastewater -- which contains higher concentrations of chemicals -- into two on-site deep wells for eventual disposal in a layer of sandy earth more than a mile underground.

Average daily discharges to the Ohio River, as reported by USS Chemicals to the U.S. EPA, are 0.79 pounds of **phenol**, 2.37 pounds of **chromium**, and smaller quantities of **zinc** and other metals. Ohio EPA records indicate several higher-than-permissible discharges of pollutants -- including **phenol** discharges of more than 20 pounds a day and excessive quantities of **chromium** -- but indicate no significant permit violations since 1980.

Most of the plant's contaminated process waters are discharged to the on-site deep wells. INFORM calculates that, based on data in Ohio EPA files, daily discharges to the wells contain from 2,000 to 12,000 pounds of **phenol**, 500 pounds of **aniline** and smaller quantities of other organic wastes.

Before discharging its process wastewaters, the plant treats them by such methods as oil-water separation, carbon adsorption, filtration and settling ponds. Specific treatment received by each wastewater stream varies according to the source, the level of contamination and the destination of the wastewater. USS Chemicals reported to INFORM that some acetone is recovered from the wastewater through the use of a stripper unit that distills a portion of the wastewater stream. Plant management was uncertain of the amount of acetone recovered, explaining that this practice is only one of many small changes that are regularly implemented throughout the plant.

USS Chemicals, the U.S. EPA and the Ohio EPA are investigating the relationship between a third type of **phenol** discharge -- known leaks from the plant's waste-storage ponds and possible leaks from deep wells -- and groundwater contamination. In January 1983 the U.S. EPA, Region V, issued an administrative order directing USS Chemicals to assess the extent of existing contamination and prevent further leaks to groundwater. The company is now complying with that order. Plant management told INFORM that extensive tests have been conducted, which revealed no leaks in the deep wells. Additionally, the torn liner in the leaking BPA waste storage pond has been replaced. USS Chemicals told INFORM that all leaks were promptly repaired and that there are no outstanding problems with any of them.

Reduction

INFORM has not learned of any waste reduction practices that have affected the Haverhill plant's wastewater stream. The plant manager told INFORM that the company had explored the possibility of reducing or recovering the large quantities of **phenol** discharged into Haverhill's wells. However, no source reduction opportunities were found because the chemical equilibrium of USS Chemicals' reactions makes the generation of some **phenol** waste unavoidable. Additionally, although the Technical Center is continually evaluating technologies which could be used to recover **phenol** from the wastewater, the company currently considers deep well injection to be environmentally safe compared to such other disposal alternatives as landfilling.

USS Chemicals predicts that deep well injection costs will increase as monitoring and testing requirements are tightened by new state and federal regulations. (Plant officials note that the Ohio EPA pays particularly close attention to the 17 deep wells in the state.) If forced by regulatory changes to give up deep well injection, USS Chemicals would probably turn to adsorption as a recovery option. The company expressed concern, however, that they may be at an economic disadvantage if the Ohio EPA alone imposed the switch to a more costly option, while competitors in other states were able to pursue deep well injection or other less costly waste management options.

SOLID WASTES

Generation

The Haverhill plant generates a wide variety of wastestreams, which contain **phenol, aniline** and other hazardous constituents. Hazardous solid wastes include about two million pounds of **phenolic** sludges annually (produced by on-site wastewater treatment) as well as process residues and contaminated filters. USS Chemicals recently sent almost nine million pounds of phenol-contaminated sludge off-site to be disposed of in a hazardous waste landfill; these wastes were formerly stored in an on-site lagoon, which USS Chemicals decided to close since it would be too costly to modify it to comply with new regulations.

Current **phenol**/BPA sludge generation at the plant is considerably lower than it previously was, since the discontinuation of the use of diatomaceous earth filters as part of its treatment process. USS Chemicals found that solid particles shedding from these filters contributed substantially to sludge generation and were, in fact, increasing rather than decreasing the total solids content of the wastewater. By removing the filters, their contribution to sludge generation has been eliminated. The quantity of **phenol** in the wastewater remains unaffected by this change. Sludge generation is currently 1.3 million pounds per year, compared to earlier generation of about 2 million pounds per year.

Only recently has the Ohio EPA determined the status of two of USS Chemicals' solid waste streams, which were difficult to classify under existing federal and state regulations.

-- USS Chemicals uses 25 drums (about 16,000 pounds) a year of DOWCLENE solvent (a mixture of the chlorinated organic solvents **trichloroethane** and **perchloroethylene**) as a degreaser and, until recently, burned the waste in an on-site boiler. Although both **trichloroethane** and **perchloroethylene** are regulated as hazardous wastes, federal law does allow some exemptions for mixtures of hazardous materials, so the company did not report this practice as hazardous waste disposal. USS Chemicals stopped burning the solvent in mid-1983, when the Ohio EPA classified DOWCLENE wastes as hazardous. USS Chemicals told INFORM that it now stores the used solvent on-site and intends to send it off-site to be recycled.

-- Since 1976 the company has burned **phenolic** residues as supplemental fuel (now at the rate of 175,000 gallons, or approximately one million pounds, per month). Because the waste is used for this purpose, USS Chemicals considers it exempt from hazardous waste regulation. In August 1983 the Ohio EPA declared the residues a hazardous waste and ordered USS Chemicals to discontinue burning them. However, the same day, the agency -- acknowledging that without the option of burning the wastes as fuel the plant could not operate -- rescinded the order and allowed USS Chemicals to continue burning **phenolic** residues, pending further decisions on the status of these wastes. USS Chemicals told INFORM that the stack tests it conducted, which were required by the Ohio EPA, revealed no indication of hazardous emissions from the burning of these wastes.

Reduction

Plant officials told INFORM that three practices recently adopted at the plant have affected its generation of hazardous solid waste -- by preventing a two million pound DPA waste stream and by reducing the waste from its **phenol** and **aniline** operations.

USS Chemicals originally planned to burn, as a fuel supplement, the low-grade waste DPA, produced as a by-product of **aniline** manufacture. Unspecified reaction modification opportunities identified by U.S. Steel's Technical Center now enable the plant to use existing equipment to manufacture a grade of DPA with the commercial specifications which allow it to be marketed as a product. As a result, the plant no longer needs to burn the DPA and has eliminated a potential waste stream of that chemical. (A small, unspecified quantity of DPA is still burned as a component of the **phenolic** residues wastestream discussed above.) Plant management told INFORM that identifying a potential market for the DPA was also an important factor in the implementation of this waste reduction practice. However, company officials were uncertain as to which of these two factors -- identification of the process modifications or identification of a DPA market -- played a greater role in the implementation of this waste reduction practice.

The effect of two other reported changes on hazardous solid waste is difficult to assess. Plant officials say that Haverhill's newer Phenol II unit is more efficient and generates less waste per pound of **phenol** produced than does the Phenol I unit. However, USS Chemicals was not able to provide details that would allow a direct comparison of waste generated by the two **phenol** processes or that could, therefore, document the waste reduction impact of this change, because the wastes from these two processes are combined and treated before they are measured.

USS Chemicals also calls its new **aniline** process, which relies on a highly unusual chemical pathway, less waste-intensive than conventional **aniline** manufacture. Haverhill is the only plant in this country that uses the Halcon process, which produces **aniline** from **phenol** and ammonia. Conventional **aniline** manufacture uses **nitrobenzene** and hydrogen as raw materials. The company has provided no information that would allow for an analysis of the impact of this practice.

OBSERVATIONS

USS Chemicals told INFORM that it considers its use of mass balances to account for the fate of materials it handles in its **phenol** operations to be quite sophisticated -- it has reduced its level of uncertainty to about one to 1.5 percent. USS Chemicals explains that this small level of error can still represent a substantial quantity of material, more than one million pounds. For example, plant management told INFORM that they were unaware, until 1983, of a 400,000 pound per year **cumene** leak from a pressure control vent in the Phenol I process area. The lost material represented less than 1/10 of one percent of their total **cumene** use (700 million pounds), far too small a quantity to be detected by routine material accounting methods. They added that while they doubt the existence of other losses of this magnitude, they cannot be certain that such leaks do not occur.

SOURCES OF INFORMATION

SOURCE	LOCATION
1. USS Chemicals' Air Permit files, 1983	Portsmouth Air Pollution Agency (PAPCA) City Health Department 728 Second Street Portsmouth, OH 45662 (614) 353-5153
2. Correspondence from USS Chemicals to U.S. Environmental Protection Agency (U.S. EPA), Region V, February 22, 1978	PAPCA (see #1)
3. Information supplied by USS Chemicals to INFORM	
4. "Human Exposure to Atmospheric Concentrations of Selected Chemicals," prepared by Systems Applications, Inc. (Contract #68-02-3066) March 1983*	U.S. Environmental Protection Agency (U.S. EPA) Office of Air Quality Planning and Standards Research Triangle Park, NC 27111 (919) 541-5315

---

* This report gives 1978 estimates of air emissions for individual plants. The U.S. EPA considers these figures unreliable due to the estimation techniques used.

5. INFORM's calculations based on data contained in Ohio Environmental Protection Agency (OEPA) files, 1979-1984 average flow and concentration

Ohio Environmental Protection Agency (OEPA)
Southeast District Office
Office of Water Pollution Control
2195 Front Street
Logan, OH 43138
(614) 385-8501

6. Wastewater Discharge Permit Application filed to OEPA, 1981

U.S. EPA, Region V
Water Division
230 South Dearborn
Chicago, IL 60604
(312) 886-6112

7. Compliance Monitoring Inspection Report, July, 1980

OEPA
(see #5)

8. Wastewater Discharge files

OEPA
(see #5)

9. U.S. EPA agreement with USS Chemicals, 1975

OEPA, Wastewater Section
361 E. Broad Street
Columbus, OH 43215
(614) 466-2390

10. Generator Hazardous Waste Annual Report, 1982

OEPA
Division of Waste Management
Permits and Manifest Records
361 E. Broad Street
Columbus, OH 43215
(614) 466-1598

11. Generator Hazardous Waste Annual Report, 1981

OEPA
(see #10)

12. Correspondence from
    OEPA, Enforcement
    Section to U.S. EPA,
    Region V, 1983

    OEPA
    Southeast District Office
    Division of Hazardous
      Materials Management
    2195 Front Street
    Logan, OH 43138
    (614) 385-8501

13. USS Chemicals' closure
    plan for the Northwest
    Lagoon, March 1983
    revision

    OEPA
    (see #12)

14. United States Steel
    Corporation, 1983
    Annual Report

    Secretary
    U.S. Steel Corporation
    Room 1538
    600 Grant Street
    Pittsburgh, PA 15230
    (412) 433-1121

15. USS Chemicals'
    "Hazardous Waste
    Analysis Plan," 1982

    OEPA
    (see #12)

USS CHEMICALS

WASTE REDUCTION

Type of Waste	G-Generation D-Disposal Q-Quantity (year)	Waste Reduction Practice	Results	Sources of Information
HAZARDOUS				
Aniline-manufacturing solid wastes	G - ? D - ? Q - ? (1984)	Use of a non-conventional chemical process PS (1982)	Process generates less waste than conventional aniline manufacture	(9)
Cumene	Air emissions from the cumene process; 290,000 lbs/yr (1983)	Installation of resin adsorption systems to prevent loss of cumene vapors EQ (1981)	715,000 lbs/yr (80%) reduction of emissions from the Phenol I unit valued at $178,750/yr	(2)
	Emissions from a pressure control vent in the Phenol I unit; Q - ? (1984)	Installation of a surplus condenser to prevent the loss of cumene vapors EQ (1983)	Recovery of 400,000 pounds of cumene in first year of use; $100,000 annual savings vs. $5,000 cost of installing condenser	(3)
Phenol-manufacturing waste	G - ? D - ? Q - ? (1984)	Use of a newer, more efficient phenol unit PS (1981)	New process generates less waste than the original phenol unit	(3)
NON-HAZARDOUS				
Acetone	Air emissions from storage tanks; 2,000 lbs/yr (1978)	Installation of floating roofs on both phenol units EQ (1969, 1981)	Reduced loss of acetone vapors	(2)
Diphenylamine (DPA) solid wastes	G - ? D - ? Q - ? (1984)	Unspecified modifications improved quality of DPA, allowing its sale as a commercial product PR,PS (1982)	Eliminated need to incinerate 2,000,000 lbs/yr of DPA as waste	(3)

Key to Waste Reduction Changes
PR- Product
PS- Process
EQ- Equipment
CH- Chemical Substitution
OP- Operational

WASTE GENERATION & DISPOSAL
USS CHEMICALS

Type of Waste ● Indicates Wastes Affected by Reduction	Generation or Disposal	Quantity of Waste	Sources of Information
Air Emissions			
HAZARDOUS			
Aniline	Emissions from a diphenyl-amine storage tank	5.83 lbs/yr (1983)	(1)
	Emissions from process and storage operations	? (listed as undetermined in air permit records) (1983)	(1)
● Cumene	Emissions from cumene oxidation process	290,000 lbs/yr (64% from Phenol I unit; 36% from Phenol II) (1978)	(2)
●	Emissions from a pressure control vent in the Phenol I unit	? (1984)	(3)
	Emissions from storage tanks	14,000 lbs/yr (1978)	(2)
Phenol	Plant-wide emissions	581,400 lbs/yr* (1978)	(4)
	Emissions from a storage tank in Phenol I unit	5,400 lbs/yr (1978)	(2)
NON-HAZARDOUS			
● Acetone	Emissions from storage tanks	2,000 lbs/yr (1978)	(2)
	Process emissions	2,000 lbs/yr (1978)	(2)
	Emissions from a feed tank	2,000 lbs/yr (1978)	(2)
Alpha methyl styrene (AMS)	Emissions from a storage tank	3,000 lbs/yr (1978)	(2)
Bisphenol A (BPA)	Process emissions	12 lbs/hour (1983)	(1)
Diphenylamine (DPA)	Emissions from 1 storage tank	0.23 lbs/yr (1983)	(1)
Ethyl amine	Emissions from 1 DPA storage tank	9.34 lbs/hour (1983)	(1)

* The U.S. EPA considers this figure unreliable because of the estimation techniques used.

WASTE GENERATION & DISPOSAL

USS CHEMICALS

● Type of Waste Indicates Wastes Affected by Reduction	Generation or Disposal	Quantity of Waste	Sources of Information
Wastewater Discharges			
Wastewater, including: HAZARDOUS	Discharged to the Ohio River	1 million gals/day (1981)	(6)
Aniline	"	? (1984)	(3)
Phenol	"	0.79 lbs/day (1981)	(6)
		21.7 lbs/day (1981)	(7)
Chromium	"	2.37 lbs/day (1980)	(7)
Lead	"	0.37 lbs/day (1980)	(7)
Zinc	"	0.66 lbs/day (1980)	(7)
Wastewater, including: HAZARDOUS	Injected into two on-site deep wells	119,000-202,000 gals/day (1980-1984)	(5)
Aniline	"	500 lbs/day (1980-1984)	(5)
Phenol	"	2,000 - 12,000 lbs/day (1980-1984)	(5)
Toluene	"	10 lbs/day (1980-1984)	(5)
NON-HAZARDOUS Dipheylamine (DPA)	"	8 lbs/day (1980-1984)	(5)
HAZARDOUS Phenol	Possible discharge to groundwater from leaks in on-site storage ponds	? (1975, 1983)	(8,9)

WASTE GENERATION & DISPOSAL

USS CHEMICALS

Type of Waste ● Indicates Wastes Affected by Reduction	Generation or Disposal	Quantity of Waste	Sources of Information
Solid Wastes			
HAZARDOUS			
● Aniline-manufacturing wastes	?	? (1984)	(3)
Asbestos waste	Sent off-site for disposal at CECOS/CER Co., Williamsburg, OH	400 lbs/yr (1982)	(10)
	"	2,500 lbs/yr (1981)	(11)
DOWCLENE solvent (perchloroethylene and trichloroethylene)	Recycled off-site	15,822 lbs/yr (1982)	(3, 15)
"Lab packs" (assorted research chemicals)	Off-site disposal at CECOS/CER Co., Williamsburg, OH	1,500 lbs/yr (1982)	(10)
● Phenol manufacturing wastes	?	? (1984)	(3)
Phenolic wastes, other:			
BPA process wastes containing phenol	Off-site disposal at CECOS/CER Co., Williamsburg, OH	24,960 lbs/yr (1982)	(10)
Carbon filters	"	30,800 lbs/yr (1982)	(10)
Contaminated soil from discontinued on-site lagoon*	Excavation and shipment to CECOS/CER Co., Williamsburg, OH	2,500,000 lbs/yr (1983)	(13)
Contaminated trash (clothing, packaging, etc.)	Off-site disposal at CECOS/CER Co., Williamsburg, OH	1,200 lbs/yr (1982)	(10)
		200 lbs/yr (1981)	(11)
Guard filters	"	21,500 lbs/yr (1981)	(11)
Phenol/BPA sludge from treatment plant	Stored in on-site ponds	1,300,000 lbs/yr (1984)	(3)
		2,285,000 lbs/yr (1982)	(10)
		1,890,000 lbs/yr (1981)	(11)

* USS Chemicals told INFORM that lagoon excavation was completed in 1984.

WASTE GENERATION & DISPOSAL

USS CHEMICALS

● Type of Waste Indicates Wastes Affected by Reduction	Generation or Disposal	Quantity of Waste	Sources of Information
Solid Wastes			
HAZARDOUS (cont.)			
Phenolic wastes, other (cont.)			
Phenol/BPA spill cleanup*	Off-site disposal at CECOS/CER Co., Williamsburg, OH	38,280 lbs/yr (1982)	(10)
	Off-site disposal at Chemical Waste Management Inc., Emelle, AL	481,920 lbs/yr (1982)	(10)
Process residues	Burned as fuel supplement on-site	2,100,000 gals/yr (1983)	(12)
Sludge from discontinued on-site lagoon	Excavation and shipment to CECOS/CER Co., Williamsburg, OH**	8,896,000 lbs/yr (1983)	(13)
Still bottom residues	Off-site disposal at CECOS/CER Co., Williamsburg, OH	1,760 lbs/yr (1982)	(10)
		16,000 lbs/yr (1981)	(11)
MAY CONTAIN HAZARDOUS CHEMICALS			
Polystyrene waste***	"	240 lbs/yr (1982)	(10)
		25,320 lbs/yr (1981)	(11)
Styrene sludge from tanks***	Off-site disposal at Chemical Waste Management Inc., Emelle, AL	3,520 lbs/yr (1982)	(10)
NON-HAZARDOUS			
● Diphenylamine (DPA)	?	? (1984)	(3)

* Spill cleanup was completed in 1982.
** USS Chemicals told INFORM that lagoon excavation was completed in 1984.
***Polystyrene operations were discontinued in the early 1980s.

Appendix A
List of Hazardous Chemicals

List of Hazardous Chemicals

"Hazardous wastes", as used in the context of this report, refers to any waste material containing any of the chemicals listed (a) as "hazardous" under section 112 of the Clean Air Act, (b) as "priority pollutants" under the federal Clean Water Act or (c) as "toxic" under the Resource Conservation and Recovery Act. These three lists, which overlap substantially, are lists of known or suspected toxic chemicals.

Nine of the chemicals listed here are among the 14 organic chemicals with an annual U.S. production of more than 5 billion pounds a year. These chemicals are: benzene, ethyl benzene, ethylene dichloride, ethylene oxide, formaldehyde, methanol, toluene, vinyl chloride and xylene.

Hazardous Air Pollutants

Six chemicals are listed under Section 112 of the Clean Air Act as hazardous air pollutants. The general category of "radionuclides" is also listed under Section 112, but radioactive materials are not included in the scope of INFORM's report.

       asbestos
       beryllium
       mercury
       vinyl chloride
       benzene
       arsenic

126 "Priority Pollutants"

This list contains the "priority pollutants" regulated by the Clean Water Act amendments of 1977 as toxic chemicals. Three chemicals have been removed by the U.S. Environmental Protection Agency from the original list of 129 substances.

Volatiles
acrolein
acrylonitrile
benzene
bis (chloromethyl) ether *
bromoform
carbon tetrachloride
chlorobenzene
chlorodibromomethane
chloroethane
2-chloroethylvinyl ether
chloroform
dichlorobromomethane
dichlorodifluoromethane *
1,1-dichloroethane
1,2-dichloroethane
1,1-dichloroethylene
1,2-dichloropropane
1,2-dichloropropylene
ethylbenzene
methyl bromide
methyl chloride
methylene chloride
1,1,2,2-tetrachloroethane
tetrachloroethylene
toluene
1,2-trans-dichloroethylene
1,1,1-trichloroethane
1,1,2-trichloroethane
trichloroethylene
trichlorofluoromethane *
vinyl chloride

Acid Compounds
2-chlorophenol
2,4-dichlorophenol
2,4-dimethylphenol
4,6-dinitro-o-cresol
2,4-dinitrophenol
2-nitrophenol
4-nitrophenol
p-chloro-m-cresol
pentachlorophenol
phenol
2,4,6-trichlorophenol

---
* de-listed by EPA in 1982

Pesticides
aldrin
α-BHC
β-BHC
γ-BHC
δ-BHC
chlordane
4,4'-DDT
4,4'-DDE
4,4'-DDD
dieldrin
α-endosulfan
β-endosulfan
endosulfan sulfate
endrin
endrin aldehyde
heptachlor
heptachlor epoxide
PCB-1242
PCB-1254
PCB-1221
PCB-1232
PCB-1248
PCB-1260
PCB-1016
toxaphene

Base/Neutral
acenaphthene
acenaphthylene
anthracene
benzidine
benzo (a) anthracene
benzo (a) pyrene
3,4-benzofluoranthene
benzo (ghi) perylene
benzo (k) fluoranthene
bis (2-chloroethoxy) methane
bis (2-chloroethyl) ether
bis (2-chloroisopropyl) ether
bis (2-ethylhexyl) phthalate
4-bromophenyl phenyl ether
butylbenzyl phthalate
2-chloronaphthalene
4-chlorophenyl phenyl ether
chrysene
dibenzo (a,h) anthracene
1,2-dichlorobenzene
1,3-dichlorobenzene
1,4-dichlorobenzene
3,3-dichlorobenzidine
diethyl phthalate
dimethyl phthalate
di-n-butyl phthalate
2,4-dinitrotoluene
2,6-dinitrotoluene
di-n-octyl phthalate
1,2-diphenylhydrazine
  (as azobenzene)
fluoranthene
fluorene
hexachlorobenzene
hexachlorobutadiene
hexachlorocyclopentadiene
hexachloroethane
indeno (1,2,3-cd) pyrene
isophorone
naphthalene
nitrobenzene
N-nitrosodimethylamine
N-nitrosodi-n-propylamine
N-nitrosodiphenylamine
phenanthrene
pyrene
1,2,4-trichlorobenzene

Other Toxic Pollutants:
(Metals, Cyanide, and Total Phenols)
Antimony, Total
Arsenic, Total
Beryllium, Total
Cadmium, Total
Chromium, Total
Copper, Total
Lead, Total
Mercury, Total
Nickel, Total
Selenium, Total
Silver, Total
Thallium, Total
Zinc, Total
Cyanide, Total
Phenols, Total

Asbestos

RCRA Toxic Chemicals

Over 350 toxic chemicals are listed by the U.S. Environmental Protection Agency under the Resource Conservation and Recovery Act (RCRA).

(i) This list contains toxic chemicals listed under RCRA as "hazardous constituents".

Acetaldehyde
(Acetato)phenylmercury
Acetonitrile
3-(alpha-Acetonylbenzyl)-4-hydroxycoumarin and salts
2-Acetylaminofluorene
Acetyl chloride
1-Acetyl-2-thiourea
Acrolein
Acrylamide
Acrylonitrile
Aflatoxins
Aldrin
Allyl alcohol
Aluminum phosphide
4-Aminobiphenyl
6-Amino-1,1a,2,8,8a,8b-hexahydro-8-(hydroxymethyl)-8a-methoxy-5-methylcarbamate azirino(2',3':3,4) pyrrolo(1,2-a)indole-4,7-dione (ester) (Mitomycin C)
5-(Aminomethyl)-3-isoxazolol
4-Aminopyridine
Amitrole
Antimony and compounds, N.O.S.[1]
Aramite
Arsenic and compounds, N.O.S.
Arsenic acid
Arsenic pentoxide
Arsenic trioxide
Auramine
Azaserine
Barium and compounds, N.O.S.
Barium cyanide
Benz[c]acridine
Benz[a]anthracene
Benzene
Benzenearsonic acid
Benzenethiol
Benzidine
Benzo[a]anthracene
Benzo[b]fluoranthene
Benzo[j]fluoranthene
Benzo[a]pyrene
Benzotrichloride
Benzyl chloride
Beryllium and compounds, N.O.S.
Bis(2-chloroethoxy)methane
Bis(2-chloroethyl) ether
N,N-Bis(2-chloroethyl)-2-naphthylamine
Bis(2-chloroisopropyl) ether
Bis(chloromethyl) ether
Bis(2-ethylhexyl) phthalate
Bromoacetone
Bromomethane
4-Bromophenyl phenyl ether
Brucine
2-Butanone peroxide
Butyl benzyl phthalate
2-sec-Butyl-4,6-dinitrophenol (DNBP)
Cadmium and compounds, N.O.S.
Calcium chromate
Calcium cyanide
Carbon disulfide
Chlorambucil
Chlordane (alpha and gamma isomers)
Chlorinated benzenes, N.O.S.
Chlorinated ethane, N.O.S.
Chlorinated naphthalene, N.O.S.
Chlorinated phenol, N.O.S.
Chloroacetaldehyde
Chloroalkyl ethers
p-Chloroaniline
Chlorobenzene
Chlorobenzilate
1-(p-Chlorobenzoyl)-5-methoxy-2-methylindole-3-acetic acid
p-Chloro-m-cresol
1-Chloro-2,3-epoxybutane
2-Chloroethyl vinyl ether
Chloroform
Chloromethane
Chloromethyl methyl ether
2-Chloronaphthalene
2-Chlorophenol
1-(o-Chlorophenyl)thiourea
3-Chloropropionitrile
alpha-Chlorotoluene
Chlorotoluene, N.O.S.
Chromium and compounds, N.O.S.

[1] The abbreviation N.O.S. signifies those members of the general class "not otherwise specified" by name in this listing.

Chrysene
Citrus red No. 2
Copper cyanide
Creosote
Crotonaldehyde
Cyanides (soluble salts and complexes), N.O.S.
Cyanogen
Cyanogen bromide
Cyanogen chloride
Cycasin
2-Cyclohexyl-4,6-dinitrophenol
Cyclophosphamide
Daunomycin
DDD
DDE
DDT
Diallate
Dibenz[a,h]acridine
Dibenz[a,j]acridine
Dibenz[a,h]anthracene(Dibenzo[a,h]anthracene)
7H-Dibenzo[c,g]carbazole
Dibenzo[a,e]pyrene
Dibenzo[a,h]pyrene
Dibenzo[a,i]pyrene
1,2-Dibromo-3-chloropropane
1,2-Dibromoethane
Dibromomethane
Di-n-butyl phthalate
Dichlorobenzene, N.O.S.
3,3'-Dichlorobenzidine
1,1-Dichloroethane
1,2-Dichloroethane
trans-1,2-Dichloroethane
Dichloroethylene, N.O.S.
1,1-Dichloroethylene
Dichloromethane
2,4-Dichlorophenol
2,6-Dichlorophenol
2,4-Dichlorophenoxyacetic acid (2,4-D)
Dichloropropane
Dichlorophenylarsine
1,2-Dichloropropane
Dichloropropanol, N.O.S.
Dichloropropene, N.O.S.
1,3-Dichloropropene
Dieldrin
Diepoxybutane
Diethylarsine
0,0-Diethyl-S-(2-ethylthio)ethyl ester of phosphorothioic acid
1,2-Diethylhydrazine
0,0-Diethyl-S-methylester phosphorodithioic acid
0,0-Diethylphosphoric acid, 0-p-nitrophenyl ester
Diethyl phthalate
0,0-Diethyl-0-(2-pyrazinyl)phosphorothioate
Diethylstilbestrol

Dihydrosafrole
3,4-Dihydroxy-alpha-(methylamino)-methyl benzyl alcohol
Di-isopropylfluorophosphate (DFP)
Dimethoate
3,3'-Dimethoxybenzidine
p-Dimethylaminoazobenzene
7,12-Dimethylbenz[a]anthracene
3,3'-Dimethylbenzidine
Dimethylcarbamoyl chloride
1,1-Dimethylhydrazine
1,2-Dimethylhydrazine
3,3-Dimethyl-1-(methylthio)-2-butanone-0-((methylamino) carbonyl)oxime
Dimethylnitrosoamine
alpha,alpha-Dimethylphenethylamine
2,4-Dimethylphenol
Dimethyl phthalate
Dimethyl sulfate
Dinitrobenzene, N.O.S.
4,6-Dinitro-o-cresol and salts
2,4-Dinitrophenol
2,4-Dinitrotoluene
2,6-Dinitrotoluene Di-n-octyl phthalate
1,4-Dioxane
1,2-Diphenylhydrazine
Di-n-propylnitrosamine
Disulfoton
2,4-Dithiobiuret
Endosulfan
Endrin and metabolites
Epichlorohydrin
Ethyl cyanide
Ethylene diamine
Ethylenebisdithiocarbamate (EBDC)
Ethyleneimine
Ethylene oxide
Ethylenethiourea
Ethyl methanesulfonate
Fluoranthene
Fluorine
2-Fluoroacetamide
Fluoroacetic acid, sodium salt
Formaldehyde
Glycidylaldehyde
Halomethane, N.O.S.
Heptachlor
Heptachlor epoxide (alpha, beta, and gamma isomers)
Hexachlorobenzene
Hexachlorobutadiene
Hexachlorocyclohexane (all isomers)
Hexachlorocyclopentadiene
Hexachloroethane
1,2,3,4,10,10-Hexachloro-1,4,4a,5,8,8a-hexahydro-1,4:5,8-endo,endo-dimethanonaphthalene
Hexachlorophene
Hexachloropropene
Hexaethyl tetraphosphate

Hydrazine
Hydrocyanic acid
Hydrogen sulfide
Indeno(1,2,3-c,d)pyrene
Iodomethane
Isocyanic acid, methyl ester
Isosafrole
Kepone
Lasiocarpine
Lead and compounds, N.O.S.
Lead acetate
Lead phosphate
Lead subacetate
Maleic anhydride
Malononitrile
Melphalan
Mercury and compounds, N.O.S.
Methapyrilene
Methomyl
2-Methylaziridine
3-Methylcholanthrene
4,4'-Methylene-bis-(2-chloroaniline)
Methyl ethyl ketone (MEK)
Methyl hydrazine
2-Methyllactonitrile
Methyl methacrylate
Methyl methanesulfonate
2-Methyl-2-(methylthio)propionaldehyde-o-(methylcarbonyl) oxime
N-Methyl-N'-nitro-N-nitrosoguanidine
Methyl parathion
Methylthiouracil
Mustard gas
Naphthalene
1,4-Naphthoquinone
1-Naphthylamine
2-Naphthylamine
1-Naphthyl-2-thiourea
Nickel and compounds, N.O.S.
Nickel carbonyl
Nickel cyanide
Nicotine and salts
Nitric oxide
p-Nitroaniline
Nitrobenzene
Nitrogen dioxide
Nitrogen mustard and hydrochloride salt
Nitrogen mustard N-oxide and hydrochloride salt
Nitrogen peroxide
Nitrogen tetroxide
Nitroglycerine
4-Nitrophenol
4-Nitroquinoline-1-oxide
Nitrosamine, N.O.S.
N-Nitrosodi-N-butylamine
N-Nitrosodiethanolamine
N-Nitrosodiethylamine
N-Nitrosodimethylamine
N-Nitrosodiphenylamine
N-Nitrosodi-N-propylamine
N-Nitroso-N-ethylurea
N-Nitrosomethylethylamine
N-Nitroso-N-methylurea
N-Nitroso-N-methylurethane
N-Nitrosomethylvinylamine
N-Nitrosomorpholine
N-Nitrosonornicotine
N-Nitrosopiperidine
N-Nitrosopyrrolidine
N-Nitrososarcosine
5-Nitro-o-toluidine
Octamethylpyrophosphoramide
Oleyl alcohol condensed with 2 moles ethylene oxide
Osmium tetroxide
7-Oxabicyclo[2.2.1]heptane-2,3-dicarboxylic acid
Parathion
Pentachlorobenzene
Pentachloroethane
Pentachloronitrobenzene (PCNB)
Pentacholorophenol
Phenacetin
Phenol
Phenyl dichloroarsine
Phenylmercury acetate
N-Phenylthiourea
Phosgene
Phosphine
Phosphorothioic acid, O,O-dimethyl ester, O-ester with N,N-dimethyl benzene sulfonamide
Phthalic acid esters, N.O.S.
Phthalic anhydride
Polychlorinated biphenyl, N.O.S.
Potassium cyanide
Potassium silver cyanide
Pronamide
1,2-Propanediol
1,3-Propane sultone
Propionitrile
Propylthiouracil
2-Propyn-1-ol
Pryidine
Reserpine
Saccharin
Safrole
Selenious acid
Selenium and compounds, N.O.S.
Selenium sulfide
Selenourea
Silver and compounds, N.O.S.
Silver cyanide
Sodium cyanide
Streptozotocin
Strontium sulfide
Strychnine and salts
1,2,4,5-Tetrachlorobenzene
2,3,7,8-Tetrachlorodibenzo-p-dioxin (TCDD)

Tetrachloroethane, N.O.S.
1,1,1,2-Tetrachloroethane
1,1,2,2-Tetrachloroethane
Tetrachloroethene (Tetrachloroethylene)
Tetrachloromethane
2,3,4,6-Tetrachlorophenol
Tetraethyldithiopyrophosphate
Tetraethyl lead
Tetraethylpyrophosphate
Thallium and compounds, N.O.S.
Thallic oxide
Thallium (I) acetate
Thallium (I) carbonate
Thallium (I) chloride
Thallium (I) nitrate
Thallium selenite
Thallium (I) sulfate
Thioacetamide
Thiosemicarbazide
Thiourea
Thiuram
Toluene
Toluene diamine
o-Toluidine hydrochloride
Tolylene diisocyanate
Toxaphene
Tribromomethane

1,2,4-Trichlorobenzene
1,1,1-Trichloroethane
1,1,2-Trichloroethane
Trichloroethene (Trichloroethylene)
Trichloromethanethiol
2,4,5-Trichlorophenol
2,4,6-Trichlorophenol
2,4,5-Trichlorophenoxyacetic acid (2,4,5-T)
2,4,5-Trichlorophenoxypropionic acid (2,4,5-TP) (Silvex)
Trichloropropane, N.O.S.
1,2,3-Trichloropropane
0,0,0-Triethyl phosphorothioate
Trinitrobenzene
Tris(1-azridinyl)phosphine sulfide
Tris(2,3-dibromopropyl) phosphate
Trypan blue
Uracil mustard
Urethane
Vanadic acid, ammonium salt
Vanadium pentoxide (dust)
Vinyl chloride
Vinylidene chloride
Zinc cyanide
Zinc phosphide

[FR Doc. 80-14307 Filed 5-16-80; 8:45 am]

(ii) This list contains toxic chemicals listed under RCRA as "discarded commercial chemical products". It overlaps substantially with, but is not identical to, the list in (i).

Acetaldehyde
Acetone (I)
Acetonitrile (I,T)
Acetophenone
2-Acetylaminoflourene
Acetyl chloride (C,T)
Acrylamide
Acetylene tetrachloride
Acetylene trichloride
Acrylic acid (I)
Acrylonitrile
AEROTHENE TT
3-Amino-5-(p-acetamidophenyl)-1H-1,2,4-triazole, hydrate
6-Amino-1,1a,2,8,8a,8b-hexahydro-8-(hydroxymethyl)8-methoxy-5-methylcarbamate azirino(2',3':3,4) pyrrolo(1,2-a) indole-4, 7-dione (ester)
Amitrole
Aniline
Asbestos
Auramine
Azaserine
Benz[c]acridine
Benzal chloride
Benz[a]anthracene
Benzene
Benzenesulfonyl chloride (C,R)
Benzidine
1,2-Benzisothiazolin-3-one, 1,1-dioxide
Benzo[a]anthracene
Benzo[a]pyrene
Benzotrichloride (C,R,T)
Bis(2-chloroethoxy)methane
Bis(2-chloroethyl) ether
N,N-Bis(2-chloroethyl)-2-naphthylamine
Bis(2-chloroisopropyl) ether
Bis(2-ethylhexyl) phthalate
Bromomethane
4-Bromophenyl phenyl ether
n-Butyl alcohol (I)
Calcium chromate
Carbolic acid
Carbon tetrachloride
Carbonyl fluoride
Chloral
Chlorambucil
Chlordane

Chlorobenzene
Chlorobenzilate
p-Chloro-m-cresol
Chlorodibromomethane
1-Chloro-2,3-epoxypropane
CHLOROETHENE NU
Chloroethyl vinyl ether
Chloroethene
Chloroform (I,T)
Chloromethane (I,T)
Chloromethyl methyl ether
2-Chloronaphthalene
2-Chlorophenol
4-Chloro-o-toluidine hydrochloride
Chrysene
C I 23060
Creosote
Cresols
Crotonaldehyde
Cresylic acid
Cumene
Cyanomethane
Cyclohexane (I)
Cyclohexanone (I)
Cyclophosphamide
Daunomycin
DDD
DDT
Diallate
Dibenz[a,h]anthracene
Dibenzo[a,h]anthracene
Dibenzo[a,i]pyrene
Dibromochloromethane
1,2-Dibromo-3-chloropropane
1,2-Dibromoethane
Dibromomethane
Di-n-butyl phthalate
1,2-Dichlorobenzene
1,3-Dichlorobenzene
1,4-Dichlorobenzene
3,3'-Dichlorobenzidine
1,4-Dichloro-2-butene
3,3'-Dichloro-4,4'-diaminobiphenyl
Dichlorodifluoromethane
1,1-Dichloroethane
1,2-Dichloroethane
1,1-Dichloroethylene
1,2-trans-dichloroethylene

NOTE: The chemicals listed as hazardous because they are ignitable (I), corrosive (C) or reactive (R) but not considered toxic (T) are not considered "hazardous wastes" in the context of INFORM's report unless they appear on other lists in this appendix.

Tetrachloroethane, N.O.S.
1,1,1,2-Tetrachloroethane
1,1,2,2-Tetrachloroethane
Tetrachloroethene (Tetrachloroethylene)
Tetrachloromethane
2,3,4,6-Tetrachlorophenol
Tetraethyldithiopyrophosphate
Tetraethyl lead
Tetraethylpyrophosphate
Thallium and compounds, N.O.S.
Thallic oxide
Thallium (I) acetate
Thallium (I) carbonate
Thallium (I) chloride
Thallium (I) nitrate
Thallium selenite
Thallium (I) sulfate
Thioacetamide
Thiosemicarbazide
Thiourea
Thiuram
Toluene
Toluene diamine
o-Toluidine hydrochloride
Tolylene diisocyanate
Toxaphene
Tribromomethane

1,2,4-Trichlorobenzene
1,1,1-Trichloroethane
1,1,2-Trichloroethane
Trichloroethene (Trichloroethylene)
Trichloromethanethiol
2,4,5-Trichlorophenol
2,4,6-Trichlorophenol
2,4,5-Trichlorophenoxyacetic acid (2,4,5-T)
2,4,5-Trichlorophenoxypropionic acid (2,4,5-TP) (Silvex)
Trichloropropane, N.O.S.
1,2,3-Trichloropropane
0,0,0-Triethyl phosphorothioate
Trinitrobenzene
Tris(1-azridinyl)phosphine sulfide
Tris(2,3-dibromopropyl) phosphate
Trypan blue
Uracil mustard
Urethane
Vanadic acid, ammonium salt
Vanadium pentoxide (dust)
Vinyl chloride
Vinylidene chloride
Zinc cyanide
Zinc phosphide

[FR Doc. 80-14307 Filed 5-16-80; 8:45 am]

(ii) This list contains toxic chemicals listed under RCRA as "discarded commercial chemical products". It overlaps substantially with, but is not identical to, the list in (i).

Acetaldehyde
Acetone (I)
Acetonitrile (I,T)
Acetophenone
2-Acetylaminoflourene
Acetyl chloride (C,T)
Acrylamide
Acetylene tetrachloride
Acetylene trichloride
Acrylic acid (I)
Acrylonitrile
AEROTHENE TT
3-Amino-5-(p-acetamidophenyl)-1H-1,2,4-triazole, hydrate
6-Amino-1,1a,2,8,8a,8b-hexahydro-8-(hydroxymethyl)8-methoxy-5-methylcarbamate azirino(2',3':3,4) pyrrolo(1,2-a) indole-4, 7-dione (ester)
Amitrole
Aniline
Asbestos
Auramine
Azaserine
Benz[c]acridine
Benzal chloride
Benz[a]anthracene
Benzene
Benzenesulfonyl chloride (C,R)
Benzidine
1,2-Benzisothiazolin-3-one, 1,1-dioxide
Benzo[a]anthracene
Benzo[a]pyrene
Benzotrichloride (C,R,T)
Bis(2-chloroethoxy)methane
Bis(2-chloroethyl) ether
N,N-Bis(2-chloroethyl)-2-naphthylamine
Bis(2-chloroisopropyl) ether
Bis(2-ethylhexyl) phthalate
Bromomethane
4-Bromophenyl phenyl ether
n-Butyl alcohol (I)
Calcium chromate
Carbolic acid
Carbon tetrachloride
Carbonyl fluoride
Chloral
Chlorambucil
Chlordane

Chlorobenzene
Chlorobenzilate
p-Chloro-m-cresol
Chlorodibromomethane
1-Chloro-2,3-epoxypropane
CHLOROETHENE NU
Chloroethyl vinyl ether
Chloroethene
Chloroform (I,T)
Chloromethane (I,T)
Chloromethyl methyl ether
2-Chloronaphthalene
2-Chlorophenol
4-Chloro-o-toluidine hydrochloride
Chrysene
C I 23060
Creosote
Cresols
Crotonaldehyde
Cresylic acid
Cumene
Cyanomethane
Cyclohexane (I)
Cyclohexanone (I)
Cyclophosphamide
Daunomycin
DDD
DDT
Diallate
Dibenz[a,h]anthracene
Dibenzo[a,h]anthracene
Dibenzo[a,i]pyrene
Dibromochloromethane
1,2-Dibromo-3-chloropropane
1,2-Dibromoethane
Dibromomethane
Di-n-butyl phthalate
1,2-Dichlorobenzene
1,3-Dichlorobenzene
1,4-Dichlorobenzene
3,3'-Dichlorobenzidine
1,4-Dichloro-2-butene
3,3'-Dichloro-4,4'-diaminobiphenyl
Dichlorodifluoromethane
1,1-Dichloroethane
1,2-Dichloroethane
1,1-Dichloroethylene
1,2-trans-dichloroethylene

NOTE: The chemicals listed as hazardous because they are ignitable (I), corrosive (C) or reactive (R) but not considered toxic (T) are not considered "hazardous wastes" in the context of INFORM's report unless they appear on other lists in this appendix.

Dichloromethane
Dichloromethylbenzene
2,4-Dichlorophenol
2,6-Dichlorophenol
1,2-Dichloropropane
1,3-Dichloropropene
Diepoxybutane (I,T)
1,2-Diethylhydrazine
O,O-Diethyl-S-methyl ester of phosphorodithioic acid
Diethyl phthalate
Diethylstilbestrol
Dihydrosafrole
3,3'-Dimethoxybenzidine
Dimethylamine (I)
p-Dimethylaminoazobenzene
7,12-Dimethylbenz[a]anthracene
3,3'-Dimethylbenzidine
alpha,alpha-Dimethylbenzylhydroperoxide (R)
Dimethylcarbamoyl chloride
1,1-Dimethylhydrazine
1,2-Dimethylhydrazine
Dimethylnitrosoamine
2,4-Dimethylphenol
Dimethyl phthalate
Dimethyl sulfate
2,4-Dinitrophenol
2,4-Dinitrotoluene
2,6-Dinitrotoluene
Di-n-octyl phthalate
1,4-Dioxane
1,2-Diphenylhydrazine
Dipropylamine (I)
Di-n-propylnitrosamine
EBDC
1,4-Epoxybutane
Ethyl acetate (I)
Ethyl acrylate (I)
Ethylenebisdithiocarbamate
Ethylene oxide (I,T)
Ethylene thiourea
Ethyl ether (I,T)
Ethylmethacrylate
Ethyl methanesulfonate
Ethylnitrile
Firemaster T23P
Fluoranthene
Fluorotrichloromethane
Formaldehyde
Formic acid (C,T)
Furan (I)
Furfural (I)
Glycidylaldehyde
Hexachlorobenzene
Hexachlorobutadiene
Hexachlorocyclohexane
Hexachlorocyclopentadiene
Hexachloroethane
Hexachlorophene
Hydrazine (R,T)
Hydrofluoric acid (C,T)
Hydrogen sulfide
Hydroxybenzene
Hydroxydimethyl arsine oxide

4,4'-(Imidocarbonyl)bis(N,N-dimethyl)aniline

Indeno(1,2,3-cd)pyrene
Iodomethane
Iron Dextran
Isobutyl alcohol
Isosafrole
Kepone
Lasiocarpine
Lead acetate
Lead phosphate
Lead subacetate
Maleic anhydride
Maleic hydrazide
Malononitrile
MEK Peroxide
Melphalan
Mercury
Methacrylonitrile
Methanethiol
Methanol
Methapyrilene
Methyl alcohol
Methyl chlorocarbonate
Methyl chloroform
3-Methylcholanthrene
Methyl chloroformate
4,4'-Methylene-bis-(2-chloroaniline)
Methyl ethyl ketone (MEK) (I,T)
Methyl ethyl ketone peroxide (R)
Methyl iodide
Methyl isobutyl ketone
Methyl methacrylate (R,T)
N-Methyl-N'-nitro-N-nitrosoguanidine
Methylthiouracil
Mitomycin C
Naphthalene
1,4-Naphthoquinone
1-Naphthylamine
2-Naphthylamine
Nitrobenzene (I,T)
Nitrobenzol
4-Nitrophenol
2-Nitropropane (I)
N-Nitrosodi-n-butylamine
N-Nitrosodiethanolamine
N-Nitrosodiethylamine
N-Nitrosodi-n-propylamine
N-Nitroso-n-ethylurea
N-Nitroso-n-methylurea
N-Nitroso-n-methylurethane
N-Nitrosopiperidine
N-Nitrosopyrrolidine
5-Nitro-o-toluidine
Paraldehyde
PCNB
Pentachlorobenzene
Pentachloroethane
Pentachloronitrobenzene
1,3-Pentadiene (I)
Perc
Perchlorethylene
Phenacetin
Phenol

Phosphorous sulfide (R)
Phthalic anhydride
2-Picoline
Pronamide
1,3-Propane sultone
n-Propylamine (I)
Pyridine
Quinones
Reserpine
Resorcinol
Saccharin
Safrole
Selenious acid
Selenium sulfide (R,T)
Silvex
Streptozotocin
2,4,5-T
1,2,4,5-Tetrachlorobenzene
1,1,1,2-Tetrachloroethane
1,1,2,2-Tetrachloroethane
Tetrachloroethene
Tetrachloroethylene
Tetrachloromethane
2,3,4,6-Tetrachlorophenol
Tetrahydrofuran (I)
Thallium (I) acetate
Thallium (I) carbonate
Thallium (I) chloride
Thallium (I) nitrate
Thioacetamide

Thiourea
Toluene
Toluenediamine
o-Toluidine hydrochloride
Toluene diisocyanate
Toxaphene
2,4,5-TP
Tribromomethane
1,1,1-Trichloroethane
1,1,2-Trichloroethane
Trichloroethene
Trichloroethylene
Trichlorofluoromethane
2,4,5-Trichlorophenol
2,4,6-Trichlorophenol
2,4,5-Trichlorophenoxyacetic acid
2,4,5-Trichlorophenoxypropionic acid alpha, alpha, alpha- Trichlorotoluene
TRI-CLENE
Trinitrobenzene (R,T)
Tris(2,3-dibromopropyl) phosphate
Trypan blue
Uracil mustard
Urethane
Vinyl chloride
Vinylidene chloride
Xylene

Appendix B
Research Method and Information Sources

Research Method and Sources of Information

INFORM's research on hazardous chemical use and waste reduction practices at each study plant came from a review of federal, state and local government agencies' environmental files and interviews at the plants themselves. All federal and state government files generally were of some value in helping to define chemicals used and wastes discharged by plants.

Corporate interviews, however, were the primary and in most cases, the only way that INFORM could learn about hazardous waste reduction initiatives -- their causes, scope and results. Of the 29 plants in the study, 13 granted such interviews. These included 11 of the 18 large plants in the study but only two of the 11 smaller ones.

Master List

The master list of organic chemical plants used to select the 29 plants to be studied was compiled from environmental files maintained by federal and state government agencies and commercial industrial directories (both state and national). The list of references at the end of this appendix contains the primary sources used.

Information on types of products manufactured at the plants on INFORM's master list was then sought in order to identify plants whose operations were predominantly organic chemical manufacturing. Table B-1 shows the types of references supplying this information. The

Table B-1. Data Sources Used for Selection of Study Plants

Type of Data Sought	Commercial Directories State	Commercial Directories National	New Jersey Industrial Survey	Literature Reviews	Phone Contacts
Plant Size					
# of Employee	XX	X	XX		
Sales ($)	XX	X			
Type of Product					
Summary		X	XX	XX	
Specific Products	XX				
Ownership*	XX	X		XX	XX
Other**	XX	X		XX	XX

XX=Primary Sources
X=Secondary Sources

* parent company, plant subsidiary/branch/division information, whether public or private

** age of facility, physical size of plant, research facility location, national rank based on sales, management structure

state industrial directories were the most useful. Other commercial directories mainly contain data of a general corporate nature and on larger plants and were not very helpful in supplying plant-specific data.

Annual reports for all of the domestic, publicly-owned plants (a few reports from privately-owned or foreign companies were also obtained) along with calls to plant officials and INFORM's general literature search, yielded further information about ownership changes, new product lines, production changes, etc.

Government Files

For each of the 29 plants selected for the study, INFORM used government files to get an understanding of the use and ultimate fate of hazardous chemicals at the plants and to form a basis for questions during the plant interviews (see Table B-2). These files are not necessarily physically separate entities and each state or local agency keeps the files according to its own organizational structure with no uniformity across states. The main sources consulted were:

-- air and water permit files containing copies of the permits regulating plant waste discharges, descriptions of the plant, and its discharges, and sometimes descriptions of its processes.

-- air emission inventories, computerized data bases listing types and quantities of chemicals discharged to the atmosphere by plants.

-- water quality data files containing periodic analyses by the plant, a consulting firm or a government office of a plant's wastewater discharges. This data was part of a permit application, an inspection report or periodic sampling required by a government agency and not the U.S. EPA STORET system, for example.

-- files maintained by individual sewage treatment plants containing discharge records/permits/ correspondence of industrial plants discharging to

Table B-2. Government Files Containing Information on INFORM Study Plants

Type of Data File	CALIFORNIA				NEW JERSEY				OHIO			
	US EPA Regional Office	State Office	Regional Offices	Local Offices	US EPA Regional Office	State Office	Regional Offices	Local Offices	US EPA Regional Office	State Office	Regional Offices	Local Offices
AIR EMISSIONS												
Air Permit Files			XX		X	X	XX		X			XX
Air Emissions Inventory			XX			X				XX		X
Inspection/Litigation Reports			X		X		XX		X			XX
Correspondence Files			X		X	X	XX		X			XX
Special Reports/Data		XX										
WASTEWATER DISCHARGES												
National Pollution Discharge Elimination System (NPDES) Files	X		XX		XX	XX			XX	XX	XX	
Water Quality Data Files			XX		XX	XX			X	XX	XX	
Sewage Treatment Plant Surveys/Permit Files	X			XX	X	XX				X	X	
Inspection/Litigation Reports			XX		XX	XX			X	XX	XX	XX
Correspondence Files	X		XX		XX	XX			X	XX	XX	
SOLID WASTES												
RCRA Notification Forms/Part A Applications	XX				XX	XX			XX	XX	XX	
RCRA Part B Applications					XX	XX						
Treatment, Storage and Disposal Annual Reports						XX				XX		
Generator's Hazardous Waste Annual Reports						XX				XX		
Inspection/Litigation Reports	XX				XX	XX	X		XX	XX	XX	
Correspondence Files	XX				XX	XX	X		XX	XX	XX	
INDUSTRIAL WASTE SURVEY			X			XX					X	

XX = Primary Source   X = Secondary Source

the sewage treatment plant. Summaries of this type of data were often found at other government offices.

-- correspondence files, including all company-to-government agency correspondence. They can include information about a plant's permit, inspection reports, waste reduction information, monitoring results, etc.

-- inspection and litigation reports, usually found in correspondence files. These detail agency inspections of the facilities (or portions of facilities) under their jurisdiction. Reports on litigation involving the plant can appear in correspondence files or in separate enforcement files.

-- Resource Conservation and Recovery Act (RCRA) plant Notification Forms and Part A Applications, containing waste quantities and basic waste handling information.

-- RCRA Part B Applications, available only for those plants requested by the government agency to submit this information, containing extensive waste type and quantity data, plant descriptions, process information, and groundwater data (if applicable).

-- RCRA Annual Reports, including annual waste generation and disposal quantities. These reports vary by state; only New Jersey's reports included off-site disposal quantities.

The Industrial Survey conducted by the New Jersey Department of Environmental Protection in 1978 was the only information source that contained comparable data on air, water and solid waste discharges from individual plants. Both Ohio and California have data bases called industrial waste surveys, but these only cover wastes regulated by RCRA. The New Jersey Industrial Survey is chemical specific (for 155 hazardous chemicals) and covers all types of emissions without regard to regulatory boundaries. Each plant reports the total quantity of the chemical brought into the plant, manufactured there and disposed of as waste. These "input/output" figures collected by the state of New Jersey are reported in a comprehensive fashion on a single form and are critical to the understanding of waste reduction both for an individual plant and in comparing the different plants' management of their wastes.

Plant Interviews

The government files contained a vast amount of information on wastes generated by the study plants but almost none on waste reduction practices. The primary source of information on waste reduction was interviews of managers at the study plants.

INFORM developed a questionnaire that sought information in 10 areas and was used as the basis for company interviews. The following topics were covered:

1. Plant ownership, size, volume of production and sales;

2. The type of products and percentage of production in organic chemicals;

3. The management structure for production and environmental decision-making;

4. The categories and amounts of toxic wastes generated;

5. Specifics on where responsibility for waste handling and regulatory filing fall;

6. Current practices for managing hazardous wastes and any changes that had occurred over the past decade;

7. The types of waste reduction practices in use;

8. The impact of waste reduction in terms of costs of implementation, economic savings and reductions in waste generation;

9. Information on recycling, reuse or recovery practices;

10. Assessment of the factors that encourage, inhibit or otherwise affect waste reduction, including technological considerations, economic factors and regulatory influences;

A profile of each study plant was drafted containing information from all sources on the plant's processes, waste generation and waste reduction practices. Plants were sent draft profiles of the data collected from public files and/or interviews for their review before publication. The 13 cooperating plants responded with corrections, additional descriptions of waste reduction practices or confirmation of the accuracy of the profiles. The completed profiles of the 29 study plants are included in Part II of this report.

Primary Sources Used for Identifying and Selecting Plants

A. Federal Government Sources

Printout of California Hazardous Waste Generators with a SIC Code 28. From: U.S. Environmental Protection Agency, Region IX, Toxics and Waste Management Division, 215 Fremont Street, San Francisco, CA 94105.

Printout of California National Pollutant Discharge Elimination System Permit Holders with a SIC Code 28. From: U.S. Environmental Protection Agency, Region IX, Administrative Services Division, 215 Fremont Street, San Francisco, CA 94105.

Printout of California Prevention of Significant Deterioration of Air Quality Program Permit Holders with a SIC Code 28. From: U.S. Environmental Protection Agency, Region IX, Air Management Division, 215 Fremont Street, San Francisco, CA 94105.

Printout of New Jersey Hazardous Waste Generators with a SIC Code 28. From: U.S. Environmental Protection Agency, Region II, Information Systems Branch, 26 Federal Plaza, New York, NY 10278.

Printout of New Jersey Hazardous Waste Manifest information for SIC Code 28. From: U.S. Environmental Protection Agency, Region II, Information Systems Branch, 26 Federal Plaza, New York, NY 10278.

Printout of Ohio Hazardous Waste Generators with a SIC Code 286. From: U.S. Environmental Protection Agency, Region V, 230 South Dearborn Street, Chicago, IL 60604.

Printout of California, New Jersey and Ohio Organic Chemical Manufacturers. From: The Organic Chemical Producers Data Base, U.S. Environmental Protection Agency, Industrial Pollution Control Division, 5555 Ridge Avenue, Cincinnati, OH 45268.

B. State Government Sources

Printout of Hazardous Waste Generators. From: California Department of Health Services, Toxic Substances Control Division, 2151 Berkeley Way, Berkeley, CA 94704.

Printout of Hazardous Waste Generators. From: New Jersey Department of Environmental Protection, Office of Hazardous Waste, 32 East Hanover Street, Trenton, NJ 08625.

Printout of Hazardous Waste Manifest Waste Pickups. From: New Jersey Department of Environmental Protection, Office of Hazardous Waste Substances, 32 East Hanover Street, Trenton, NJ 08625.

Printout of the Industrial Survey Data Base. From: New Jersey Department of Environmental Protection, Office of Science and Research, 190 West State Street, Trenton, NJ 08625.

Air Emissions Inventory System: Point Source Report for 1980, Ohio Environmental Protection Agency, Air Division, 361 Broad Street, Columbus, OH 43215.

Printout of National Pollutant Discharge Elimination System Permit Holders. From: Ohio Environmental Protection Agency, Industrial Wastewater Section, 361 Broad Street, Columbus, OH 43215.

Printout of Off-Site Hazardous Waste Generators. From: Ohio Environmental Protection Agency, Hazardous Materials Management Division, 361 Broad Street, Columbus, OH 43215.

C. Non-Government Sources

Chemical Week 1983 Buyers Guide. McGraw-Hill, Inc., 1221 Avenue of the Americas, New York, NY 10020.

California State Industrial Directory 1982. MacRae's Blue Book Inc., 817 Broadway, New York, NY 10003.

1983 California Manufacturers Register. Times Mirror Press, 1115 South Boyle Avenue, Los Angeles, CA 90023.

Million Dollar Directory 1982, Volume I. Dun & Bradstreet Inc., 3 Century Drive, Parsippany, NJ 07054.

Standard & Poor's Register of Corporations, Directors and Executives, 1982, Volume 3. Standard & Poor's, Inc., 25 Broadway, New York, NY 10004.

New Jersey State Industrial Directory 1982. McRae's Blue Book, Inc., 817 Broadway, New York, NY 10003.

Ohio Industrial Directory 1980. Harris Publishing Co., 2057-2 Aurora Road, Twinsburg, OH 44087.

Ohio Manufacturer's Directory 1983. Manufacturer's News Inc., 4 East Huron Street, Chicago, IL 60611.

Ohio State Industrial Directory 1982. McRae's Blue Book, Inc., 817 Broadway, New York, NY 10003.

Kline Guide to the Chemical Industry, 1980, 4th Edition, Charles A. Kline & Co. Inc., 330 Passaic, Fairfield, NJ 07006.

1980 Directoy of Chemical Producers, Stanford Research Institute, Palo Alto, CA.

Local telephone directories for California, New Jersey and Ohio.

Annual Reports and 10K Reports for individual companies.

Primary Sources Used for Information on Individual Study Plants

A. Air Emissions from Individual Plants

Bay Area Air Quality Management Division, 939 Ellis Street, San Francisco, CA 94109.

South Coast Air Quality Management District, 9150 East Flair Drive, El Monte, CA 91731.

New Jersey Department of Environmental Protection, Bureau of Air Pollution Control, Trenton, NJ 08625.

New Jersey Department of Environmental Protection, Newark Field Office, 1100 Raymond Boulevard, Newark, NJ 07102.

Ohio Environmental Protection Agency, Southeast District Office, Division of Air Pollution Control, 2195 Front Street, Logan, OH 43138.

Ohio Environmental Protection Agency, Central District Office, Air Division, 361 East Broad Street, Columbus, OH 43215.

Ohio Environmental Protection Agency, North East District Office, 2110 East Aurora Road, Twinsburg, OH 44087.

Southwestern Ohio Air Pollution Control Agency, 2400 Beekman Street, Cincinnati, OH 45214.

Toledo Environmental Services Agency, 26 Main Street, Toledo, OH 43605.

Portsmouth Air Pollution Control Agency, 740 Second Street, Portsmouth, OH 45662.

B. Air Inventories

California Air Resources Board, 1102 Q Street, Sacramento, CA 95812.

Ohio Environmental Protection Agency, Central District Office, Air Division, 361 East Broad Street, Columbus, OH 43215.

C. Wastewater Discharge Permit/Compliance Reports

Regional Water Quality Control Board, 1111 Jackson Street, Oakland, CA 94607.

New Jersey Department of Environmental Protection, Division of Water Resources, 1474 Prospect Street, Trenton, NJ 08625.

U.S. Environmental Protection Agency, Region V, Water Division, 230 South Dearborn Street, Chicago, IL 60604.

Ohio Environmental Protection Agency, Southwest District Office, Industrial Wastewater Section, 7 East Fourth Street, Dayton, OH 45402.

Ohio Environmental Protection Agency, Southeast District Office, Division of Air Pollution Control, 2195 Front Street, Logan, OH 43138.

U.S. Environmental Protection Agency, Region V, Eastern District Office, 25089 Center Ridge Road, Westlake, OH 44145.

D. Sewage Treatment Plant User Surveys and/or Discharge Permits

Union Sanitary District, 4057 Baine Avenue, Fremont, CA 94536.

City of Richmond Water Pollution Control Plant, 601 Canal Boulevard, Richmond, CA 94804.

West Contra Costa Sanitary District, 2910 Hilltop Drive, Richmond, CA 94806.

Los Angeles County Sanitation District, 1955 Workman Mill Road, Whittier, CA 90607.

U.S. Environmental Protection Agency, Region II, Permits Administration Branch, 26 Federal Plaza, New York, NY 10278

New Jersey Department of Environmental Protection, Division of Water Resources, 1474 Prospect Street, Trenton, NJ 08625.

Ohio Environmental Protection Agency, Central District Office, Pretreatment Division, 361 East Broad Street, Columbus, OH 43215.

Toledo Environmental Services Agency, 26 Main Street, Toledo, OH 43605.

City of Cincinnati-Metropolitan Sewer District, 1600 Gest Street, Cincinnati, OH 45204.

E. Hazardous Waste Reports

U.S. Environmental Protection Agency, Region IX, 215 Fremont Street, San Francisco, CA.

U.S. Environmental Protection Agency, Region II, Permits Administration Branch, 26 Federal Plaza, New York, NY 10278.

New Jersey Department of Environmental Protection, Division of Waste Management, 32 East Hanover Street, Trenton, NJ 08625.

U.S. Environmental Protection Agency, Region V, Waste Management Branch, 230 South Dearborn Street, Chicago, IL 60604.

Ohio Environmental Protection Agency, Central District Office, Division of Hazardous Waste Management, 361 East Broad Street, Columbus, OH 43215.

Ohio Environmental Protection Agency, Southwest District Office, 7 East Fourth Street, Dayton, OH 45402.

Ohio Environmental Protection Agency, Southeast District Office, Division of Hazardous Materials Management, 2195 Front Street, Logan, OH 43138.

F. Industrial Waste Survey

California Department of Health Services, Toxic Substances Control Division, 2151 Berkeley Way, Berkeley, CA 94704.

New Jersey Department of Environmental Protection, Office of Science and Research, 190 West State Street, Trenton, NJ 08625.

Ohio Environmental Protection Agency, Southeast District Office, Division of Hazardous Materials Management, 2195 Front Street, Logan, OH 43138.

Authors' Biographies

David J. Sarokin worked extensively on hazardous waste and toxic chemical issues in New York as an Environmental Analyst for the State Assembly's Environmental Conservation Committee (1981). His experience includes siting of hazardous waste facilities, control of abandoned dumpsites and statewide data collection on toxic substances information. Prior to this, he was a staff researcher at the California Institute of Technology (1976-1978), studying environmental problems associated with coastal erosion in southern California. He holds a Masters of Science degree in Marine Environmental Science from the State University of New York at Stony Brook.

Warren R. Muir is President of Hampshire Research Associates, Inc. He served from 1978-1981 as the Director of the Environmental Protection Agency's Office of Toxic Substances with the responsibility for implementing the federal Toxic Substances Control Act. As the senior staff scientist with the President's Council on Environmental Quality from 1971-1978, he was instrumental in formulating national policy objectives for the regulation of toxic substances.

Dr. Muir has served as an advisor and consultant to many national and international environmental organizations, including the National Academy of Sciences, the Federal Interagency Testing Committee, the World Health Organization, and the United Nation's Director General's Programme Advisory Committee. He holds a Ph.D. in chemistry from Northwestern University.

Catherine G. Miller is experienced in the use of technical information for making environmental public policy. She has worked for the Environmental Protection Agency since 1971, and is the author of half a dozen EPA studies, including: assessments of the economic benefits of municipal wastewater treatment and multi-media pollution control programs and the policy uses of air quality models and exposure assessment modeling. As an EPA operations research analyst she developed management and budget analyses of operational and policy issues including data information systems.

Sebastian R. Sperber is a law student at Columbia University and has been an integral part of INFORM's Waste Reduction research since the project's inception in June of 1982. He has travelled widely throughout New Jersey, California and Ohio in pursuit of the raw data needed on each of the chemical plants under study, and has spoken before a number of university groups about the problems of hazardous wastes and INFORM's research in this area.

Other Related INFORM Publications

TOXIC WASTES:

<u>Tracing a River's Toxic Pollution: A Case Study of the Hudson</u> (1985). The first complete inventory of point source discharges of selected toxic chemicals into a major U.S. river basin. $12.00.

<u>A Directory of 82 Organic Chemical Plants in New York</u> (1984). The only up-to-date, accurate, and comprehensive list of chemical manufacturers of this state. $15.00.

<u>A Directory of 95 Organic Chemical Plants in Ohio</u> (1984). The only up-to-date, accurate, and comprehensive list of chemical manufacturers of this state. $15.00.

<u>A Directory of 84 Organic Chemical Plants in California</u> (1983). The only up-to-date, accurate, and comprehensive list of chemical manufacturers of this state. $15.00.

<u>A Directory of 181 Organic Chemical Plants in New Jersey</u> (1983). The only up-to-date, accurate, and comprehensive list of chemical manufacturers of this state. $15.00.

INFORM's Board of Directors

James B. Adler
President
Adler & Adler
 Publishing Co.

Michael J. Feeley
Manager
Brown Brothers Harriman
 & Co.

Jane R. Fitzgibbon
Senior Vice President
Ogilvy & Mather Advertising

C. Howard Hardesty, Jr.
Chief Executive Officer
Purolator Courier Corp.

Timothy L. Hogen
President
T.L. Hogen Associates

Lawrence S. Huntington
Chairman of the Board
Fiduciary Trust Company
 of New York

Martin Krasney

Dr. Jay T. Last

Charles A. Moran
Senior Vice President
Manufacturers Hanover
 Trust Company

Kenneth F. Mountcastle, Jr.
Senior Vice President
Dean Witter Reynolds, Inc.

Linda Stamato
Associate Director
Institute of Judicial
 Administration

Frank T. Thoelen
Audit Partner
Arthur Andersen & Co.

Grant P. Thompson
Senior Associate
The Conservation Foundation

Edward H. Tuck
Partner
Shearman & Sterling

Joanna D. Underwood
Executive Director
INFORM

Frank A. Weil
Partner
Wald, Harkrader & Ross

Anthony Wolff
Writer and Photographer